Biopesticides

Biopesticides

Botanicals and Microorganisms for Improving Agriculture and Human Health

Timothy O. Adejumo, Ralf T. Vögele (Eds.)

Logos Verlag Berlin

Bibliographic information published by Die Deutsche Bibliothek
Die Deutsche Bibliothek lists this publication in the Deutsche Nationalbibliografie;
detailed bibliographic data is available in the Internet at http://dnb.ddb.de.

ISBN 978-3-8325-5264-0

Logos Verlag Berlin GmbH
Georg-Knorr-Str. 4, Geb"aude 10
D-12681 Berlin
Germany

Tel.:   +49 (0)30 / 42 85 10 90
Fax: +49 (0)30 / 42 85 10 92
http://www.logos-verlag.com

PREFACE

Conventional pesticides are responsible for an extensive pollution of the environment, serious health hazards due to a presence of residues in food and fodder, development of resistance in targeted insect populations, a decrease in biodiversity, and outbreaks of secondary pests. Despite the public health risks, application of pesticides is on the increase in developing and developed countries to prevent crop damage, and to increase food and agricultural productivity. However, there is poor monitoring and lack of internal regulations and control in some governments. The WHO reported roughly about three million cases of pesticide poisoning annually, resulting in 220,000 deaths worldwide, and that 99% of the deaths are associated with pesticides in developing countries, where only 25% of the world's production of pesticides is used.

Biopesticides however, have readily available sources, are effective and easily biodegradable, exhibit various modes of action, are cheaper, and inherently less toxic to humans and the environment, they do not leave harmful residues, and are usually more specific to target pests. The use of biopesticides is markedly safer for the environment and users, and more sustainable than the application of chemicals, and are therefore used as potential alternatives to synthetic pesticides, especially as components in Integrated Pest Management (IPM) strategies.

The book '**Biopesticides: Botanicals and Microorganisms for Improving Agriculture and Human Health**' is a collection of articles, up to date reviews and research contributions by 20 eminent researchers from both developed and developing countries. It emphasises the current issues of importance and the progress made in the fields of agricultural, environmental and soil microbiology, plant pathology and ethnobotany, and aims to bring together all available and relevant information on biopesticides. It comprises 12 Chapters on emerging issues on biopesticides from important and useful botanicals to beneficial microorganisms that show great potential in both agriculture and human health.

The book will be of immense help to both the undergraduate and postgraduate students, biologists and agriculturists, who would like to broaden their knowledge and gain substantial experience about biopesticides in agriculture and health, this will enable them to contribute significantly in making the world a safer and healthier place.

Timothy Olubisi Adejumo
November, 2019

DEDICATION

This book is dedicated to the Alexander von Humboldt Stiftung/Foundation for sponsoring research fellowships and research awards, to promote academic cooperation between excellent scientists and scholars from abroad and from Germany.

Alexander von Humboldt
Stiftung / Foundation

ACKNOWLEDGEMENTS

I am forever grateful to Almighty God for the multitude of mercy obtained that made me a beneficiary of the Georg Forster Research Fellowship of the Alexander von Humboldt (AvH) Stiftung/Foundation, Germany and member of Humboldt family. This opportunity indeed has brought a significant turning point to my academic career. Great appreciation goes to the AvH; through their sponsorships of a book grant, renewed research stays on four (4) occasions at Göttingen, Detmold and Stuttgart in Germany, with the recent sponsorship of my first Ph.D. student and Junior researcher, Dr. Olukayode O. Orole. All these, together with the sponsorship of the Humboldt-Kolleg titled "Commercialisation of research-based University research efforts for national development" at Adekunle Ajasin University, Akungba-Akoko, Nigeria in 2018, and the publishing of this book are hereby greatly acknowledgcd.

The completion of this book titled 'Biopesticides: Botanicals and Microorganisms for Improving Agriculture and Human Health' has taught me a great lesson in life that virtually nothing is impossible to a motivated, focused and courageous person, who puts his trust in the invisible God. There were great discouragements from the onset and along the way of execution of this project, but the word 'Only be thou strong and very courageous' sustained me throughout. I will like to express my gratitude to many people who provided support, contributcd chapters to share their knowledge with the scientific community, and those that read through, offered comments, and assisted in the editing and proof reading. Thanks also to my numerous students who have been with me over the years, whose names were not mentioned.

Special thanks to my lovely wife, Prof. (Mrs) Arinpe G. Adejumo; my precious children: Toluwani, Ooreoluwa and IbukunOluwa for their encouragements and understanding, especially when I was away for half of a year during this recent research fellowship.

ABOUT THE EDITORS

Timothy Olubisi Adejumo is a Professor of Agricultural/Environmental Microbiology & Biotechnology at Adekunle Ajasin University, Akungba-Akoko, Nigeria. His research interests are molecular diagnostics, mycotoxicology, phytopathology, soil microbiology, biopesticides and microbial ecology. He holds B.Sc and M.Sc degrees in Microbiology (Ilorin) and Ph.D in Agriculture: Mycology and Plant Pathology (Ibadan). He was former Dean, Faculty of Science (2017-2019), Head of Department of Microbiology (2008-2010, 2014-2016), Acting Director, Centre for Research and Development (2010-2014); Director, Linkage and International Programmes Office (2015- 2016) and Director, Centre for Space, Energy and Environmental Research (2016-2017). He is a Georg Forster Research Fellow of Alexander von Humboldt (AvH) Foundation, Germany; with Renewed Research Fellowships in 2010, 2013, 2017 and 2019, where he was a Visiting Research Professor at Georg-August University, Göttingen, Max Rübner Institute, Detmold and University of Hohenheim, Stuttgart, Germany. A seasoned researcher, former Research Associate and Graduate Research Fellow of International Institute of Tropical Agriculture (IITA), Ibadan. He is currently serving as Pioneer Ambassador scientist and member of International Board of Ambassadors of the International Society for Microbial Ecology, The Netherlands. He is the Founder, Pioneer Chairman and Member of Board of Trustees of Nigerian Society for Microbial Ecology. He serves as External Assessor and Postgraduate External Examiner of many Universities, and he is happily married with children.

Ralf Thomas Voegele is a German Biologist, specialising in Phytopathology and Microbiology. He is Chair of Phytopathology, Managing Director of the Institute of Phytomedicine and Dean of the Agricultural Science Faculty at the University of Hohenheim, Stuttgart, Germany. He studied biology at the University of Constance (1983-1989), and obtained his doctorate in biology (Dr. rer. nat.) in 1993. He carried out two postdoctoral research stays, the first at the College of Biological Science of the University of Guelph (1993-1995) supported by a postdoctoral research grant from the Deutsche Forschungsgemeinschaft, the second at McMaster University in Hamilton, Canada (1996-1997). He habilitated from the University of Constance in 2007 and was awarded the *venia legendi* for Phytopathology and Microbiology. 2010 he was called to the Chair of Phytopathology at the University of Hohenheim. The same year, he was elected Managing Director of the Institute of Phytomedicine (re-election 2012 and 2017), and in 2015 he was elected Dean of the Faculty of Agricultural Sciences (re-election 2019). He is married and has one daughter.

CONTRIBUTORS

1. **Timothy Olubisi Adejumo**, *Dept. of Microbiology, Adekunle Ajasin University, Akungba-Akoko, Nigeria.*
2. **Ralf Thomas Voegele**, *University of Hohenheim, Faculty of Agricultural Sciences, Institute of Phytomedicine, Dept. of Phytopathology, (360), Germany.*
3. **Olubukola Oluranti Babalola**, *Food Security and Safety Niche Area, Faculty of Natural and Agricultural Sciences, North-West University, Mmabatho 2745, South Africa.*
4. **Olanrewaj*u* F. Olotuah**, *Dept. of Plant Science and Biotechnology, Adekunle Ajasin University, Akungba-Akoko, Nigeria.*
5. **Sabine Gruber**, *Institute of Crop Production (340a), University of Hohenheim, Germany.*
6. **Sabine Zikeli**, *Centre for Organic Farming (309), University of Hohenheim, Germany.*
7. **Festus A. Olajubu**, *Dept. of Microbiology, Adekunle Ajasin University, Akungba-Akoko, Nigeria.*
8. **Omena Bernard Ojuederie**, *Food Security and Safety Niche Area, Faculty of Natural and Agricultural Sciences, North-West University, Mmabatho 2745, South Africa.*
9. **Chinenyenwa Fortune Chukwuneme**, *Food Security and Safety Niche Area, Faculty of Natural and Agricultural Sciences, North-West University, Mmabatho 2745, South Africa*
10. **Oluwaseyi Samuel Olanrewaju**, *Food Security and Safety Niche Area, Faculty of Natural and Agricultural Sciences, North-West University, Mmabatho 2745, South Africa*
11. **Modupe Ayilara**, *Food Security and Safety Niche Area, Faculty of Natural and Agricultural Sciences, North-West University, Mmabatho 2745, South Africa.*
12. **Olukayode O. Orole**, *Dept. of Microbiology, Federal University, Lafia, Nasarawa State, Nigeria.*
13. **Tope Taofeek Adegboyega**, *Food Security and Safety Niche Area, Faculty of Natural and Agricultural Sciences, North-West University, Mmabatho 2745, South Africa.*
14. **Oludare T. Osuntokun**, *Dept. of Microbiology, Adekunle Ajasin University, Akungba-Akoko, Nigeria.*
15. **Olatunde Philip Ayodele**, *Dept. of Agronomy, Faculty of Agricultural Sciences, Adekunle Ajasin University, Akungba-Akoko, Nigeria.*
16. **Jialu Xu,** *Institute of Crop Production (340a), University of Hohenheim, Germany.*

17. **Regina G. Belz**, *Institute of Agricultural Sciences in the Tropics (490f), University of Hohenheim, Germany.*
18. **Emmanuel Okrikata**, *Dept. of Biological Sciences, Federal University Wukari, Taraba State, Nigeria.*
19. **Victor I. Gwa**, *Dept. of Crop Production and Protection, Faculty of Agriculture and Agricultural Technology, Federal University Dutsin-Ma, PMB 5001, Katsina State, Nigeria.*
20. **Johanna Bitzenhofer**, *University of Hohenheim, Faculty of Agricultural Sciences, Institute of Phytomedicine, Dept. of Phytopathology, (360), Germany.*

TABLE OF CONTENTS

General Issues

CHAPTER 1

Introduction:
Potential of Botanicals and Microorganisms as Biopesticides

by
Timothy Olubisi Adejumo
Dept. of Microbiology, Adekunle Ajasin University,
P.M.B. 001, Akungba-Akoko, Ondo State, Nigeria.
Email: timothy.adejumo@aaua.edu.ng

Abstract

Current agricultural practices depend heavily on chemical inputs (such as fertilizers, pesticides, herbicides among others) which cause a deleterious effect on the nutritional value of farm product and health of farm workers and consumers. The increasing awareness of health challenges as a result of consumption of poor quality crops has led to a quest for new and improved technologies of improving both the quantity and quality of crop without jeopardizing human health. A reliable alternative to the use of chemical inputs are botanicals and microbial inoculants that can act as biofertilizers, bioherbicide, biopesticides, and biocontrol agents. Plants and microorganisms are the major sources of biopesticides due to the high components of bioactive compounds and antimicrobial agents. Biopesticides are biological control or natural pest management agents that are based on beneficial microorganisms (bacteria, fungi, viruses and protozoa, beneficial nematodes i.e. entomopathogenic nematodes, parasitoids, predators) and biologically-based active ingredients such as botanical products and pheromones. Many of these biological agents and their products (such as *Bacillus thuringiensis* (*Bt*-toxin), *Azadirachta indica* (neem plant extract) are used against various plant pathogens and pests in Nigeria, Kenya, Botswana, Ghana and other African countries. Biopesticides are applied in the same manner as chemical pesticides like gammalin, actellic dusts and DDT powder. Although, the synthetic pesticides have been found to have positive effect on diseases and pests, but harmful effect on man and the ecosystem.

Keywords: biopesticides, botanical pesticides, microbial pesticides, indigenous knowledge, human health

Introduction

The use of chemical fungicides is considered as the most effective method of plant disease management and often practised worldwide. However, repeated use of certain chemical

fungicides has led to the appearance of fungicide- resistant pathotypes of several pathogens. In recent years, there has been considerable pressure by consumers to reduce or eliminate chemical fungicides in food products. The use of chemical fungicides for the
management of plant diseases has its limitations due to their carcinogenic, teratogenic
properties, high and acute residual toxicity, hormonal imbalance, slow and long degradation period, environmental pollution and deterioration of food quality and adverse effects
on human health (Brent and Hollomon, 1998; Dubey *et al.*, 2007; Kumar *et al.* 2007). Their
uninterrupted and indiscriminate use has not only led to the development of resistant
strains, but the accumulation of toxic residues on food grains used for human consumption
has also led to health problems (Sharma and Meshram, 2006).

Agrochemicals are commonly used in agricultural production to control or prevent diseases, pests and weeds in order to maintain high quality of agricultural products and eliminate or reduce yield losses (Alori and Babalola, 2018). With this, food is produced at
reduced costs and farmers therefore get higher profit from their farm but serious concerns
were being raised about health risks resulting from residues in drinking water and food and
from occupational exposure (Alori and Fawole, 2017). Suyal *et al.* (2016) reiterated that
heavy doses of chemical fertilizer, although leading to self-reliance in food production,
causes harmful impacts on living organisms and also depreciate the environment. The
chemical contaminates the food produced and goes further to alter the normal body functions of the consumer (Sayre, 2009). Baker *et al.* (2002), reported 75% of pesticide residues
in conventionally grown produce. Water supplies are polluted by toxic insecticides, herbicides, and chemical fertilizers used (Alori and Fawole, 2017). Plants and microorganisms
are the major sources of biopesticides due to the high components of bioactive compounds
and antimicrobial agents (Nefzi *et al.*, 2016).

Numerous studies have documented the antifungal and antibacterial effects of extracts and
oils from plants (Canillac and Mourey 2001, Adejumo and Langenkämper, 2010;
Adejumo, 2012 and Orole *et al.* 2016). The examination of indigenous local herbs and
plant materials has also been reported from different parts of the world. Higher plants contain essential oils and a wide spectrum of secondary metabolites such as phenols, flavonoids, quinones, tannins, alkaloids, saponins and sterols. Such plant-derived chemicals
may be exploited for their different biological properties, because of their natural/plant
origin, they are biodegradable and do not usually leave toxic residues or by-products. Medicinal plants constitute an effective source of both traditional and modern medicine (Abd
El-Ghani, 2016). Plants have been shown to have genuine utility and about 80% of the
rural population depends on them as primary health care (Akinyemi, 2000). Plants have
been used as sources of remedies for the treatment of many diseases since ancient times
and people of all continents especially Africa have this old tradition. Despite the remarkable progress in synthetic organic medicinal products of the twentieth century, over 25% of

the prescribed medicines in industrialized countries are derived directly or indirectly from plants (Newman *et al.*, 2000). However, plants used in traditional medicine are still under-studied (Kirby, 1996). In developing countries, notably in West Africa, new drugs are not often affordable. Thus, up to 80% of the population uses medicinal plants as remedies (Kirby, 1996; Hostellmann and Marston, 2002).

Botanical medicine has been used throughout history similar to the way modern pharma-ceuticals are used today (to improve human health). Fossil evidence suggests that plants were used medicinally in prehistoric times (Robert, 1976). Although, the first use of plant-derived pharmaceuticals may be difficult to pinpoint, the use of botanicals for human health is the basis for the modern pharmaceutical industry (Newman and Cragg, 2007 and Raskin *et al.*, 2002). Botanicals and the purification of their active ingredients have played a significant role in the history of mankind. The purification of an early botanical from opium poppy (*Papaver somniferum*) was described in great detail in 1806 in a report sum-marizing 57 studies (Boussel *et al.*, 1982). The botanical was very effective for pain and insomnia, leading to a world social crisis when its addictive nature was realized in the mid 19th century (Boussel *et al.*, 1982).

For botanicals to be reliable for research purposes and consumer products, they must be standardized with sufficient quality controls to ensure consistent composition, safety, and potency. This includes uniform cultivation of source plants with controls to monitor for contamination from other species, pesticides, and environmental toxins. The active com-ponents of botanicals must be identified by activity-guided fractionation with the use of *in vitro* assays that require little test material, followed by validation *in vivo*. Concentrations of active compounds within the botanicals can then be accurately measured to ensure the delivery of a dependable dose in the final product. The use of bioenhancing agents may be considered for compounds with poor bioavailability. Standardization of botanical thera-peutics can only be achieved when the active compounds are identified and biological ac-tivity is confirmed, thus ensuring a consistent product.

A careful selection of microbes and intelligent design of test assays are the key steps in developing new technologies for effective utilization of microorganisms for sustainable agriculture, environmental protection, and human and animal health. Several microbial ap-plications are widely known in solving major agricultural (crop productivity, plant health protection, and soil health maintenance) and environmental issues (bioremediation of soil and water from organic and inorganic pollutants). Wastewater treatment and recycling of agricultural and industrial wastes are other important uses of microbial technology. It is expected that microbes in combination with developments in electronics, software, digital imaging, and nanotechnology will play a significant role in solving global problems of the twenty-first century, including climate change. These advances are expected to enhance sustainability of agriculture and the environment.

The global challenge is to secure high and quality yields and to make agricultural produce environmentally compatible. Eco-friendly alternative has been claimed to be the need of the hour, despite many years of effective control by the conventional synthetic agrochemicals due to human health and environmental concerns threatening its continued use. Therefore, there is a need to develop biopesticides which are effective, biodegradable and do not leave any harmful effect on environment (Nicholson, 2007).

What are Biopesticides?

Biopesticides are defined as compounds that are derived from natural organisms or substances such as animals, plants, bacteria, and certain minerals, including their genes or metabolites to manage agricultural pests by means of specific biological effects rather than as broader chemical pesticides (Sporleder et al., 2013, https://www.epa.gov/ingredients-used-pesticide-products/what-are-biopesticides). They are products and by-products of naturally occurring substances such as insects, nematodes, microorganisms, plants as well as semiochemicals (Gasic and Tanovic, 2013). According to FAO definition, biopesticides include biocontrol agents that are passive agents, in contrast to biocontrol agents that actively seek out the pest, such as parasitoids, predators, and many species of entomopathogenic nematodes. Biopesticides differ in their modes of action from conventional chemical pesticide considerably; their modes of action are almost always specific. The rationale behind replacing conventional pesticides with biopesticides is that the latter are more likely to be selective and biodegradable" (http://www.fao.org/biotech/spec-term-n.asp?id_glo¼ 4875&id_lang¼TERMS_E). Based on the nature and origin of the active ingredients, biopesticides fall into several categories such as botanicals, antagonists, compost teas, growth promoters, predators and pheromones (Semeniuc et al., 2017). Using biopesticides efficiently therefore requires specific user knowledge on the agent and the target pest for optimizing application time, field rates, and application intervals (Sporleder et al., 2013).

Classification of Biopesticides

Biopesticides can be classified as three major classes as follow:

- **Biochemical pesticides**: pesticides based on naturally occurring substances that control pests by non-toxic mechanisms, in contrast to chemical pesticides that contain synthetic molecules that directly kill or inactivates the pest. Biochemical pesticides fall into different biologically functional classes, including plant extracts, substances that interfere with mating, such as insect sex pheromones, as well as various scented plant extracts that attract insect pests to traps. Botanical pesticides that have already been commercialized include neem, pyrethrum, cotton and tobacco. Other sources of botanical pesticides include garlic, euphorbia, citrus, pepper among others (Lengai and Muthomi, 2018). According to Ragunath *et al.*, 2014, Biochemical pest control agents include four (4) general biologically functional classes:

1. **Semiochemicals**: These are chemicals emitted by plants or animals that modify the behaviour of receptor organisms of like or different kinds. They include pheromones, allomones, and kairomones. Pheromones are substances emitted by a member of one species that modify the behaviour of others within the same species. Allomones are chemicals emitted by one species that modify the behaviour of a different species to the benefit of the emitting species. Kairomones are chemicals emitted by one species that modify the behaviour of a different species to the benefit of the receptor species.

2. **Hormones**: These are biochemical agents synthesized in one part of an organism and translocated to another where they have controlling, behavioural, or regulating effect.

3. **Natural Plant Regulators**: These are chemicals produced by plants that have toxic, inhibitory, stimulatory, or other modifying effects on the same or other species of plants. Some of these aretermed "plant hormones" or "phytohormones."

4. **Enzymes**: In this regard, enzymes are protein molecules, which are the instrument for expression for gene action and catalyze biochemical reactions.

- **Microbial pesticides and other entomopathogens**: pesticides that contain microorganisms, like bacteria, fungi, or virus or protozoan, which attack specific pest species, or entomopathogenic nematodes as active ingredients. Microbial pesticides can control different kinds of pests, although each separate active ingredient is relatively specific for its target pest(s). These include biofungicides (*Trichoderma*), bioherbicides (*Phytopthora*) and bioinsecticides (*Bacillus thuringiensis*), the most widely used microbial pesticides. Each strain of this bacterium produces a different mix of proteins and, specifically, kills one or a few related species of insect larvae. While some Bt's control moth larvae found on plants, other Bt's are specific for larvae of flies and mosquitoes. The target insect species are determined by whether the particular Bt produces a protein that can bind to a larval gut receptor, thereby causing the insect larvae to starve. Biopesticides exhibit different modes of action against pathogens such as hyperparasitism, competition, lysis and predation. Microbial biopesticides include bacteria species such as *Pseudomonas, Bacillus, Xanthomonas, Rahnella and Serratia* or fungi such as *Trichoderma, Verticillium* and *Beauveria* species (Kachhawa, 2017). Plant growth promoting *rhizobacteria* (PGPR) protect plants from biotic and abiotic stresses and they also enhance plant growth and enhance formation of root hairs. The most common species of PGPR include *Agrobacterium, Ensifer, Microbacterium, Bacillus, Rhizobium, Pseudomonas, Chryseobacterion and Rhodococcus* (Abbamondi *et al.*, 2016). They colonize the environment around the plant roots, fix nitrogen, increase phosphate solubilisation and result in general increase in plant yield. Species of *Pseudomonas* and *Bacillus* have been used as biofertilizers with reports showing increase in plant growth, yield and phosphorous and zinc content in fruits and soils. Natural enemies including predators, pathogens and some insects are also used as biopesticides in management of insect pests. Parasitoids, wasps, beetles, lace wings, bugs and lady birds are used in management of destructive pests such as boll worms (*Helicoverpa armigera*) in important crops such as cotton. Compost teas are filtrates of compost extracts and are similarly used as biopesticides (Ghorbani *et al.*, 2005). Species of *Trichoderma, Bacillus, Pseudomonas, Beauveria* have been commercialized as microbial pesticides (Lengai and Muthomi, 2018).

- **Plant-Incorporated Protectants (PIPs)**:
  Plant-Incorporated-Protectants (PIPs). This class of biopesticides consists of genetically modified plants (or insecticidal transgenic crops) that produce chemicals (pesticides) that act as protection against pest infestation. In general, PIPs are typically extracted from the transgenes (protein-based cytotoxins) of the insect pathogenic bacteria *Bacillus thuringiensis* (Bt) (All, 2017; http://www.epa.gov/opp00001/biopestici des/). In principle, PIPs, also termed semi-chemical pesticides, are also widely used for pest control. This is due to the minimal impact these class of biopesticides exert on humans and the environment (All, 2017, Walia *et al.*, 2017). Consequently, significant research and scientific resources are dedicated to PIPs as natural pest control agents.

Advantages of Biopesticides

Some of the benefits of botanical and microbial biopesticides over synthetic pesticides include the following:

- Lack of polluting residues: The residues of biopesticides are not toxic to humans or other animals, and they can be applied even when a crop is almost ready for harvest. They are much less harmful to the ecosystem than the synthetic pesticides. The organisms (i.e. biocontrol agents) are essentially nontoxic and non pathogenic to wild life, humans, and other organisms not closely related to the target pest. They pose fewer risks than conventional pesticides. The safety offered by microbial insecticides is their greatest strength (Usta, 2013).
- High level of safety to non-target organisms: The toxic action of microbial insecticides (biopesticides) is often specific to a single group or species of insects. This specificity means that most biopesticides do not directly affect beneficial insects (including

predators or parasites of pests) in treated areas. Biopesticides generally affect only the target pest and closely related organisms, in contrast to broad spectrum conventional pesticides that may affect organisms as different as birds, insects and mammals. Biopesticides are usually inherently less toxic than conventional pesticides.

- In some cases, pathogenic microorganism can become established in a pest population or its habitat. They provide control during subsequent pest generations or seasons.

- Biopesticides help to reduce the risk of pest resistance. There is evolution of resistance in the pest population to most of the synthetic pesticides used in combating plant. Plant pests and diseases.

- Biopesticides often are effective in very small quantities and often decompose quickly, resulting in lower exposures and largely avoiding the pollution problems caused by conventional pesticides.

- When used as a component of Integrated Pest Management (IPM) programs, biopesticides can greatly reduce the use of conventional pesticides, while crop yields remain high.

- It requires less data and less time to register than conventional pesticides.

- Most microbial pesticides replicate in their target hosts and persist in the environment due to horizontal and vertical transmission, which may cause long-term suppression of pest populations even without repeating the application.

- The use of biopesticides is markedly safer for the environment and users, and more sustainable than the application of chemicals, hence their use as alternatives to chemical pesticides as components in Integrated Pest Management (IPM) strategies.

Biopesticides in Sustainable Agricultural Production

The availability of source materials of biopesticides makes them inexpensive to attain since they are found within the natural environment, and some of them are used for other purposes like food and feed. Biopesticides are safe products both for the applicant and the consumer since they have no toxicity (Damalas and Koutroubas, 2015). Biopesticides can therefore be suitably incorporated in integrated pest management (IPM), which helps to reduce the amount of chemical pesticides in management of crop pests. Natural products decompose quickly which makes them safer for use in the environment. Pesticides from natural sources have very short re-entry intervals which guarantee safety for the applicant. Biopesticides are also used in the decontamination of agricultural soils through introduction of important microbial species. They provide advantages as safe environment and healthy food for human consumption, however, there are factors that limit their full adoption as pest and disease management options (Ghorbani et al., 2005).

Biopesticides of botanical Origin

Botanical pesticides are first-generation pesticides being used in traditional agriculture for more than a century. These are higher plant origin pesticides which can directly or indirectly kill or reduce the target pest population. They serve as important alternatives to minimize the use of synthetic pesticides, because they possess an array of properties including toxicity to the pest, repellency, antifeedancy and insect growth regulatory activities against pests of agricultural importance.

They can either be plant extracts or essential oils. They are obtained from plants parts such as leaves, barks, flowers, roots, rhizomes, bulbs, seeds, cloves or fruits which are either fresh or dried. Dried plant parts are preferred as this reduces water concentration resulting in higher yield of active ingredient (Chougule and Andoji, 2016). The active compounds in plants include phenols, quinones, alkaloids, steroids, terpenes, alcohols and saponins (Mizubuti *et al.,* 2007). Different plant families have varied antimicrobial bioactive compounds which include oil components such as α- and β-phillandrene, limonene, camphor, linalool, β-caryophyellene and linalyl acetate depending on the plant family (Ali *et al.,* 2017). The most common and already commercialized botanical pesticides are derived from neem (*Azadirachta indica*), pyrethrum (*Chrysanthemum cinerariifolium*), sabadilla (*Schoenocaulon officinale*) and tobacco (*Nicotiana tabacum*).

Advantages of botanicals over synthetic pesticides

Bhagat *et al.*, 2014 and Prakash *et al.*, 2014 highlighted the advantages of botanicals over synthetic pesticides are as follows:

- Possess low mammalian toxicity and thus constitute least/no health hazards and environmental pollution
- Practically, no risk of developing pest resistance to these products, when used in natural forms
- Less hazards to non target organisms and pest resurgence
- Promote sustainable agriculture: It does not cause ill effect on the crop plants, soil health and environment. No adverse effect on plant growth, seed viability and cooking quality.
- Reduce crop losses: Several plant diseases/plant pathogens can be effectively managed by reducing disease incidence and related losses in the crop plants.
- Eco-friendly and organic farming: It is eco-friendly in nature, does not cause ecological imbalance and suitably fit in any agroecosystem. It suitably fits in the organic farming system.
- Biodegradable: It rapidly degrades under the exposure of sunlight.
- Integrated disease management: It can be suitably incorporated in the framework of integrated disease management.

- Cheaper and easily available: It is relatively cheaper than conventional chemical fungicides and easily available. Thousand seventy-five species of higher plants have been found to possess pesticidal property against insects, mites, nematodes, molluscs, birds and rodent pests of agricultural importance. Some of the botanicals like neem, bel, ocimum, senwar, pyrethrum, tobacco, karanj, mahua, cymbopogon and sweet flag have already attained the status of potential pesticides of plant origin against field pests including phytonematodes and plant mites and also against insect pests in storage ecosystems.

Microorganisms as Sources of Biopesticides

Microorganism-based biocontrol agents form the bulk of commercialized biopesticides and they include bacteria, viruses, fungi, nematodes and protozoa. Lengai and Muthomi, 2018 reported that up to 175 reported microbial based biopesticide active agents have been used in the management of pathogens, weeds, insects and nematodes. Majority of the microbial biopesticides are used to manage soil borne pathogens (Vinale *et al.*, 2008). Bacterial species that have been utilised as biopesticides include *Bacillus, Pseudomonas, Burkholderia, Xanthomonas, Enterobacter, Streptomyces, Serratia* and these are either obligate facultative or crystalliferous. Fungi used as biopesticides include species of *Trichoderma, Beauveria, Metarhizium, Paecilomyces, Fusarium, Pythium, Penicillim and Verticillium. Steinernama and Heterarhabditis* are nematode species used to make biopesticides (Kachhawa, 2017). The mechanisms of action exhibited by microorganisms against plant pathogens include hyperparasitism, competition, secretion of volatile compounds, antibiosis and parasitism. The rhizosphere is usually concentrated with various classes of important microorganisms. Other rich sources of microorganisms include hay, manure, cow shed, as well as straw.

Modes of Action of Biopesticides

Each type of biopesticide exhibits varied modes of action.

Microbial pesticides act on pathogens by antagonism, hyper parasitism, antibiosis and predation (Lengai and Muthomi, 2018). Hyperparasitism has been found to be one of the most reported modes of action on many biocontrol agents. The antagonist kills the pathogen or its propagules, while some attack the sclerotia or the hypha of the fungal pathogen. *Pasteuria penetrans* is a biocontrol agent that parasitizes on root-knot nematodes *of Meloidogyne spp.*, while species of genus *Trichoderma* exhibit predation mode of action by producing enzymes that directly kill cell walls of pathogens and colonize the environment therein.

By antibiosis mechanism, some microorganisms produce compounds that kill other microorganisms, mostly common with the following bacterial species: *Pseudomonas, Agrobacterium, Bacillus, Burkholderia, Pantoea* as well as fungus *Trichoderma spp.* Sufficient

25

quantities of antibiotics need to be produced for enhanced biocontrol. Some microbial species like *Bacillus cereus* produce multiple compounds that could suppress more than two pathogens making it to be effective in crop disease management.

Other microorganisms like *Lysobacter and Myxobacteria* produce lytic enzymes which hydrolyze compounds leading to suppression of pathogen. *Beauveria bassiana* inhibits chitin development in insects by conidia attaching to the body of insects. After germination, the hypha penetrates through the cuticle and grows throughout the insect body and eventually killing it (Prasad and Syed, 2010).

Botanical pesticides inhibit growth of pathogens, modify their cellular structures and morphology and exhibit neurotoxicity on insects. They also repel insects, suppress oviposition and feeding. Ngegba *et al* (2018) reported that extracts of neem (*Azadirachta indica*) and Mexican sunflower (*Tithonia diversifolia*) inhibited growth of rotting disease pathogens of tomato, *Aspergillus niger, Fusarium oxysporum and Geotrichum candidium* by up to 100%.

Semiochemicals such as female sex pheromones are used to lure the male insect pests which are then sterilized thereby decreasing their effectiveness. Upon mating with the sterile male insects, the females lay unfertilized eggs thereby reducing harmful insect populations. Some bioactive compounds cause partitioning of fungal cell membranes making them permeable leading to leakage of cell contents, while others cause separation of cytoplasmic membrane that leads to damage of the intracellular components and swelling of cells and eventual death. The compounds allicin in garlic (*Allium sativum*) bulbs cause suffocation of the pest due to effects on receptors of neurotransmitters (Baidoo and Mochiah, 2016).

Predators mainly kill the prey through parasitization or injection of toxic substances which eventually kill the prey. Natural enemies predate on insect pests which balances their population in the ecosystem. The mechanisms used by predators to lure insects include scents and other attractants. Some of these scents, called pheromones, have been commercialized and are being used in the management of important crop pests such as *Tuta absoluta*.

Conclusion

Synthetic pesticides are considered more effective than biopesticides in managing crop pests. Their effectiveness sometimes has nonetheless no much significance in managing a particular population of pest as would the biopesticides. Our world needs effective, environmentally smart agricultural technologies that are safe for people and protect our natural resources (Parliman, 2001). Biopesticides are revolutionizing farming practices around the world, improving productivity for organic crops, making conventional harvests safer, reducing the environmental impact of agriculture and ensuring that consumers are not ingesting chemicals on their food (Business Daily, 2013). It has been reported that biopesticides in other instances perform better than synthetic pesticides when applied in the right regimes, concentrations and appropriate frequencies (Shah *et al.*, 2013).

Sustainable farming starts with a healthy soil, which results in a healthy plant. The healthy soil concept is now being trumpeted everywhere and being adopted by conventional farmers. Biopesticides play a critical role in ensuring optimal soil health as the foundation for sustainable agriculture and food and production (Maksymiv, 2015). Conventional farmers also find benefits from the innovations in biopesticides, especially around harvest time and they are increasingly adopting biopesticides to eliminate synthetic chemical residues on the crops they grow. Farmers can spray biopesticides right up to harvest and then export without any residue issues (Damalas and Koutroubas, 2015). While traditional chemical pesticides often lose their efficiency as pests build up resistance, biopesticides have managed to thwart the natural ability of insects to adapt and develop resistance, leading to healthier crops, better food, wine and cannabis products for consumers and a cleaner environment (Plata-Rueda *et al.*, 2017).

Ragunath *et al.*, 2014 reported that Biopesticides is a key components of integrated pest management (IPM) programs, which is receiving much practical attention as a means to reduce the load of synthetic chemical products for controlling plant diseases. In most cropping systems, biological pesticides should not necessarily be viewed as wholesale replacements for chemical control of plant pests and diseases, but rather as a growing category of efficacious supplements that can be used as rotation agents to retard the onset of resistance to chemical pesticides and improve sustainability. In organic cropping systems, biopesticides can represent valuable tools that further supplement the rich collection of cultural practices that ensure against crop loss to diseases.

According to Lengai and Muthomi (2018), despite the many challenges facing the adoption of biopesticides, they still remain suitable alternatives to conventional pesticides. The use of synthetic chemicals has raised numerous concerns due to their negative effects on the environmental, human health, natural enemies and ecosystem balance. Some of the active ingredients of synthetic pesticides have been found to be carcinogenic thus posing a threat to human life. Biopesticides offer better alternative to synthetic pesticides due to their low toxicity, biodegradability and low persistence in the environment. The base materials for biopesticides are readily available and inexpensive. Data on toxicity levels, chemistry, active compounds and their compatibility with other methods of pests and disease management is needed to aid in formulation and commercialization. Globally, researchers have conducted studies on effectiveness of natural plant protection products with significant results being from *in vitro* experiments. Isman and Grieneisen, 2014 observed a rapidly growing publications on botanical insecticides, but much of the data is limited in its' reproducibility and thereby does not provide a basis for comparison with existing or future studies. Unfortunately, the studies do little to advance knowledge, except to add another species to the list of potentially useful plants. Lengai and Muthomi (2018) reported that there are studies on effectiveness of biopesticides under controlled environments and field conditions with varying results. Further research is therefore recommended to close the gaps in formulation of biopesticides. Stable products under field conditions will be a guarantee of utter effectiveness of biopesticides in crop pest management. Scientists and researchers should make greater efforts to investigate the utility of plant extracts for crop protection in field trials, in collaboration with local farmers, engineers in the government and industry because such studies should prove more valuable than laboratory-only studies, as well as providing stable, durable formulations of biopesticides.

Future Prospects

Kumar (2013) suggested that recombinant DNA technology being deployed for enhancing efficacy of biopesticides should continue, in addition to the continuous search for new biomolecules and improving of efficiency of known biopesticides. Also, fusion protein is being designed to develop next-generation biopesticides. This technology allows selected toxins (not toxic to higher animals) to be combined with a carrier protein which makes them toxic to insect pests when consumed orally, while they were effective only when injected into a prey organism by a predator (Fitches *et al.*, 2004). Several other innovative approaches are being applied to develop biopesticides as effective, efficient and acceptable pest control measure among the farmers and common man. Many biopesticides target a single pest species, but it is always desirable to have biopesticide that can control a range of pest species. Biological pesticides are expected to provide predictable performance, and they must do so in an economically viable manner for their better acceptability and adaptability.

Damalas, and Koutroubas (2018) recommended the co-operation between the public and private sectors to facilitate the development, manufacturing, and sale of this environmentally friendly alternative, as the discovery of new substances and research on formulation and delivery would boost commercialization and use of biopesticides. While new substances could serve as a promising option for use in pest control, more field research is required to assess the efficacy on specific pest problems in various cropping systems. Furthermore, microencapsulation based on nanotechnology could improve the residual action of biopesticides, and this could increase their field use.

References

Abbamondi, G.R., Giuseppina, T., Nele, W., Sofie, T., Wouter, S., Panagiotis, G., Carmine, I., Wesley, M.R., Barbara, N. and Jaco, V. (2016) Plant Growth-Promoting Effects of Rhizospheric and Endophytic Bacteria Associated with Different Tomato Cultivars and New Tomato Hybrids. Chemical and Biological Technologies in Agriculture, 1, 1-10.

Abd El-Ghani, M.M. (2016) Traditional medicinal plants of Nigeria: an overview. *Agric. and Biol. J. North America.* ISSN Online: 2151-7525, doi:10.5251/abjna.2016.7.5. 220.247.

Adejumo T.O. (2012) Evaluation of botanicals as biopesticides on the growth of *Fusarium verticillioides* causing rot diseases and fumonisin production of maize. *Journal of Microbiology and Antimicrobials.* 4 (1), 23-31.

Adejumo, T.O. and G. Langenkämper (2010) Potential of some botanicals as biopestides against *Fusarium verticillioides*, the causal agent of ear and stalk rot disease and fumonisin production in maize. 3rd Biopesticide Conference, CCS Agricultural University, Hisar, India. 20-22 Oct., 2010.

Akinyemi, B. (2000) Recent concept in plaque formation. *J. Clin. Pathol.* 30:13-16.

All, J. (2017), Insecticidal Transgenic Crops (PIP – Plant Incorporated Protectants). *UGA Extension Special Bulletin ed. Entomologist* 28, 193–195. UGA Extension, Georgia, USA.

Ali, A.M., Mohamed, D.S., Shaurub, E.H. and Elsayed, A.M. (2017) Antifeedant Activity and Some Biochemical Effects of Garlic and Lemon Essential Oils on *Spodoptera littoralis* (Boisduval) (Lepidoptera: Noctuidae). Journal of Entomology and Zoology Studies, 3, 1476-1482.

Alori, E.T. and Babalola, O.O. (2018) Microbial Inoculants for Improving Crop Quality and Human Health in Africa. *Front. Microbiol.* 9 (2213): 1-12.

Alori, E. T. and Fawole, O. B. (2017) "Microbial inoculants-assisted phytoremediation for sustainable soil management," in Phytoremediation: Management of Environmental Contaminants, Switzerland, eds A. A. Ansari, S. S. Gill, G. R. Lanza, and L. Newman (Berlin: Springer International Publishing), doi:10.1007/978-3-319-52381-1_1.

Baidoo, P.K. and Mochiah, M.B. (2016) Comparing the Effectiveness of Garlic (*Allium sativum* L.) and Hot Pepper (*Capsicum frutescens* L.) in the Management of the Major Pests of Cabbage *Brassica oleracea* (L.). Sustainable Agriculture Research, 2, 83-91. https://doi.org/10.5539/sar.v5n2p83.

Baker, B., Benbrook, C., Groth, E. III, and Benbrook, K. (2002) Pesticide residues in conventional IPM-growth and organic foods: insights from three data set. *Food Addit. Contam.*19,427–446.

Bhagat, S., A. Birah, R. Kumar, M. S. Yadav and C. Chattopadhyay (2014) Plant Disease Management: Prospects of Pesticides of Plant Origin. Chapter 7. 119-130. In: Advances in Plant Biopesticides, Dwijendra Singh (Ed.). 401pp. ISBN 978-81-322-2005-3 ISBN 978-81-322-2006-0 (eBook). DOI 10.1007/978-81-322-2006-0. Springer New Delhi Heidelberg New York Dordrecht London.

Boussel, P., Bonnemain, H., Bove, F.J. (1982) History of pharmacy and pharmaceutical industry. New York, NY: Thieme-Stratton, Inc.

Brent, K.J., Hollomon, D.W. (1998) Fungicide resistance: the assessment of risk. FRAC, Global Crop Protection Federation, Brussels 2:1 – 48.

Business Daily (2013) Chemical Ban Hits Vegetable Exports to the EU Market. 14 February 2013.

Canillac N, Mourey A (2001) Antibacterial activity of the essential oil of *Picea excelsa* on *Listeria*, *Staphylococcus aureus* and coliform bacteria. Food Microbiol 18:261–268.

Chougule, P.M. and Andoji, Y.S. (2016) Antifungal Activity of Some Common Medicinal Plant Extracts against Soil Borne Phytopathogenic Fungi *Fusarium oxysporum* Causing Wilt of Tomato. International Journal of Development Research, 3, 7030-7033.

Damalas, C.A. and Koutroubas, S.D. (2015) Farmers' Exposure to Pesticides: Toxicity Types and Ways of Prevention. Toxics, 1, 1-10.

Damalas, C.A. and Koutroubas, S.D. (2018) Current Status and Recent Developments in Biopesticide Use. Agriculture 8, 13: 2-6. doi:10.3390/agriculture8010013 www.mdpi.com/journal/agriculture

Dubey, S.C., Suresh, M., Singh, B. (2007) Evaluation of *Trichoderma* species against *Fusarium oxysporum* f.sp. *ciceris* for integrated management of chickpea. Biol Control 40:118–127.

Fitches E, Edwards MG, Mee C, Grishin E, Gatehouse AM, *et al.* (2004) Fusion proteins containing insect-specific toxins as pest control agents: snowdrop lectin delivers fused insecticidal spider venom toxin to insect haemolymph following oral ingestion. J Insect Physiol 50: 61–71.

Gasic, S. and Tanovic, B. (2013) Biopesticide Formulations, Possibility of Application and Future Trends. Journal Pesticides and Phytomedicine (Belgrade), 2, 97-102.

Ghorbani, R., Wilcockson, S. and Leifert, C. (2005) Alternative Treatments for Late Blight Control in Organic Potato: Antagonistic Micro-Organisms and Compost Extracts for Activity against *Phytophthora infestans*. Potato Research, 48, 181-189.

Hostellmann, K. and Marston, A. (2002) Twenty years of research into medicinal plants: results and perspectives. *Phytochem. Rev.* 1: 275-285.

http://www.fao.org/biotech/spec-term-n.asp?id_glo¼4875&id_lang¼TERMS_E.

https://www.epa.gov/ingredients-used-pesticide-products/what-are-biopesticides.

http://www.epa.gov/opp00001/biopesticides/

Isman, M.B. and M.L. Grieneisen (2014) Botanical insecticide research: many publications, limited useful data. Trends in Plant Science 19 (3): 140-145.

Kachhawa, D. (2017) Microorganisms as a Biopesticides. Journal of Entomology and Zoology Studies, 3, 468-473.

Kirby, G.C. (1996) Medicinal plants and the control of parasites. *Trans. Roy. Soc. Trop. Med. Hyg.* 90: 605609.

Kumar, R., Mishra, A.K., Dubey, N.K., Tripathi, Y.B. (2007) Evaluation of *Chenopodium ambrosioides* oil as a potential source of antifungal, anti-aflatoxigenic and antioxidant activity. Int J Food Microbiol 115:159–164.

Kumar, S. (2013) The Role of Biopesticides in Sustainably Feeding the Nine Billion Global Populations. J Biofertil Biopestici 4: e114. doi:10.4172/2155-6202.1000e114

Lengai, G. and Muthomi, J. (2018) Biopesticides and Their Role in Sustainable Agricultural Production. *Journal of Biosciences and Medicines*, **6**, 7-41. doi: 10.4236/jbm. 2018.66002.

Maksymiv, I. (2015) Pesticides: Benefits and Hazards. Journal of Vasyl Stefanyk Precarpathian National University, 1, 70-76.

Mizubuti, G.S.E., Junior, V.L. and Forbes, G.A. (2007) Management of Late Blight with Alternative Products. Pest Technology, 2, 106-116.

Nefzi, A., Abdallah, B.A.R., Jabnoun-Khiareddine, H., Saidiana-Medimagh, S., Haouala, R. and Danmi-Remadi, M. 2016) Antifungal Activity of Aqueous and Organic Extracts from *Withania somnifera* L. against *Fusarium oxysporum* f.sp. *radicis lycopersici*. Journal of Microbial and Biochemical Technology, 8, 144-150.

Newman, D.J., Cragg, G.M. (2007) Natural products as sources of new drugs over the last 25 years. *J. Nat. Prod.* 3:461–77. 3.

Newman, D.J., Cragg, G. and Snader, K.M. (2000) The influence of natural products upon drug discovery. *Nat. Prod. Rep.* 17: 175-285.

Ngegba, P.M., Kanneh, S.M., Bayon, M.S., Ndoko, E.J. and Musa, P.D. (2018) Fungicidal Effect of Three Plants Extracts in Control of Four Phytopathogenic Fungi of Tomato (*Lycopersicum esculentum* L.) Fruit Rot. International Journal of Environment, Agriculture and Biotechnology, 1, 112-117. https://doi.org/10.22161/ijeab/3.1.14.

Nicholson, G.M. (2007) Fighting the global pest problem: Preface to the special Toxicon issue on insecticidal toxins and their potential for insect pest control, Toxicon (49): 413–422.

Orole, O.O., Adejumo, T.O. and Orole, R.T. (2016) Antifungal Activity of Nine Medicinal Plants against *Aspergillus* species from Cocoa Beans (*Theobroma cacao*). *Journal of Agriculture and Ecology Research International* 7(2), 1-11.

Parliman, D.J. (2001) Soil Analyses for 1,3-Dichloropropene (1,3-DCP), Sodium N-Methyldithiocarbamate (Metam-Sodium), and Their Degradation Products near Fort Hall,

Idaho, September 1999 through March 2000. Water-Resources Investigations Report 01-4052. PAN., 2011.

Plata-Rueda, A., Martínez, L.C., Santos, M.H., Fernandes, F.L., Wilcken, C.F., Soares, M.A., Serrao, J.E. and Zanuncio, J.C. (2017) Insecticidal Activity of Garlic Essential Oil and Their Constituents against the Mealworm Beetle, Tenebrio molitor Linnaeus (Coleoptera: Tenebrionidae). Scientific Reports, 7, 46406.

Prakash, A., J. Rao, J. Berliner, S.S. Pokhare, T. Adak and K. Saikia (2014) Botanical Pesticides for the Management of Plant Nematode and Mite Pests Chapter 6 (Pages 89-118). In: Advances in Plant Biopesticides, Dwijendra Singh (Ed.). 401pp. ISBN 978-81-322-2005-3 ISBN 978-81-322-2006-0 (eBook). DOI 10.1007/978-81-322-2006-0. Springer New Delhi Heidelberg New York Dordrecht London.

Prasad, A. and Syed, N. (2010) Evaluating Prospects of Fungal Biopesticide *Beauveria bassiana* (Balsamo) against *Helicoverpa armigera* (Hubner): An Eco-Safe Strategy for Pesticidal Pollution. Asian Journal of Experimental Biological Sciences, 3, 596-601.

Ragunath, P.K., Abhinand, P.A., and Archanna, K.(2014) Relevance of Bioinformatics in Biopesticide Management: A Comparative Comprehensive Review. 345-356. In: K. Sahayaraj (Ed.), Basic and Applied Aspects of Biopesticides 384pp. ISBN 978-81-322-1876-0 ISBN 978-81-322-1877-7 (eBook). DOI 10.1007/978-81-322-1877-7 Springer New Delhi Heidelberg New York Dordrecht London.

Raskin, I., Ribnicky, D.M., Komarnytsky, S, Ilic, N., Poulev, A., Borisjuk, N., Brinker, A., Moreno, D.A., Ripoll, C., Yakoby, N., O'Neal, J.M., Cornwell, T., Pastor, I., Fridlender, B. (2002) Plants and human health in the twenty-first century. *Trends Biotechnol.* 20 (12):522–531.

Riaz, T., Khan, S.N. and Javaid, A. (2008) Antifungal Activity of Plant Extracts against *Fusarium oxysporum*—The Cause of Corm-Rot Disease of Gladiolus. Mycopath, 1&2, 13-15.

Robert, J.C. (1976) Pharmaceutical history. Roanoke, VA: Certified Medical Representatives Institute, Inc. 2.

Semeniuc, C.A., Pop, C.R. and Rotar, A.M. (2017) Antibacterial Activity and Interactions of Plant Essential Oil Combinations gainst Gram-Positive and Gram-Negative Bacteria. Journal of Food and Drug Analysis, 25, 403-408.

Sayre, L. (2009) The Hidden Link between Factory Farms and Human Illness. Finland, MN: Mother Earth News, Organic Consumers Association. Available at: http://www.motherearthnews.com/Natural-Health/Meat-Poultry-HealthRisk.aspx

Shah, J.A., Inayatullah, M., Sohail, K., Shah, S.F., Shah, S., Iqbal, T. and Usman, M. (2013) Efficacy of Botanical Extracts and a Chemical Pesticide against Tomato Fruit-Worm, *Helicoverpa armigera*. *Sarhad Journal of Agriculture* 1, 93-96.

Sharma K, Meshram NM (2006) Bioactivity of essential oils from *Acorus calamus* Linn. and *Syzygium aromaticum* Linn. against *Sitophilus oryzae* Linn. in stored wheat. Biopestic Int 2:144–152.

Sporleder, M. and Lacey, L.A. (2013) Biopesticides. Chapter 16 - *Insect Pests of Potato*. Pages 463-497. https://doi.org/10.1016/B978-0-12-386895-4.00016-8.

Suyal, D. C., Soni, R., Sai, S., and. Goel, R (2016) "Microbial inoculants as biofertilizer," in Microbial Inoculants in Sustainable Agricultural Productivity, ed. D.P. Singh (NewDelhi: Springer India), 311–318. doi:10.1007/978-81-3222647-5_18.

CHAPTER 2

Indigenous Traditional Knowledge: Application, Documentation, Advantages and Deficiencies

by

Olukayode Olugbenga Orole[1] and Timothy Olubisi Adejumo[2],

[1]Department of Microbiology, Federal University, Lafia, Nasarawa State, Nigeria.
[2]Dept. of Microbiology, Adekunle Ajasin University, P.M.B. 001,
Akungba-Akoko, Ondo State, Nigeria.
Corresponding Email*: orolekayode@gmail.com*

Abstract

"The acquisition and practice of beliefs, laws, passed down generational lines, and modified through experiences, observations, and information made indigenous traditional knowledge (ITK) important. ITK as it is popularly referred to has allowed continuity and survival of indigenous families and traditional societies providing means of beating the odd in farming, dealing with environmental challenges, maintaining diverse life around the local communities, and in economic life of the people. While ITK is local and peculiar to poor people living in a particular place in their day to day living, it is adopted in other places to make survival possible. It is presently confronted with myriad of challenges such as loss resulting from youth migration to urban centers, introduction of new technology, disrespect from people who are not the originators, and unacceptance because of it unscientific nature. It is believed that documentation will help place ITK in the right position and place it should be, and its adoption along with scientific knowledge will go a long way in helping to provide solutions to a host of challenges that modern man presently faces, as the local knowledge is rich covering every area of human endeavor.

Keywords: Indigenous knowledge, Sustainable development, Communities, Generations, Documentation

a) Introduction

Indigenous traditional knowledge (ITK) is acquired knowledge for a given people, society, or community from generation to generation. It is unique mostly to poor communities, and it is indigenous and local in nature originating naturally (Altieri, 1995). Mahapatro *et al.* (2017) defined ITK "as the assemblage of awareness and understanding of various facts which people have developed over a large span of time and continue to expand it". Accumulation of these understanding through practical experiences and observations help the indigent people to survive through time, as the accumulated knowledge affords them the skills for surviving in their local under-developed domains (Rajasekaran, 1993). The indigenous communities are characteristically poor and mostly share religious belief, kinship, songs, taboos and others. The unique nature of ITK is seen in its specificity to a particular community or society, repetitive though unfixed in attribute, community ownership guiding community decision making processes, and dynamic and reproducibility. ITK is attributed to poor communities of the world where the local indigenous knowledge shapes their understanding of agriculture, health care, wildlife and forestry management, biodervisity and preservation (Sharma, 2015; Anaeto *et al.*, 2013; Asiabaka, 2010; Warren, 1991). Pandey *et al.* (2017) opined that a flow of locally acquired knowledge is necessary for preservation, development, and sustainability of indigenous wisdom.

ITK has been variously described by terms like focal ecology, indigenous technical knowledge ethnology, indigenous knowledge, rural knowledge customary laws and knowledge of the land (Kyasiimire, 2010; Altieri, 1995). It shapes conservation and other day-to-day communal activities of the poor population. ITK is important because it helps in transmitting knowledge needed to sustain agricultural and the economic facet of life of the local society (Fernandez, 1994). The way of life of the indigenous people in managing their daily living is basically dependent on their belief, traditions, experiences, and acquired knowledge overtime. These understanding defines how they practice agriculture, conserve natural resources, practice traditional healing, and manage natural disasters issues so as to be able to survive. UNEP (2009) describes ITK as being local and indigent to a particular community, comprising accumulated knowledge resulting from acquired skills, belief and societal practices passed from generation to generation, transmission of which sharpens and fine-tunes it as a requirement for surviving and achieving a stable livelihood.

ITK originated from farmers, community leaders, elders, folklores, songs, myths, poetry, stories, languages, beliefs amongst others (Satapathy *et al.*, 2002). World Conservation Strategy of International Union in 1980 recognized the position of ITK, followed by the World Commission on Environment in 1987 and the United Nations conference on Environment and Education in 1992. The three bodies accorded ITK a position of recognition, as it afforded the traditional societies a united front as regards the management of the resources around them for survival. The bodies and others variously appreciated the contributions of ITK in societies and localities around the world. The indigenous knowledge is influenced by belief system, spirituality, brotherhood, experiences, wisdom of the people and their leaders which ultimately is dependent on the resources available to the community, so as to achieve and ensure quality minimal livelihoods for the survival and continuation of the local people. Passage of the knowledge from one generation to another refines and fine-tunes it to the extent that it becomes a code, a direction that guide their way of life (Pushpangadan *et al.*, 2002). Transmission of the knowledge is encouraged by its friendly, economic, and socially accepted nature, and suited to specific problem solving attributes (Pandey *et al.*, 2017; Sharma, 2015).

Traditional indigenous knowledge has over time generated a people who have been able to sustainably co-exist in harmony with nature, managing the limited available resources to advance their local development. Though, ITK is derogatorily characterized as the knowledge of the poor and the marginalized populations, it has found prospects where modern scientific knowledge is been found wanton, especially in the areas of sustainable agriculture and biodiversity management (Gorjestani, 2000). The knowledge while being indigent and local in nature is a part of the peoples live. These bodies of skills, tradition, beliefs, and culture cannot be separated from their life; it is a way of life passed from father to children and grandchildren. World Bank (1997) concluded that in being able to ensure survival of the local poor, ITK has greatly contributed and improved the global knowledge economy by granting assets to the traditional people for investment, food production, shelter provision, and other survival management skills, through the acquisition of local knowledge adapted from generation to generation, and modified by varied environmental conditions and cultural values.

The memory houses information and skills needed for survival hence dissemination and transmission of ITK would be through activities such as stories, songs, folklore, proverbs, dances, myths, cultural values, beliefs, rituals, community laws, local languages, agricultural practices, equipment, materials, plant species and animal breeds as noted by Grenier (1998). Continuity, passage, and dissemination of ITK is dependent on the interest and ability of young generation to acquire, store, memorise, trust, and practice the activities entrenched in such belief system and respect it (Atteh, 1989). Traditional indigenous knowledge has over time generated a people who have been able sustainably co-exist in harmony with nature, managing the limited available resources to advance their local development. In a study that investigated the nature and types of indigenous knowledge being used by rural women, the extent of use as well as the domains of use in sustainable development, Olatokun and Ayanbode (2008) revealed that the majority of the rural women were farmers and illiterates, but have vast knowledge of traditional medicine. There was an extensive use of oral ITK in various domains: culture transfer and preservation, food security, saving and lending money, population control, childcare and many other things, but its greatest impact was in the area of food production.

Characteristics peculiar to Indigenous traditional knowledge
Attributes peculiar to ITK defines its uniqueness and nature. These characteristics as defined by Warren (1991) and Raseroka (2002) are listed;

1. Specific to the local community where it is generated.

   ITK are generated through experiences, observations, local and spiritual teachings, facts common to people located in the same community. These accumulates and though slow in acquisition, systematically guides the people in their daily activities and sustainability for many generations and become specific to these people who are the originator of such knowledge. The knowledge is local and unique to a given culture or society

2. Expression of ITK is usually in the form of teaching though admonition, songs, folk-lore, stories, myths, legend, ritual, training, paintings, drawings and ceremonies. Other means of acquiring the knowledge are through personal experiences and observations of the dynamic environment.

3. Transmission and passage can be orally, imitation, or through demonstrations. For the knowledge to be transmitted from generation to another, it requires series of practices and repetition for mastery to become a way of life. Transmission of the knowledge is important as it focuses on different segment of the community and individual live stages such as in sustainable agricultural practices, economic decision taken, and management of ecological challenges. In transmitting or passing the knowledge from generation to generation, the locals make concerted effort to protect the knowledge from being hijacked, desecrated, and abused. Such knowledge is respected and kept away from visitors. Passage and transmission of ITK is from generation to generation, in many societies by word of mouth (Louise, 1998).

4. ITK is tacit in nature. It is difficult to transfer it from the local people who respect and believe in it to another person. It is difficult to transmit some of such knowledge through writing and by verbalizing because they are usually in languages and terms that are incomprehensible and meaningless to reasoning.

5. The knowledge is garnered through experiences and experimentations, thus they are not theoretical in nature.

6. ITK should be cost-effective, participatory, and sustainable, and environmentally friendly (Vanek, 1989; Hansen and Erbaugh, 1987).

7. It should be applicable for local-level decision-making in different areas of living in the rural communities.

8. ITK has value for the culture in which it originated.

The nature of ITK is that observations and experiences over long period of time are accumulated and fine-tuned so that a range of options the community considers the most adaptable and appropriate for its continuation and survival is transferred to the next generation. This is important because of changing environmental conditions to include ecology, value system, sociocultural belief of the people, and their relationships (Hammersmith, 2007). ITK is easy to adopt, friendly and dynamic; as it is an assemblage of the cultural practices and tradition, of the local communities appropriate and relevant in solving individual and societal problems of the local communities and help them to evolve.

The prospects of ITK is advancing better and productive living cannot be over flogged. Lately, many bodies have keyed into the need to respect and give relevance to ITK along with scientific knowledge to enhance better living for all, while tolerating the beliefs of traditional societies in helping to solve the many challenges we are faced with. ITK has great prospects in helping to better understand climatic events which will go a long way in aiding subsistence farmers, especially in developing countries which will greatly sustain the economy in such places (Lwoga et al., 2010; Hart and Mouton, 2007; Nyong et al., 2007).

Indigenous traditional knowledge is different from scientific knowledge. While ITK is a cumulative of series of experiences generated from generations and experimental observations made up of ideas, perceptions, and wisdom fine-tuned through trial and error method (Louise, 1998), unique to a people who respect and practice it (Aluma, 2010; Thrupp 1989). It is also empirical; Scientific knowledge is a systemic, theoretical, and analytical knowledge acquired through learning (Ellen and Harris, 1996). It is also objective and could be evaluated.

b) Some areas where traditional knowledge is applicable

1. Integrated Pest Management
Apart from the economic implications of pests to the traditional indigent man and his household, pest causes and transmit diseases, destroy crop yield, and can infest living and non-living host creating nuisance sight. To control and manage pest the people use various indigenous methods depending on the pest in question. Based on observation and experiences, the locals know what to use to control termite (neem), lice (soap), rats (traps), birds in the field and so on (Azoro et al., 2002; Okwuanga, 1994).

2. Biodiversity conservation
Differences are noticed between organisms at the species level within an ecosystem. The observed differences resulting from genetic variations creates a diversity in the level of organism to organisms and the environment. Type and species distribution of organisms varies within the designated environment, and based on ITK, management of ecological constituents of a community is appropriately undertaken to protect and conserve such

resources. ITK is unique to traditional people and location specific, hence conservation of natural resources and biodiversity of the living component of such locality is under the confines of the knowledge believed, respected, and practiced by the indigenous people. In this locations or territories, concentration of biodiversity is usually high and better pro-tected. Galloway McLean (2010) and Tauli-Corpuz *et al.* (2009) both agreed that protec-tion and conservation of biodiversity cannot be the exclusive right of government, alluding to the role and contributions of indigenous people at conserving biodiversity.

3. Environmental impact assessment

After years of accumulation of knowledge through experiences, observations, stories, songs, teachings, and spiritual beliefs, the traditional man can reliably describe the climate of his region. Based on visible indicators in the environment, they are better prepared for a possible climate change that can adversely alter their food production and sustainability of same. Observation of climate change and other environmental parameters can dictate type of farming system to engage in in the next planting season, decision to move from one place to another location to avert a looming natural disaster, type of crop to plant, type of clothing to wear which all serve to ensure safety of life and continuity from one gener-ation to another. ITK dictate decisions to take when environmental changes were observed, and also guide in the build-up to implementation of such decisions.

4. Animal husbandry and wildlife management

ITK has trained the local communities in the act of breeding animals, preparing traditional fodder and forage crops, and nomadic rearing of animals (Ghosh and Sahoo, 2011). The acquired knowledge is also handy in wildlife management of different species of animals. The belief that some animals are messengers and some spirit is respected.

5. Agriculture

Survival from generation to generation and migration from place to place are as a result of viable and practical traditional knowledge of the agricultural systems. This is employed in the production of food needed for sustenance of the traditional family and their community. Tables 1 and 2 show the plants used as biopesticides. Understanding of the environment by farmers, guides and dictates when to plant, apply manure, weed, apply insecticides, fungicides and when to harvest. They know the type of crop to plant on an arable land, and whether to do crop rotation, mixed cropping (Lwoga *et al.*, 2010; Nyota and Mapara, 2008). They have in-depth knowledge of animal husbandry, animal diseases, and the effect of climate change on their crops. This knowledge determines their farming practices, con-servation techniques, animal husbandry and management of agroforestry, which is from generation to generation. ITK as aided in the sustenance of generations of people who have

Table 1: Plant Species with biopesticidal potential.

S/No.	Common Name	Species	Family	Plant Parts
1.	Bead vine, Coral bead plant, Coral bean or Crab's eye	*Abrus precatorius L.*	Fabaceae	L, S
2.	Garlic	*Allium sativum L.*	Alliaceae	L
3.	Cashew	*Anacardium occidentale L.*	Anarcadiaceae	L
4.	African custard-apple, Wild custard apple or Wild soursop	*Annona senegalensis* Pers.	Asteraceae	S, B
5.	Sweet wormwood, Sweet annie, Sweet sagewort, Annual mugwort or Annual wormwood	*Artemisia annua L.*	Asteraceae	L, B
6.	Neem, Nimtree or Indian Lilac	*Azadirachta indica* A. Juss.	Meliaceae	L,B,R, F
7.	Desert date	*Balanites aegyptiaca* Linn Bel.	Zypophyllaceae	R
8.	Black-jack, Beggar-ticks, Cobbler's pegs	*Bidens pilosa L.*	Asteraceae	L
9.	Hemp, Grass, Hashish, Mary Jane pot, Marijuana	*Cannabis sativa L.*	Cannabaceae	L, S, F
10.	Chili pepper, bird pepper, tabasco pepper	*Capsicum frutescens L*	Solanaceae	F
11.	Pawpaw, Papaya	*Carica papaya L.*	Caricaceae	R, B
12.	Mums or Chrysanths	*Chrysanthemum coccineum* Wild.	Asteraceae	L, F
13.	Horsewood	*Clausena anisata*	Rutaceae	L, R
14.	Dalbergia	*Dalbergia saxatilis*	Fabaceae	L, B
15.	Pepper fruit	*Dannettia tripetala*	Annonaceae	L
16.	Tasmanian Blue Gum, Eurabbie, Blue Gum or Blue Eucalyptus	*Eucalyptus globules* Labill	Myrtaceae	L, B
17.	Gamhar, Gmelina, Gumhar, Malay beechwood, Malay bush beech, Malay bush-beech or Snapdragon	*Gmelina arborea* Juss.	Verbenaceae	L
18.	Chan, Chinese mint, Horehound, Hyptis or Mint weed	*Hyptis sauvcolens* Poit.	Labiatae	Shoot
19.	Barbados nut, Black vomit nut, Curcas bean or Physic nut	*Jatropha curcas L*	Euphorbiaceae	sap, F, S, B
20.	African mahogany, Dry zone mahogany	*Khaya senegalensis* A. Juss	Mcliaccac	S, B
21.	Plum mango	*Lannea acida*	Anacardiaceae	B
22.	Henna Plant	*Lawsonia inermis*	Lythraceae	L
23.	Chinaberry, Persian lilac, Ale-laila, paraiso	*Melia azadarach L*	Meliaceae	L, R, B
24.	Button grass	*Mitracarpus scaber* Zucc.	Rubiaceae	Shoot
25.	Tobacco	*Nicotiana tabacum L.*	Solanaceae	L
26.	African basil, Basilic, Basilic sauvage	*Ocimum gratissimum L.*	Liminaceae	L
27.	Clapperton's parkia	*Parkia clappentoniana* Keay.	Mimosaceae	S, B
28.	Belhambra, Packalacca	*Phytolacca dodecandra* L'Herit.	Phytolaceae	L, F
29.	West African black pepper, Ashanti pepper, Benin pepper	*Piper guineense* Schum &Thonn.	Piperaceae	F
30.	Camel's foot	*Piliostigma thonningii*	Caesalpiniaceae	R, B
31.	African mesquite, iron tree	*Prosopis africana* Linn.	Mimosaceae	S, B
32.	Chickenspike	*Spenoclea zeylanica* Gearth	Sphenocleaceae	Shoot
33.	Aztec marigold, chinchilla, Dwarf marigold, False marigold, Khaki bush, Khaki-bush, Little marigold, Marigold	*Tagetes minuta L.*	Asteraceae	L
34.	Fish bean, Fish-poison bean, or Vogel's tephrosia	*Tephrosa vogelii* Hook	Fabaceae	L
35.	Bitter leaf plant, little ironwood	*Vernovia amygdalina L*	Asteraceae	L

Source: Okwute, S.K. (2012). Slightly modified. L = Leaf; B = Bark; S = Seed; R = Root; F = Fruit

41

Table 2: Some examples of botanical insecticides and field pests of crops.

	Plant Name	Product/Trade Name	Group/Mode of Action	Targets
1.	*Lonchocarpus* spp. *Derris eliptica*	Rotenone	Insecticidal	Aphids, bean leaf beetle, cucumber beetles, leafhopper, red spider mite
2.	*Chrysanthemum cinerariaefolium*	Pyrethrum/Pyrethrins	Insecticidal	Crawling and flying insects such as cockroaches, ants, mosquitoes, termites
3.	*Nicotiana tabaccum*	Nicotine	Insecticidal Antifungal	Aphids, thrips, mites, bugs, fungus, gnat, leafhoppers
4.	*Azadirachta indica*	Azadirachtin/Neem Oil Neem cake Neem powder Bionimbecidine (GreenGold)	Repellent Antifeedant Nematocide Sterilant Anti-fungal	Nematodes, Sucking and Chewing Insects (Caterpillars, Aphids, Thrips, Maize Weevils)
5.	Citrus trees	d-Limonene Linalool	Contact poison	Fleas, Aphids, Mites, Paper Wasp, House Cricket.
6.	*Shoenocaulon officinale*	Sabadilla dust	Insecticidal	Bugs, blister beetles
7.	*Ryania speciosa*	Ryania	Insecticidal	Caterpillars, thrips, beetles, bugs, aphids
8.	Adenium *obesum (Heliotis sp)*	Chacals Baobab(Senegal)	Insecticidal	Cotton pests, particularly the larvae of ballworm

Source: Okwute, S.K. (2012)

been able to survive as a result of the capacity to meet their needs as a result of their understanding of the acquired knowledge passed down to them (Akullo, 2007). Based on the large amount of knowledge acquired and accumulated over the years, indigenous people have been able to meet their demand for food and meat supply efficiently. Knowledge of the time to clear land, sow, harvest, weed, type of farming method to engage in had greatly helped in maintaining the clan and the community. They fully understand what is needed to be done after a bumper harvest, and techniques to apply in the preservation of their crops and harvest (Sundaramari et al., 2011). Based on the acquired knowledge, the farmers can easily convert their excess harvest and other agricultural materials into value-added products for their use. Owolabi and Okunlola (2015) in his study established that farmers were aware of indigenous methods, and that majority of them utilized the knowledge in controlling pests and diseases in their cocoa farms, as they regard it as simple to use, affordable, cost effective, sustainable and compatible with their culture. Recently, Mobolade et al. (2019) reported that farmers use efficient post-harvest handling traditional storage containers that are comparatively cheap, eco-friendly and impart high shelf life to stored food grains, and that these could be applied in modern storage areas with minor modifications. The traditional wisdom and methods of storage can protect commodities from insect infestation for substantially longer periods.

6. Rural development and agroforestry

Understanding of the available resources within and around the local community is learnt from elders who have wealth of indigenous knowledge. Development activities such as building work, constructions of mills and bridges, canoe and other such works requiring use of wood and type of wood are supervised by them. They determine the type of wood and tools that best suit the work at hand. They are also adept at determining when to replace and how and when to do maintenance work (Manna et al., 2011).

7. Traditional medicine and health

The use of herbal plant materials to include leaves, stems, roots, flowers, bark of the plants date back to time immemorial. Traditional communities understand the importance of different herbal concoctions in the treatment of different ailments and infections. Based on acquired traditional knowledge, the quantity to apply, what should constitute the herbal remedy, dosage system, when to use, and associate spiritual connotation are transmitted from generation to generation. Traditional herbal remedies reduce cost of treatment, and presently herbal materials are compounded for use in the treatment of infections caused by resistant microorganisms. Traditional medicine is applied as a curative and preventive alternatives in the management of diseases.

8. Social-economic activities

The social life of the indigenous people is defined by their value system which is embedded in their traditions and believe in the afterlife. The community leaders, elders, and parents are majorly responsible for instilling discipline in their children and enforcing rules and regulations in the community. The community leaders are also managers and monitor activities in their traditional community setting, based on passed down spiritual guidelines. Trading and other economic activities are guided by rules and fear of the consequences of flouting community's regulation, so as to be able to relate with other communities with the intention of building stronger economic and protective ties, which can preserve and sustain the community and its available natural resources.

c) Loss of Indigenous Traditional Knowledge

Indigenous traditional knowledge is lost in various ways from the traditional societies from where they evolve (Lwoga et al., 2010). ITK of a people had been reported transferred to other communities where it had been manipulated and used without approval from the indigenous community. The loss and inadequate appreciation of ITK erodes the importance that would have been attached to the indigenous knowledge. Shrinkage of resource rich areas under indigenous communities, and increasing population with the attendant colonization of barren lands, coupled with deforestation and desertification result in the declining knowledge of the indigenous people. This calls for urgent need for application and integration of the knowledge and acceptance. Despite the diverse nature of indigenous knowledge in agriculture management as encompassing sowing, harvesting, crop rotation, storage and transportation, use of plant parts for the control of diseases and infections, animal breeding, fish farming, soil and water conservation, pesticide and insecticide use, management of natural resources, and maternal care which are all common to the indigenous and local communities is been lost.

The dynamic nature of traditional knowledge make it prone to losses in part or as a whole. Loss of ITK is reportedly due to adoption of the knowledge by people who are not the custodians or originators and introduction of foreign technologies. ITK of different communities and people with separate cultures from which the knowledge was transmitted are ill-suited to preserve the knowledge (Ellen and Harris, 1996, Agarwal, 2004). Such communities adopt the knowledge for the benefit it offers, while neglecting the spiritual and cultural attachments.

Migration from local communities to urban areas in order to secure better ways of life has continued to deal blows to ITK. While the productive youth population that are supposed to uphold the local belief system and ITK move to urban centres, the disadvantaged remaining elders die with their wealth of knowledge, thus creating memory lapses which leads to decline in accumulated knowledge. UNFPA (2007) reported that about half of the world's population now reside in urban centres particularly in developing countries, and projected that by 2030, over 78% of the people alive would be resident in cities where ITK will not be needed for survival (Kumar, 2010). Global warming and the changing natural environment, cultural changes, intrusion of foreign technologies, and disappearing local communities are all contributing to the loss of ITK (Ellen and Harris, 1996).

International bodies such as World Conservation Strategy of International Union and Conservations of Natural Resources, World Commission on Environment and Development, and United Nations Conference on Environment and Education variously appreciated the contributions of ITK in societies and localities around the world. The indigenous knowledge is influenced by belief system, spirituality, brotherhood, experiences, wisdom of the people and their leaders which ultimately is dependent on the resources available to the community so as to achieve and ensure quality minimal livelihoods for the survival and continuation of the local people (Rangnekar, 1994).

d) Documentation of Indigenous Tradition Knowledge
Indigenous traditional knowledge are not recorded until lately. Gathering of the scattered local knowledge and documenting them is important for sustenance and development of a particular community (Brokensha, 1990), and cannot be overstressed as the knowledge is a requisite for maintaining resources available to such a community, and for the provision

of needs necessary for the continual survival of the community. Documenting ITK negates that of scientific knowledge because i) large volume of the knowledge cannot be validated, ii) the most common and practised of the ITK are the one documented (Pandey *et al.*, 2017). Despite opposition to the knowledge advanced by traditional communities and local people as a result of the spiritual attachment such knowledge connotes and the unscientific reasoning associated with indigenous knowledge, acceptance and recognition of the ITK has increased over the time as a result of political or economic will of proponent of local knowledge.

Opposition recorded against the indigenous tradition knowledge was basically to the word "knowledge". The class of people who disagree with documenting ITK argued that the knowledge transmitted from many generation down was obtained from spiritual teaching, beliefs, and adherence to the same believing that religious and spiritual teachings are mostly unscientific and a threat to modern scientific knowledge as the knowledge could not be verified nor tested (Howard and Widdowson, 1997; Widdowson and Howard, 2002). Unfortunately, the recognition accorded ITK presently is because proponents after careful study believe that documentation will provide subtle, thorough, and more environmentally friendly ways of managing the limited available resources (Ellis, 2005; Parlee *et al.*, 2005; Berkes *et al.*, 2003).

WIPO (2017) define ITK documentation as primarily a process in which ITK is identified, collected, organized, registered or recorded in some way, as a means to dynamically maintain, manage, use, disseminate and/or protect ITK according to specific goals. Documentation of ITK is important as a result of the contributions recorded by local communities in the application of their individual community acquired knowledge in fostering food production (Tabuti and van Damme, 2012) and helping ensuring food security over time. To protect and encourage ITK, Lwoga *et al.* (2010) recommended documentation in permanent form that could be easily accessed such as in the form reports, inventories, maps, matrices and decision trees; audio-visuals such as photos, films, videos or audio cassettes as well as dramas, stories, songs, drawings, seasonal pattern charts, daily calendars (IIRR, 1996).

Other reasons adduced to protecting ITK are associated with the economic significance of such knowledge. ITK has in-built value system of respect that binds and make members of the communities loyal to the community, which inadvertently creates a sustaining economic front that promotes survival. The economic knowledge of the poor locals though simple, is reliable and could be tapped into in the areas of health, food production, and biodiversity conservation to mention a few, is laden with rich resource needed for further development (Dutfield, 2004). The capacity of ITK to ensure sustainable management of available resources to the traditional communities is never in doubt, hence survival from generation to generation. The knowledge is rich in a lot of areas such as in the management and conservation of natural environment to be amenable to the need of the people. Hamwey (2004) explained that these are made possible through respect for the beliefs, laws and rules guiding the communities which dictate the life style of the people.

A host of challenges and problems have cropped up with respect to documentation. Ngulube (2002) described the concern of issues related to methodology, access, intellectual property rights and the media and formats in which to preserve the knowledge as a topical one. Others are the problem of generational gap when there is no one to transfer the knowledge to or memories fail or in the event of death. Distorted and fragmented transmission of the knowledge is another problem documentation faces. These could result from gender dynamics, politics, power, culture, conflicts, resistance, religious beliefs and government policies (Mudege, 2005).

Conclusion

Indigenous traditional knowledge (ITK) has been viable in sustaining indigenous and local communities and societies for centuries, contributing fine-tuned, long-term understanding of the changing environment. The knowledge acquired from past generations have been able to economically and socially create close knit communities with values and beliefs from place to place, and country to country who are resilient and have been able to adapt to ecological and sociocultural environments they are presently boxed in by technological advancement. While ITK increases unity of purpose among the locals, it reduces and prevents inter-community conflicts which should be appreciated. Incorporating local and traditional acquired knowledge into present day scientific knowledge would go a long way in complementing modern scientific knowledge and make problem solving easier and faster.

ITK is disappearing as a result of urban migration and the adoption of modern lifestyles which presently does not respect nor appreciate the accumulation of wisdom, traditional practices and values which was passed down by several past generations. ITK over time enhances survival and longevity because it meets local needs and aspirations of the traditional people economic wise, hence the capacity of the people to build resilient communities. Appreciation and appropriate recognition of the traditional knowledge system to encompass community ownership and diversity of knowledge will help build societies that can better cope with the range of challenges that modern man is confronted with.

There is the need for well funded effective extension system and extensive empirical documentation of indigenous practices of controlling pests and diseases of crops by scientists, experts, Non-Governmental Organizations and relevant organizations to enhance utilization by future farmers. Such indigenous methods can be used in combination with scientific methods by farmers to promote sustainable agriculture.

References

Akullo, D., Kanzikwera, R., Birungi, P., Alum, W., Aliguma, L., & Barwogeza, M. (2007) Indigenous Knowledge in Agriculture: a case study of the challenges in sharing knowledge of past generations in a globalized context in Uganda.Durban, South Africa. http://WWW. ifla. Org/iv/ifla73/index. htm.

Altieri, M.A. (1995) Agroecology: The Science of Sustainable Agriculture. 2nd Edition.

Aluma, J.R. (2010) Integration of indigenous knowledge (IK) agriculture and health development processes in Uganda. 1-14.

Anaeto, F.C., Asiabaka, C.C., Nnadi, F.N., Aja, O.O., Ajaero, J.O., Ukpongson, M.A., Ugwoke, F.O. (2013) Integrating indigenous knowledge system in extension education: The potential for sustainable agricultural development in Nigeria. Research J Agricultural Engineering Management.;2:332-4

Asiabaka, C.C. (2010) Scaling up agricultural technologies for food security and poverty reduction: whose knowledge counts. The farmer or the scientist? Proc16th Inaugural Lec, Federal University of Technology, Owerri; 2:2-14.

Atteh, O.D. (1989) Indigenous local knowledge as key to local-level development: possibilities, constraints and planning issues in the context of Africa. Seminar on reviving local self-reliance: challenges for rural/regional development in Eastern and Southern Africa. Unpublished

Azoro, C., Okoro, C.C., Ijebor, J., Garba, M.I., Uduozue, I.M., Okonkwo, C.I., Idris, O.M. (2002) The modern engineer in society. Alpha Graphics Press, New Bern;86-104.

Berkes, F., Colding, J., Folke, C. (2003) Navigating social–ecological systems: building resilience for complexity and change. Cambridge University Press, Cambridge, UK

Brokensha, D. (1990) Indigenous knowledge system and development. Lanham, MD: University Press of America

Dutfield, G. (2004) Developing and Implementing National System for Protecting Traditional Knowledge: Experience in Selected Developing Countries (United Nations Conference on Trade and Development: Protecting and Promoting Traditional Knowledge: system, national experiences and international dimensions). New York and Geneva: United Nations, p. 141

Ellen, R., and Harris, H. (1996) Concepts of indigenous environmental knowledge in scientific and development studies literature – a critical assessment. Draft paper. East-West Environmental Linkages Network Workshop 3. International Development Research Centre, Canterbury

Ellis, S.C. (2005) Meaningful consideration? A review of traditional knowledge in environmental decision making. Arctic 58(1):66 – 77.

Fernandez, M.E. (1994) Gender and indigenous knowledge. Indigenous Knowledge Development Monitor.;2:6.

Galloway McLean, K. (2010) Advance Guard: Climate Change Impacts, Adaptation, Mitigation and Indigenous Peoples – A Compendium of Case Studies. United Nations University – Traditional Knowledge Initiative, Darwin, Australia. Available at: http://www. unutki.org/news.php?doc_id=101&news_id=92

Ghosh, K., and Sahoo, B. (2011) Indigenous traditional knowledge. Orissa Review; 65-70.

Gorjestani, N. (2000) Indigenous knowledge for development: Opportunities and challenges, Proc VIth UNCTA Science Congr, (Geneva); 4:7-11.

Grenier, L. (1998) Working with Indigenous Knowledge: A Guide to Researchers. Ottawa: Internation Development Research Center.

Hammersmith, J.A. (2007) Converging Indigenous and Western Knowledge Systems: Implications for Tertiary Education.Unpublished Doctoral Thesis. Pretoria: University of South Africa (UNISA)

Hamwey, R. (2004) Traditional Knowledge and the Environment: Statement by the United Nations environment program (United Nations Conference on Trade and Development: Protecting and Promoting Traditional Knowledge: system, national experiences and international dimensions). New York and Geneva: United Nations.

Hansen, D. O., and Erbaugh, J.M. (1987) "The Social Dimension of Natural Resources Management" in D. D. Southgate and J. F. Disinger (eds), Sustainable Resources Development in the Third World, Westview Press, Boulder, pp81-94.

Hart, T., and Mouton, J. (2007) Indigenous knowledge and its relevance for agriculture: a case study in Uganda. Indilinga: African Journal of Indigenous Knowledge Systems, 4(1), 249-

Howard, A., and Widdowson. F. (1997) "Traditional Knowledge Advocates Weave a Tangled Web", Policy Options, April: 46-48

International Institute of Rural Reconstruction (IIRR) (1996) Recording and using indigenous knowledge: a manual. International Institute of Rural Reconstruction, Silang, Carite, Philippines.

Kumar, K.A. (2010) Local Knowledge and Agricultural Sustainability: A Case Study of Pradhan Tribe in Adilabad District. Centre for Economic and Social Studies, 1-38

Kyasiimire, S. (2010) The role of indigenous knowledge in conservation of Uganda's National Parks (Bwindi). Kampala: Royal Geographical Society London: IT Publications.

Louise, G. (1998). Working with indigenous knowledge. A guide for researchers. International Development Research Centre (IDRC), Ottawa, Canada.

Lwoga, E. T., Ngulube, P., & Stilwell, C. (2010) Understanding indigenous knowledge: Bridging the knowledge gap through a knowledge creation model for agricultural development. SA Journal of Information Management, 12(1), 8-pages.

Mahapatro, G.K., Debajyoti, C., and Gautam, R.D. (April, 2017) Indian Indigenous Traditional Knowledge (ITK) on termites: Eco-friendly approaches to sustainable management. Indian Journal of Traditional Knowledge Vol. 16 (2), April 2017, pp. 333-340

Manna, K.R., Das, K.A., Krishna, D.R.S., Karthikeyan, M., Singh, D.N. (2011) Fishing crafts and gear in river Krishna. Indian J Traditional Knowledge; 2:491-97.

Mobolade A.J., N. Bunindrob, D. Sahoob, Y. Rajashekarb (2019) Traditional methods of food grains preservation and storage in Nigeria and India. Annals of Agricultural Sciences 64: 196–205.

Mudege, N. N. (2005), Knowledge production and dissemination in land resettlement areas in Zimbabwe: the case of Mupfurudzi, Ph.D. Thesis, Wageningen University, Wageningen, available at: http://library.wur.nl/wda/dissertations/dis3817.pdf (accessed 23 June 2009).

Ngulube, P. (2002) "Managing and Preserving indigenous knowledge in the knowledge management Era: Challenges and opportunities for information professionals. "Information Development18(2):95-102

Nyong, A.; Adesina, F.; Elasha, B.O. (2007) "The value of indigenous knowledge in climate change mit-igation and adaptation strategies in the African Sahel", in Mitigation and Adaptation Strategies for Global Change, Vol. 12, pp. 787–797.

Nyota, S., and J. Mapara. (2008) Shona Traditional children's games and play songs as indigenous ways of knowing, in I.M. Zulu (ed), Journal of Pan African Studies, Vol. 2, Number 4 pp 184-202. Online: http://www.jpanafrican.com South Africa (UNISA).

Okwuanga, J.A. (1994) Case study of indigenous knowledge in national and international development. Workshop on Indigenous Knowledge System in Agricultural Research and Extension for ADP Field Staff at Igbariam NRCRI Sub-station.

Okwute, S.K. (2012). Plants as Potential Sources of Pesticidal Agents: A Review, Pesticides - Advances in Chemical and Botanical Pesticides, Dr. R.P. Soundararajan (Ed.). Chapter 9, pp. 207–232, InTech, Rijeka, Croatia.

Olatokun, W.M., and O.F. Ayanbode (2008) Use of indigenous knowledge by rural women in the development of Ogun state. Indilinga– African J. Indigenous Knowledge Systems 7 (1): 47-63.

Owolabi, K.E. and J. O. Okunlola (2015) Farmers' Utilization of Indigenous Knowledge Techniques for the Control of Cocoa Pests and Diseases in Ekiti State, Nigeria. Asian Journal of Agricultural Extension, Economics & Sociology 4(3): 247-258.

Pandey, V., Mittal, R., and Sharma, P. (2017) Documentation and Application of Indigenous Traditional Knowledge (ITK) for Sustainable Agricultural Development. Asian Journal of Agricultural Extension, Economics & Sociology, 15(3): 1-9, Article no.AJAEES.31481

Parlee B, Berkes, F, Teetl'it Gwich'in Renewable Resources Council (2005) Health of the land, health of thepeople: a case study on Gwich'in berry harvesting from northern Canada. EcoHealth 2: 127-137.

Pushpangadan, P.S., Rajasekharan, George, V. (2002) Indigenous Knowledge and benefit sharing - A TBGRI experiment In IK strategies for Kerala, (NSE Publication, Thiruvananthapuram); 274-79.

Rajasekaran, B. (1993) A framework for incorporating indigenous knowledge system into development in India. PhD Thesis, Iowa State University, Ames, Iowa.

Rangnekar, S. (1994) Studies on knowledge possessed by women related to livestock production. Interaction; 12:103-11.

Raseroka, H.K. (2002) From Africa to the world- the globalisation of IKSs: setting the scene. In Snyman, R. (ed.) SCECSAL 2002: From Africa to the world- the globalisation of indigenous knowledge systems. Proceedings of the 15th Standing Conference of Eastern, Central and Southern African Library and Information Associations, 15-19 April, Caesars Gauteng Conference Centre, South Africa. Pretoria: LIASA: 1-5.

Satapathy, C.S., Veeraswami, Satapathy, B. (2002) Indigenous technical knowledge: Method of documentation and rationalization. Indian J Traditional Knowledge; 12:14-23.

Sharma, S. (2015) Indigenous use of medicinal plants for respiratory problems in Punjab. M.Sc. thesis, Punjab Agricultural University, Ludhiana; 2015

Sundaramari, M., Ganesh, S., Kannan, G.S., Seethalakshmi, M., Gopalsamy. (2011) Indigenous grain storage structures of South Tamil Nadu. Indian J Traditional Knowledge; 2:380-83.

Tabuti, J. R., & Van Damme, P. (2012) Review of indigenous knowledge in Uganda: Implications for its promotion. Afrika Focus, 25(1), 29-38.

Tauli-Corpuz, V., R. de Chavez, E. Baldo-Soriano, H. Magata, C. Golocan, M.V. Bugtong, L. Enkiwe-Abayao, and J. Cariño. (2009) Guide on Climate Change and Indigenous Peoples. Second Edition. Baguio City, Philippines: Tebtebba Foundation. Available at: http://www.tebtebba.org/index.php/content/160-2ndedition-of-guide-on-climate-change-and-indigenous-peoples-now-released

Thrupp, L. (1989) 'Legitimizing Local Knowledge: "Scientized Packages" orEmpowerment for Third World People', in D. Warren, J. Slikkerveer and S. Titilola (eds.) Indigenous Knowledge Systems: Implications for Agriculture andInternational Development, Studies in Technology and Social Change, No. 11.Ames, Iowa: Technology and Social Change Program, Iowa State University.

UNEP (2009) Indigenous Knowledge in Disaster Management in Africa. Nairobi, Kenya.

UNFPA (2007) State of World Population 2007; Unleashing the Potential of Urban Growth.

Vanek, E. (1989) "Improved Attitude towards Indigenous Knowledge Systems: The Case of the   World   Bank" in D. M. Warren; L. J. Slikkerveer and S. O. Titilola (eds), Indigenous Knowledge Systems: Implications for Agricultural and international Development, Studies in Technology and Social Change, No 11, Iowa State University, Iowa, pp162-170.

Warren, D.M. (1991) Indigenous knowledge systems and development. J Ext Sys.;4:45-56.

Widdowson, F., and Howard. A. (2002) "The Aboriginal Industry's New Clothes", Policy Options, March: 30-34.

World Bank, Knowledge and Learning Center (1997) Indigenous knowledge for development: A framework for action. Unpublished report.

World Intellectual Property Organization (WIPO) (2017) Documenting Traditional Knowledge – A Toolkit. WIPO: Geneva.

CHAPTER 3

Botanicals in Plant Disease Control

Victor I. Gwa
Department of Crop Production and Protection,
Faculty of Agriculture and Agricultural Technology,
Federal University Dutsin-Ma, PMB 5001,
Katsina State, Nigeria
Emails: *igwa@fudutsinma.edu.ng & gwavictor9@gmail.com*

Abstract

Botanical pesticides play an important role in the management of plant diseases, and serve as the most important alternative in the field of crop protection. Plant diseases develop as a result of interaction of susceptible crop, virulent pathogen and favorable environmental condition both in the field and in store. These diseases are controlled by botanical pesticides which contain more than thousands of biochemical compounds, believed to have different antimicrobial activities on various pathogens. Solvents used for the extraction of biologically active compounds are acetone, chloroform, ethanol, ether, methanol and water. The active compounds are extracted from different parts of the plant such as the bulbs, tubers, rhizomes, roots, leaves, stems, barks, seeds and flowers. Botanical pesticides are relatively cheap, target specific, environmentally friendly, biodegradable and does not induce resistance, compared to systemic pesticides. In view of this, botanical pesticides are therefore easily integrated with other plant disease management strategies to reduce the risks associated with the application of synthetic pesticides, to increase export quality by producing chemical-free crops. However, it is recommended that field trials be conducted to access the effectiveness of botanicals rather than only *in vitro* and green house experiments. In addition, studies on biosafety of botanical pesticides should be carried out to determine the level of toxicity on humans consuming the crops, animals feeding on them, as well as the quantity of residual chemicals in the harvested crop in order to provide food security at economically low cost without affecting nutrition, and to prolong postharvest shelf life of the produce.

Keywords: Botanical pesticides; Plant diseases; Antimicrobial; Food security; Shelf life

Introduction

The use of chemical control methods in the control of plant diseases caused by fungal pathogens and other plant pathogens may considerably reduce the effect of plant diseases, and may also be more available and quick to apply, but application of chemical pesticides may not always be the most acceptable method. Though, different types of synthetic fungicides have been used to control different types of plant diseases in different crops which have resulted to increased production and self sufficiency; there are major disadvantages with the use of chemicals such as pollution of the ground and surface water, soil, atmosphere as well as development of resistance by the pathogens (Kumar *et al.,* 2007; Malkhan *et al.,* 2012). According to Dubey *et al.* (2007) and Kumar *et al.* (2007), the use of synthetic chemicals to control post harvest pathogens biodeterioration has been restricted due to their carcinogenicity, teratogenecity, high and acute residual toxicity, hormonal imbalance, long degradation period, environmental pollution and their adverse effects on food and side effects on humans. There is also another problem of chemical residues or food poisoning in foods or crop produce harvested. The problem has necessitated researchers to develop alternatives solutions to chemical fungicides to combat diseases both in the field and after harvest, to reduce bioaccumulation and bioconcentration of toxic compounds in our food and crop produce in order to make them safer for consumption. Food safety is an important area of concern to consumers since the nature of food consumed has direct effect on the health of the individual. Improving food safety will therefore increase food security as more population will have access to quality and sufficient food, devoid of toxic chemical compounds. This necessitated the use of biological control based methods with the use of botanicals collected from different parts of the plant such as leaves, stems, fruits, seeds, roots and barks which have been proved to be environmentally safe, economically cheap and ecologically biodegradable (Gwa and Nwankiti 2017b; Gwa *et al.,* 2017; Nwankiti and Gwa, 2018). In addition, botanicals are natural in origin and have little or no adverse effect on plant physiology, and have minimum adverse effects on the physiological processes of plants and can be easily converted into ecologically phytochemical compounds. Some of these plants are known to synthesize phytochemical compounds with antimicrobial potencies and are used successfully in the control of diseases, both in humans and in crops such as yam, cassava, tomato, cowpea, rice, etc. (Bediakao *et al.,* 2007; Okigbo, *et al.,* 2009a and b; Okigbo *et al.,* 2010; Gwa and Nwankiti, 2017b; Sani and Gwa, 2018). There are different plant products such as plant extracts, gums, essential oils and resins etc that have been shown to have biological activity against phytopathogenic fungi both *in vitro* and *in vivo,* and can be used as botanical pesticides (Romanazzi, *et al.,* 2012). Plants have ability to synthesize aromatic secondary metabolites like phenols, phenolic acids, quinones, flavones, flavonoids, flavonols, tannins and coumarins (Chima, 2012; Aidah *et al.,* 2014). Botanical pesticides are natural plant products that are specific in mode of action, show limited field persistence, target specific and have a shorter shelf life, nontoxic to antagonistic microorganisms with no residual accumulation. Research has it that about 250,000 higher plant species have untapped reservoir of bioactive chemical compounds with many potential application as agrochemical and pharmaceuticals (Cowan, 1999). Studies conducted by different researchers in different parts of the world approved the potency of different plant products for the control of plant pathogenic fungal growth and mycotoxins production. Examples include *Allium sativum* and *Nicotiana tabacum* (Ijato, 2011); *Moringa oleifera, Vernonia amygdalina* (Mamkaa and Gwa, 2018), *Azadirachta indica, Zingiber officinale* (Sani and Gwa, 2018); *A. indica, P. guineense, Z. officinale, C. papaya* and *N. tabacum* (Gwa and Akombo, 2016); *A. indica, N. tabacum, Z. officinale* (Gwa *et al.*, 2018a). It is therefore very important to develop pesticides that are safe and sustainable in the environment, and the use of botanical pesticides is one of such methods especially in developing countries that have little or no resources in combating health challenges as a result of consumption of crops produced from synthetic chemicals.

Pesticides and Botanical Pesticides

The definition of pesticides may be looked at in different ways, depending on the on what the researcher has in mind. The most standard definition is the one given by Food and Agriculture Organization (FAO) of the United Nations (2002), which defined a pesticide as 'any substance or mixture of substances intended for preventing, destroying, or controlling any pest including vectors of human or animal diseases, unwanted species of plants or animals causing harm during, or otherwise interfering with the production, processing, storage, or marketing of food, agricultural commodities, wood and wood products, or animal feedstuffs, or which may be administered to animals for the control of insects, arachnids or other pests in or on their bodies'. In other words, a pesticide is any chemical substance designed to kill or retard the growth of pests that damage or interfere with the growth of crops, shrubs, trees, timber and other vegetation desired by humans. A pesticide may also be defined as any chemical substance or biological agent that interferes with the growth pattern as well as or physiological system of pests or kills the pest entirely, thus reducing the attack below economic injury level or preventing the attack of the agricultural crops or commodities completely, thereby increasing crop yield and quality and preventing economic losses.

Some plants contain chemical compounds that are toxic to insects and disease causing organisms when extracted from the plants and applied on infected crops. These insect and pathogen killer compounds derived from plants are called botanicals or botanical pesticides. Botanical pesticides are also chemicals that occur in natural and are extracted from plants. Botanical pesticides are highly degradable by micro organisms in the soil and are good alternatives to chemicals. This is because they are safer to the applicator, economically cheap, environmentally friendly to the soil and target specific. Botanicals are considered better alternative because they are easily degraded into harmless chemical compounds within few hours or days.

Mode of Action of Pesticides

Mode of action of a pesticide is defined as the specific biochemical interaction through which a pesticide produces its effect. This may include the protein, specific enzyme, or biological step affected. Some classification may be targeted at physical characteristics, chemical composition, or pest controlled. Mode of action refers to the specific biological process the pesticide interrupts (Bloomquist *et al.*, 2008). The knowledge of the mode of action (how pesticides work) of pesticide is important to enhance quality and maintain sustainability of pesticides. It is therefore very important to understand how the pests' targeted system works. Similarly, the modes of action of the pesticides to control the pest are very important in order to avoid development of resistance by the targeted pests. The development of resistance by pests can be traced to the use of pesticides with same mode of action which kills only the susceptible pests and leave those that are resistance to the entire class of pesticides with same mechanism of action. (Brown, 2005).

Classification of Pesticides

According to Chengala and Singh (2017) pesticides are generally classified based on:

1. **Target pests:** Insecticides, Herbicides, Fungicides, Rodenticides, Fumigants, Bactericides, Acarioides.
2. **Mode of action**: Systemic Pesticides and Non Systemic (contact) Pesticides
3. **Chemical nature**: Organochlorines, Organophosphates, Carbamates, Pyrethrins and Pyrethroids

The universally accepted classification of pesticides is the one given by Council on Scientific Affairs (1997) which classifies pesticides into two groups namely:

- **Synthetic pesticides:** Organochlorines, Organophosphates, Carbamates and Pyrethrins
- **Biopesticides:** Microbial Biopesticides, Plant Incorporated Protectants, Botanical Pesticides and Pheromones. Commercially available botanical pesticides are classified as follows:

Botanical pesticides: Neem based pesticides, rotenone, pyrethrum and eucalyptus essential oil.

Sources of Botanical Pesticides

Studies conducted in different parts of the world have shown that many plants possess pesticidal activities and various researchers have confirmed the potency of the different parts of plant. These plants are either used in aqueous or dried forms. They are dried under the sun to a constant weight before grinding into powder (Plates 1-4). Most of the parts of plant used include the underground parts such as bulbs, tubers, rhizomes, roots and so on compared with the other parts of the plants above ground such as the leaves, stems, barks, seeds, flowers for their phytochemical compounds which possess antimicrobial activities. There are different types of botanicals or botanical pesticides which are used in the control of different plant diseases (Table 1). Botanicals such as *A. indica, Z. officinale A. sativum, A. cepa, O. gratissimum* etc are some of the most commonly used botanical pesticides used in the management of plant diseases all over the world. These botanicals may be combined in one way or the other in the management of different plant diseases in the concept of integrated pest management (IPM) programs, which often times are generally safe for human handling and friendly to the environment compare to the hazardous synthetically produced chemical pesticides, based on the needs of researchers in order to achieve a better result.

Plate 1: Seeds of *Piper nigrum*

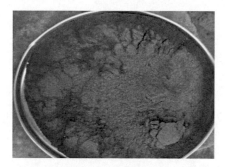

Plate 2: Ground seeds of *P. nigrum*

Plate 3: Ground leaves of *Azadirachta indica*

Plate 4: Ground leaves *Nicotiana tabacum*

Table 1: List of some botanical pesticides known to control plant diseases

Plant	Part used	Preparation	Pathogen/Diseases Controlled	Reference
Azadirachta Indica	Leaves	Aqueous	*Colletotrichum lindmuthianum*/anthracnose disease of cowpea	Amadioha and Obi (1998)
	Leaves	Aqueous	*Aspergillus flavus*/postharvest dry rot disease of yam tuber	Gwa and Akombo (2016)
	Leaves	Aqueous	*A. niger*/postharvest dry rot disease of yam tuber	Gwa and Ekefan (2018)
	Leaves	Aqueous	*Colletotrichum sp*/anthracnose disease of yam	Gwa and Nwankiti (2017b)
	Leave	Aqueous	*Fusarium oxysporum*/Plant pathogens causing disease	Hadian (2012)
	Leaves	Aqueous	*Rhizoctonia solani*/Plant pathogens causing disease	Hadian (2012)
	Leaves	Ethanol	*A. niger, A. flavus, Botrytis cenerea, Colletotrichum capsici, Phytophthora capsici*/various diseases of pepper	Zakari *et al.* (2015)
	Leaves	Aqueous	*Colletotrichum capsici, Cercospora capsici, Phytophthora capsici*/anthracnose disease of pepper	Sopialena *et al.* (2018)
	Seeds	Aqueous	*F. oxysporum, Rhizoctonia solani*/ postharvest diseases of tomato	Sani and Gwa (2018)
	Leaves	Aqueous	*Botryodiplodia theobromae*/postharvest dry rot disease of yam	Gwa *et al.* (2017)
	Leaves	Aqueous	*Penicillium expansum*/ postharvest dry rot disease of yam	Gwa *et al.* (2018a)
	Leaves	Aqueous and ethanol	*A. niger, F. oxysporum, R. stolonifer, Geotrichum candidium*/postharvest transit rot of tomato	Ijato *et al.* (2010)
	Seed	Oil	*Colletotrichum destructivum*/anthracnose disease of cowpea	Obi and Barriuso-Vargas (2013)
	Leaves	Aqueous	*Aspergillus ochraceous*/ postharvest dry rot disease of yam	Gwa *et al.* (2018b)
	Leaves	Aqueous	*A. solani*/ postharvest rot disease of yam	Suleiman (2010)
	Leaves	Aqueous	*Pyricularia grisea*/ rice blast disease	Hubert *et al.* (2015)
	Leaves	Aqucous	*F. oxysporum*/ postharvest dry rot disease of yam	Gwa (2018)
	Leaves	Aqueous and Ethanol	*A. niger, B. theobromae, F. solani, P. oxalicum*/postharvest rot disease of cassava	Okigbo *et al.* (2009a)
	Leaves	Water and Ethanol	*B. theobromae*/ postharvest rot disease of yam tuber	Nweke (2015)

	Leaves	Aqueous	*F. oxysporum, A. niger, Rhizopus sto-lonifer, P. oxalicum*/postharvest dry rot disease of yam	Taiga (2011)
	Seeds	Oil	*C. capsici/* anthracnose disease of pepper fruit	Jeyalakshmi and Seetharaman (1998)
	Seeds	Aqueous	*R. stolonifer/* wet rot disease of sweet potato	Tijjani *et al.* (2013)
	Seeds	Oil	*R. stolonifer/* postharvest wet rot of yam	Nahunnaro (2008)
Acacia nilotica	Leaves	Aqueous and Ethanol	*C. gloeosporiorides, C. orbiculare, Cladosporium sp, Curvularia sp, Fusarium sp, Pestalotiosis sp/* postharvest fungal diseases of *Piper nigrum*	Aisha (2013)
Acorus calamus	Leaves	Aqucous	*C. capsici/* anthracnose disease of pepper chilli fruits	Jeyalakshmi and Seetharaman (1998), Korpraditskul *et al.* (1999)
Aframomum melegueta	Seeds	Aqueous and Ethanol	*A. niger, B. theobromae, F. solani, P. oxalicum*/postharvest disease of cassava	Okigbo *et al.* (2009a)
Allium sativum	Bulb, Leaves	Aqueous	*Pyricularia grisea*/rice blast disease	Hubert *et al.* (2015)
	Bulb	Aqueous	*A. niger, A. flavus, F.solani, Geotricum candidum, Rhizoctonia carotae, F. oxysporum/* various postharvest diseases potato and carrot	Amaeze *et al.* (2013)
	Bulb	Aqueous and Ethanol	*A. niger, A flavus, F. oxysporum, F. solani, R. stolonifer, B. theobromae/* various postharvest rots diseases of yam	Ijato *et al.* (2011)
	Bulb	Aqueous	*F. oxysporum, R. solani/ Plant pathogens causing disease*	Hadian (2012)
Aloe barbadensis	Leaves	Aqueous	*F. oxysporuum, A. niger, R. stolonifer, P. oxalicum*/postharvest dry rot disease of yam	Taiga (2011)
Aloe vera	Bulb	Aqueous	*Pyricularia grisea*/rice blast disease	Hubert *et al.*(2015)
Alipinia galangal	Leaves, bark, Stems	Aqueous and Ethanol	*C. gloeosporiorides, C. orbiculare, Cladosporium sp,Curvularia sp, Fusarium sp, Pestalotiosis sp/* postharvest fungal diseases of *Piper nigrum*	Aisha (2013)
Anacardium occidentale	Leaves	Aqueous and Ethanol	*B. theobromae*/postharvest yam tuber rot	Nweke (2015)
Annona muricata		Aqueous	*Pyricularia grisea* /rice blast disease	Hubert *et al.* (2015)

59

Argemone mexicana	Leaves, Bark	Aqueous and Ethanol	*C. gloeosporiorides, C. orbiculare, Cladosporium sp,Curvularia sp, Fusarium sp, Pestalotiosis sp /* postharvest fungal diseases of *Piper nigrum*	Aisha (2013)
Bidens pilosa	Leaves	Aqueous	*Pyricularia grisea/*rice blast disease	Hubert *et al.* (2015)
Carica papaya	Leaves	Aqueous	*A. solani /*postharvest rot disease of yam	Suleiman (2010)
	Leaves	Aqueous	*R. nigricans, Mucor circinelloides/* soft rot of yam	Chima (2012)
	Leaves	Aqueous	*E. coli, S. aureus, C. albicans/* bacteria diseases	Subramanian *et al.* (2014)
	Leaves	Aqueous	*A. flavus/* postharvest dry rot disease of yam	Gwa and Akombo (2016)
	Leaves	Aqueous	*Colletotrichum sp/* postharvest rot of yam	Gwa and Nwankiti (2017b)
	Leaves	Aqueous	*B. theobromae/* postharvest dry rot disease of yam	Gwa *et al.* (2017)
	Leaves	Aqueous	*F. oxysporum/* postharvest dry rot disease of yam	Gwa (2018)
	Leaves	Aqueous	*A. niger /* postharvest dry rot disease of yam	Gwa and Ekefan (2018)
	Leaves	Aqueous	*P. expansum /* postharvest dry rot disease of yam	Gwa *et al.* (2018a)
	Leaves	Aqueous	*A. ochraceous /* postharvest dry rot disease of yam	Gwa *et al.* (2018b)
Chromolena odorata	Leaves	Aqueous and Ethanol	*Ceracystosis fimbriata, B. theobromae, R. stolonifer, F. solani/* postharvest rot disease of sweet potato	Anukwuorji *et al.* (2013)
	Leaves	Aqueous and Ethanol	*A. niger, F. oxysporum, R. stolonifer,Geotrichum canditum/* postharvest rot disease of tomato	Ijato *et al.* (2010)
Cymbopogon citrates	Leaves	Aqueous	*Alternaria spp/* leaf spot disease	Shafique *et al.* (2012)
	Leaves	Aqueous and Ethanol	*Ceracystosis fimbriata, B. theobromae, R. stolonifer, F. solani/* postharvest rot disease of sweet potato	Anukwuorji *et al.* (2013)
	Leaves	Aqueous	*C. destructivum/* anthracnose disease of cowpea	Obi and Barrius-Varga (2013)
Cymbopogon martini	Leaves	Aqueous	*C. capsici/* Anthracnose disease of chilli pepper	Jeyalakshmi and Seetharaman (1998)
Caesalpinia coriaria	Leaves, bark, stems	Aqueous and Ethanol	*C. gloeosporiorides, C. orbiculare, Cladosporium sp, Curvularia sp, Fusarium sp, Pestalotiosis sp/* postharvest fungal diseases *of Piper nigrum*	Aisha (2013)

60

Citrus aurantifolia	Leaves	Aqueous and Ethanol	*B. theobromae/* postharvest rot of yam tuber	Nweke (2015)
Camellia sinensis	Leaves	Aqueous	*Pyricularia grisea/* rice blast disease	Hubert *et al.* (2015)
Chrysanthemum coccineum	Leaves	Aqueous	*Pyricularia grisea/* rice blast disease	Hubert *et al.* (2015)
Coffee Arabica	Leaves	Aqueous	*Pyricularia grisea/* rice blast disease	Hubert *et al.* (2015)
Cassia alata	Leaves	Aqueous and Ethanol	*B. theobromae/* postharvest rot of yam tuber	Nweke (2015)
Colocasia esculenta	Leaves	Aqueous	*A. niger, B. theobromae/* various postharvest diseases of crops	John and Gideon (2018)
Datura stamonium	Leaf	Aqueous	*Pyricularia grisea/* rice blast disease	Hubert *et al.* (2015)
Dioscorea alata	Peel	Ethanol and Methanol	*F. oxysporum, R. stolonifer, B. theobromae, T. viride/* postharvest rot disease of yam	Okigbo *et al.* (2015)
D. rotundata	Leaf	Aqueous	*A. niger, B. theobromae/* various postharvest diseases of crops	John and Gideon (2018)
Emblica officinalis	Rhizome	Aqueous and Ethanol	*C. gloeosporiorides, C. orbiculare, Cladosporium sp, Curvularia sp, Fusarium sp, Pestalotiosis sp/* postharvest fungal diseases of *Piper nigrum*	Aisha (2013)
Euphorbia tirucalli	Leaves, stems	Aqueous and Ethanol	*C. gloeosporiorides, C. orbiculare, Cladosporium sp, Curvularia sp, Fusarium sp, Pestalotiosis sp/* postharvest fungal diseases of *Piper nigrum*	Aisha (2013)
Lantana camara	Leaf	Aqueous	*A. flavus*	Tijjani *et al.* (2013)
Manihot esculenta	Leaf	Aqueous	*A. niger, B. theobromae/* various postharvest diseases of crops	John and Gideon (2018)
Musa paradisiacal	Peel	Ash	*R. stolonifer/* postharvest rot disease of yam	Nahunnaro (2008)
Moringa oleifera	Seeds	Aqueous	*A. niger, B. theobromae/* postharvest rot of cowpea seed	Mamkaa and Gwa (2018)
	Seeds	Aqueous	*R. stolinifer/* wet rot disease of sweet potato	Tijjani *et al.* (2013)
	Seed	Aqueous	*A. flavus*	Tijjani *et al.* (2014)
Nicotiana tabacum	Leaves	Aqueous	*A. flavus/* postharvest dry rot disease of yam	Gwa and Akombo (2016)
	Leaves	Aqueous	*Colletotrichum sp/* postharvest dry rot disease of yam	Gwa and Nwankiti (2017b)
	Leaves	Aqueous	*B. theobromae /* postharvest dry rot disease of yam	Gwa *et al.* (2017)
	Leaves	Aqueous	*F. oxysporum/* postharvest dry rot disease of yam	Gwa (2018)
	Leaves	Aqueous	*A. niger/* postharvest dry rot disease of yam	Gwa and Ekefan (2018)
	Leaves	Aqueous	*P. expansum/* postharvest dry rot disease of yam	Gwa *et al.* (2018a)
	Leaves	Aqueous	*A. ochraceous/* postharvest dry rot disease of yam	Gwa *et al.* (2018b)

61

	Leaf	Aqueous	*F. oxysporum, A. niger, R. stolonifer, P. oxalicum/* postharvest dry rot disease of yam	Taiga (2011)
	Leaf	Aqueous/ Ethanol	*F. oxysporum, A. niger, R. stolonifer, F. solani, A. flavus, B. theobromae/* postharvest rot disease of yam	Ijato (2011)
	Leaf	Aqueous	*Pyricularia grisea/* rice blast disease	Hubert *et al.* (2015)
	Leaf	Ethanol, ethylacetate, Hexane, Acetone, Butanol, Aqueous	*Erwinia carotovora, Escherichia coli, Pseudomonas aeroginosa, staphylococcus aureus, salmonella typhi, Agrobacterium tumefaciens, Bacillus cereus/* control of bacteria diseases	Jehan Bakht *et al.* (2012)
Ocimum sanctum	Leaf	Aqueous	*C. capsici/* fruit rot of mango	Korpraditskul *et al.* (1999)
Ocimum gratissimum	Leaf	Aqueous, Ethanol	*Ceracystosis fimbriata, F. solani, R. stolonifer, B. theobromae/* postharvest rot disease of sweet potato	Anukwuorji *et al.* (2013)
	Leaf	Aqueous, Ethanol	*A. niger, F. oxysporum, F. solani, R. stolonifer, B. theobromae, A. flavus/* postharvest dry rot disease of yam	Ijato (2011)
	Leaf	Aqueous	*C. destructivum/* anthracnose disease of cowpea	Obi and Barriuso-Vargas (2013)
Piper nigrum	Seed	Aqueous	*A. flavus/* postharvest dry rot disease of yam	Gwa and Akombo (2016)
	Seed	Aqueous	*Colletotrichum sp/* postharvest dry rot disease of yam	Gwa and Nwankiti (2017a)
	Seed	Aqueous	*B. theobromae* / postharvest dry rot disease of yam	Gwa *et al.* (2017)
	Seed	Aqueous	*F. oxysporum/* postharvest dry rot disease of yam	Gwa (2018)
	Seed	Aqueous	*Staphylococcus aureus, B. subtilis, P. aeruginosa, E. coli, A. alternata, A. niger, A. flavus, F. oxysporum/* various pathogens of plant diseases	Shiva Rani *et al.* (2013)
	Seed	Aqueous	*A. niger* /postharvest dry rot disease of yam	Gwa and Ekefan (2018)
Piper guineense	Seed	Aqueous	*P. expansum/* postharvest dry rot disease of yam	Gwa *et al.* (2018a)
	Seed	Aqueous	*A. ochraceous/* postharvest dry rot disease of yam	Gwa *et al.* (2018b)
	Seed	Aqueous	*A. niger, A. flavus, F. solani, F. oxysporum, Geotrichum candidum, R. carotae/* postharvest rot disease of potato and carrot	Amaeze *et al.* (2013)
Tridax procumbens	Leaf	Ethanol	*A. niger, A. flavus, B. cinerea, C. capsici, P. capsici/* postharvest fruit rot of pepper	Zakari *et al.* (2015)
Tribulus terrestris	Leaf	Aqueous, Ethanol	*C. gloeosporiorides, C. orbiculare, Cladosporium sp, Curvularia sp, Fusarium sp, Pestalotiopsis sp/* postharvest fungal diseases of *Piper nigrum*	Aisha (2013)
Vernonia amygdalina	Leaf	Aqueous	*A. niger, B. theobromae/* postharvest diseases of cowpea seed	Mamkaa and Gwa (2018)

62

	Leaf	Ethanol	*A. niger, A. flavus, C. capsici, B. cinerea, P. capsici*/ postharvest fruit rot of pepper	Zakari *et al.* (2015)
	Leaf	Aqueous	*R. stolonifer*/ postharvest rot disease of yam	Nahunnaro (2008)
Xylopia aethiopica	Fruit	Oil	*C. destructivum* / anthracnose disease of cowpea	Obi and Barriuso-Vargas (2013)
	Leaf, seed	Aqueous	*F. oxysporum, A. niger, A. flavus*/ postharvest rot disease of yam	Okigbo and Nmeka (2005)
Zingiber officinale	Rhizome	Aqueous	*A. flavus*/ postharvest dry rot disease of yam	Gwa and Akombo (2016)
	Rhizome	Aqueous	*Colletotrichum sp*/ postharvest dry rot disease of yam	Gwa and Nwankiti (2017)
	Rhizome	Aqueous	*B. theobromae*/ postharvest dry rot disease of yam	Gwa *et al.* (2017)
	Rhizome	Aqueous	*F. oxysporum*/ postharvest dry rot disease of yam	Gwa (2018)
	Rhizome	Aqueous	*A. niger*/ postharvest dry rot disease of yam	Gwa and Ekefan (2018)
	Rhizome	Aqueous	*P. expansum* / postharvest dry rot disease of yam	Gwa *et al.* (2018a)
	Rhizome	Aqueous	*A. ochraceous*/ postharvest dry rot disease of yam	Gwa *et al.* (2018b)
	Rhizome	Aqueous	*F. oxysporum, R. solani*/ postharvest rot disease of tomato	Sani and Gwa (2018)
	Rhizome	Aqueous, Ethanol	*Ceracystosis fimbriata, F. solani, B. theobromae, R. stolonifer*/ postharvest rot disease of sweet potato	Anukwuorji *et al.*(2013)
	Rhizome	Aqueous, Ethanol	*C. gloeosporiorides, C. orbiculare, Cladosporium sp, Curvularia sp, Fusarium sp, Pestalotiopsis sp*/ postharvest fungal diseases of *Piper nigrum*	Aisha (2013)
	Rhizome	Aqueous, Ethanol	*A. niger, A. flavus, F. oxysporum, F. solani, R. stolonifer, B. theobromae*/ postharvest dry rot disease of yam	Ijato (2011)
	Rhizome	Aqueous	*P. grisea*/ rice blast disease	Hubert *et al.* (2012)

Choice of botanical pesticides

Botanicals or botanical pesticides are major alternative methods to the use of synthetic chemical pesticides in the management of plant pathogens and diseases. The major characteristic features of botanicals include selectivity towards target organisms, low mammalian toxicity and non persistence in the soil and bioaccumulation in the ecosystem (Grdisa and Grsic, 2013) had contributed in the research into different botanicals from different sources of plant across the world. Botanical pesticides have little or no effect on the environment and generally friendly to the applicator (Dimetry, 2014).

The major advantages of botanical pesticides are as follows:

- There is no fear of environmental pollution and hence eco-friendly.
- Botanical pesticides are designed to affect only one pest (target specific).
- Botanical pesticides are biodegradable and non persistent.
- Botanical pesticides are easy to prepare and apply.
- Botanical pesticides are low cost alternatives to synthetic chemical pesticides.
- Ideally suited for integration with most of other plant protection measures used in Integrated Pest Management (IPM) programs.
- Botanical pesticides have low mammalian toxicity

Extraction methods of plant botanical pesticides

According to Handa *et al.* (2008), extraction is the separation of medicinally active portions of plant using selective solvents through standard procedures. The main reason extraction is carried out is simply to separate the soluble plant metabolites from insoluble cellular marc (residue). The plant metabolites extracted from the initial crude extracts includes flavonoids, terpenoids, glycosides, alkaloids and phenolics. Some of these extracts initially obtained from the plants may be used directly as medicinal agents either in the form of fluid extracts or in the form of tinctures, while some of the extracts may require further processing to obtain the desired chemical compounds. According to Pandey and Tripathi (2014), there are several methods of extraction which include maceration, infusion, percolation, digestion, decoction, aqueous-alcoholic extraction by fermentation, hot continuous extraction (Soxhlet), counter current extraction, ultrasound extraction (sonication), microwave-assisted extraction, supercritical fluid extraction, and distillation techniques (water distillation, steam distillation, phytonic extraction (with hydro fluorocarbon solvents). For aromatic plants, hydro water and steam distillation, hydrolytic maceration followed by distillation, expression and effleurage (cold fat extraction) may be employed. Some of the latest extraction methods for aromatic plants include headspace trapping, solid phase micro extraction, protoplast extraction, micro distillation.

Maceration: Maceration is a technique used in wine making and has been adopted and widely used in medicinal plants research. Maceration involved soaking plant materials (coarse or powdered) in a stoppered container with a solvent and allowed to stand at room temperature for a period of minimum 3 days with frequent agitation (Handa *et al.,* 2008). The processed product is intended to soften and break the plant's cell wall which releases the soluble phytochemicals. After 3 days, the mixture is pressed or strained by filtration. In this conventional method, heat is transferred through convection and conduction, and the choice of solvents will determine the type of compound extracted from the samples (Azwanida, 2015).

Infusion: It is a dilute solution of the readily soluble components of the crude extracts. Fresh infusions are prepared by macerating the solids (coarse or powdered plant materials) for a short period of time with either cold or boiled water (Bimakr, 2010). However, the maceration period for infusion is shorter and the sample is boiled in specified volume of water (e.g. 1:4) for a defined time (Handa *et al.,* 2008).

Decoction: It is only suitable for extraction of the water soluble heat-stable compounds, hard plants materials (e.g. roots and barks) by boiling it in water for 15 minutes, cooling, straining and passing sufficient cold water through the plant materials to produce the required volume (Bismakr, 2010), which usually resulted in more oil-soluble compounds compared to maceration and infusion.

Percolation: This is the procedure used most frequently to extract active ingredients in the preparation of tinctures and fluid extracts. A percolator (a narrow, cone-shaped vessel open at both ends) is generally used. The solid ingredients are moistened with an appropriate amount of the specified menstruum and allowed to stand for approximately 4 h in a well closed container, after which the mass is packed and the top of the percolator is closed. Additional menstruum is added to form a shallow layer above the mass, and the mixture is allowed to macerate in the closed percolator for 24 h. The outlet of the percolator is then opened and the liquid contained therein is allowed to drip slowly. Additional menstruum is added as required, until the percolate measures about three-quarters of the required volume of the finished product. The marc is then pressed and the expressed liquid is added to the percolate. Sufficient menstruum is added to produce the required volume, and the mixed liquid is clarified by filtration or by standing followed by decanting (Cowan, 1999).

Digestion: This is a kind of maceration in which gentle heat is applied during the maceration extraction process. It is used when moderately elevated temperature is not objectionable and the solvent efficiency of the menstruum is increased thereby (Bimakr, 2010).

Soxhlet extraction: Soxhlet extraction is only required where the desired compound has a limited solubility in a solvent, and the impurity is insoluble in that solvent. If the desired compound has a high solubility in a solvent, then a simple filtration can be used to separate the compound from the insoluble substance. The advantage of this system is that instead of many portions of warm solvent being passed through the sample, just one batch of solvent is recycled. This method cannot be used for thermolabile compounds, as prolonged heating may lead to degradation of compounds (Sutar, 2010).

Serial exhaustive extraction: This is a method of extraction which involves successive extraction with solvents of increasing polarity from a non-polar (hexane) to a more polar solvent (methanol) to ensure that a wide polarity range of compound could be extracted (Pandey and Tripathi, 2014). Some researchers employ soxhlet extraction of dried plant material using organic solvent. This method cannot be used for thermolabile compounds as prolonged heating may lead to degradation of compounds (Das, 2010).

Sonication: This method involves the use of ultrasound with frequencies ranging from 20 kHz to 2000 kHz; this increases the permeability of cell walls and produces cavitation. If the intensity of ultrasound is increased in a liquid, then it reaches at a point at which the intramolecular forces are not able to hold the molecular structure intact, so it breaks down and bubbles are created, this process is called cavitation (Baig *et al.,* 2010). Collapse of bubbles can produce physical, chemical and mechanical effects which result in the disruption of biological membranes to facilitate the release of extractable compounds and enhance penetration of solvent into cellular materials and improve mass transfer (Cares *et al.,* 2009).

Although, the process is useful in some cases like extraction of *Rauwolfia* root, its large-scale application is limited due to the higher costs. One disadvantage of the procedure is the occasional but known deleterious effect of ultrasound energy (more than 20 kHz) on the active constituents of medicinal plants through formation of free radicals, and consequently undesirable changes in the compounds (Cowan, 1999).

Plant tissue homogenization: Plant tissue homogenization in solvent has been widely used by researchers. Dried or wet, fresh plant parts are grinded in a blender to fine particles, put in a certain quantity of solvent and shaken vigorously for 5-10 min or left for 24 h after which the extract is filtered. The filtrate may then be dried under reduced pressure, and redissolved in the solvent to determine the concentration. Some researchers however centrifuged the filtrate for clarification of the extract (Das, 2010).

Plant material

Plant based natural compounds can be obtained from different parts of the plant such as bark, flowers, fruits, leaves, roots, seeds, etc. This means that any part of the plant may contain useful active substances that can be used to inhibit pathogens that cause diseases in plants.

Choice of solvents

The phytotoxicity of the active substances found in the plant material is to some great extent determined by the type of solvent used in the extraction of the phytochemicals. Properties of a good solvent in plant extractions include:

- It should contained high preservative action.
- There should be low toxicity.
- The solvent should not cause the extract to complex or dissociate.
- It should be able to promote rapid physiologic absorption of the extract.
- There should be ease of evaporation at low heat.

Factors affecting the choice of solvent are:

- Rate of extraction
- Differences in compounds extracted
- Quantity of phytochemical to be extracted
- Toxicity of the solvent in the bioassay process
- Diversity of inhibitory compounds extracted.
- The potential health hazard of the extractants
- The ease at which the extracts will subsequently be handled after extraction

The choice of the types of solvent used depends on the targeted compounds to be extracted. It may also be influenced by what is intended with the extract. Since the phytochemicals extracted will contain some traces of residual solvent, the solvent should contain no toxic substances and should also not interfere with the bioassay.

Factors affecting variation in extraction methods

- pH of the solvent used
- Solvent-to-sample ratio
- Particle size of the plant tissues
- Temperature of the environment
- Duration of the extraction period
- Type of solvent used for extraction

The basic principle is to grind the plant material (dry or wet) finer, which increases the surface area for extraction thereby increasing the rate of extraction. Earlier studies reported that solvent to sample ratio of 10:1 (v/w) solvent to dry weight ratio has been used as ideal.

Solvents used for active component extraction

Table 2 presents the solvents used and phytochechemical compounds extracted.

Table 2: Solvents used and phytochemical compounds extracted

Acetone	Chloroform	Ethanol	Ether	Methanol	Water
Phenol	Terpenoids	Tannins	Alkaloids	Anthocyanins	Anthocyanins
Flavonols	Flavonoids	Polyphenols	Terpenoids	Terpenoids	Starches
		Polyacetylenes	Coumarins	Saponins	Tannins
		Flavonols	Fatty acids	Tannins	Saponins
		Terpenoids		Xanthoxyllines	Terpenoids
		Sterols		Totarol	Polypeptides
		Alkaloids		Quassinoids	Lectins
				Lactones	
				Flavones	
				Phenones	
				Polyphenols	

The basic parameters influencing the quality of an extract are:
- Extraction procedure
- Plant part used as starting material
- Solvent used for extraction

Effect of extracted plant phytochemical depends on:
- Particle size of plant material
- Moisture content in the extract
- Nature of the plant material
- The degree of processing
- The origin of plant

The variations in different extraction methods that will affect quantity and secondary metabolite composition of an extract depends on:
- Nature of the solvent
- Polarity
- Type of extraction method used
- Time of extraction
- Concentration of the solvent
- Temperature of the environment

Diseases of some crops

Plants are affected by different pathogenic organisms resulting in various diseases of economic importance. Some diseases affect plants, causing losses above the economic injury level, hence the need to control them. While some depending on the time of attack, duration of attack, susceptibility of plant and environmental conditions, little damages may be incurred and hence there may be no need to control such diseases. Plates 5-8 present some of the diseases of some crops.

Plate 5: Fusarium wilt disease of tomato

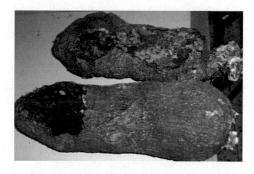

Plate 6: Dry rot tuber disease of white yam

Plate 7: Anthracnose disease of water yam

Plate 8: Leaf spot disease of white yam

Bioassay and evaluation of potency of botanical pesticides

There are several methods that are used for evaluating the antimicrobial activities of biological pesticides. Some of the methods used for screening the potency of the various botanical used in controlling different plant diseases are described below.

Poisoned food technique: This method involved poisoning the nutrient medium with the chemical substance to be tested, and then growing the fungus on the poisoned medium principle (Nene and Thapilyal, 2002). The technique uses a solid or liquid medium. In this method, the medium is prepared and poured into sterilized Petri dishes, and 5 ml of each plant extracts and chemical fungicide (used as poison) at different concentrations are poured into Petri dishes containing 15 ml of the medium separately (Nene and Thapilyal 2002) mixed well and allowed to solidify. The medium thus prepared is called poisoned food. The solidified medium is then inoculated centrally at the point of intersection of the two perpendicular lines drawn at the bottom of the plate (Amadioha and Obi, 1999), with 5 mm diameter discs as inoculums which is obtained from one-week-old culture of the fungus to be tested and grown on Potato Dextrose Agar (PDA) plates containing the media (Vadashree et al., 2013; Gwa and Nwankiti, 2017a; Gwa and Ekefan, 2018). The procedure for the control experiment is the same, except that 5 ml of sterile distilled water is added to PDA instead of plant extracts or chemical (poison). The treatments and control experiments are completely randomized (Gomez and Gomez 1984) and incubated at ambient room temperature (30 ±5°C) for 120 hours (Gwa and Nwankiti, 2018; Gwa and Richard, 2018; Gwa et al., 2018a; Gwa et al., 2018b). Measurements of radial growth of the mycelia for both the poisoned and control plates are done at 24 hr interval (Gwa and Richard, 2018; Gwa et al., 2018a; Gwa et al., 2018b), and fungitoxicity is calculated as percentage growth inhibition of the tested fungus using the formula described by Korsten and De Jager, (1995). The efficacy of plant extracts using poisoned food technique have been thoroughly researched by several workers (Yadav and Thrimurty, 2006; Sharma and Tripathi, 2006; Gwa and Nwankiti, 2018; Gwa and Richard, 2018; Sani and Gwa, 2018; Gwa et al., 2018a; Gwa et al., 2018b).

Slide germination method: This technique uses spore suspension of the test pathogen prepared from fresh culture using sterile acidified water. Spore concentration is adjusted to about 20-25 spores per microscopic field at low magnification. A drop of spore suspension is placed on a clean grease free cavity slide, and then one drop of double concentration of the test chemical is added to it. Both the drops are mixed thoroughly in cavity using inoculation/dissecting needle. The cavity slides are placed in moist chambers where conditions are favourable for the growth of the test pathogen and incubated at desired temperature for about 5 to 10 days. At the end of the incubation period, the germinated and ungerminated spores in 10-20 microscopic fields at low magnification are counted to determine percent inhibition of spore germination. Slide germination technique was used by Raghuvansi et al. (2006) to test the efficacy of leaf extracts of some plants against spore germination of *Colletotrichum falcatum,* the red rot pathogen of sugarcane. The authors reported that leaf extracts of neem, bhang and shisham gave a significant inhibition in conidial germination of *C. falcatum.* In same vein, Maji et al. (2005) used slide germination method for *in vitro* screening of twenty plant species that inhibited the growth of three fungal pathogens: *Peridiopsora mori* (Brown rust), *Pseudocercospora mori* (Black leaf spot) and *Phyllactinia corylea* (Powdery mildew).

Well or cup diffusion method: In this method, a sterile cork borer is used to make wells in the pathogen seeded plates. The test compounds are introduced into the wells under strict aseptic conditions; plates are incubated at required temperature for the growth of test pathogen and zone of inhibition if any is determined. Microbial growth is determined by measuring the diameter of the zone of inhibition. Satish *et al.* (1999) and Nair and Chanda (2005) had used well or cup diffusion method to determine the antibacterial activity of aqueous and methanol extracts of different plant species.

Filter paper disc or zone of inhibition method: This method uses a sterile filter paper disc which is impregnated with an antimicrobial agent and is placed on a pathogen inoculated agar plate. These plates are kept for incubation and are looked for zones of inhibition. The presence of a clear zone of inhibition surrounding the disk indicates the inhibitory (antimicrobial) activity of the material under study. Antibacterial activity of aqueous and methanol extracts of different plant species using agar disk diffusion method was studied by Bauer *et al* (1996) and Salie *et al.* (1996).

Methods of application of botanicals

Foliar application: Foliar application of botanicals is a commonly used method of application of botanicals for the control of different plant diseases including bacterial, fungal and viral diseases. Research conducted by Nagarajan *et al.* (1990) showed that a lot of botanicals posses' antiviral compounds which when sprayed on tobacco leaf plant early in the season before flowering gave a good protection against tobacco mosaic virus (TMV) infection.

Spraying of crude leaf extract of *Azadirachta indica* was most potent in reducing bean common mosaic virus infectivity under field conditions (Tripathi and Tripathi, 1982). In similar studies, Ogwulumba *et al.* (2008) used *C. papaya* leaf extracts to reduce incidence of foliar myco-pathogens of groundnut *(Arachis hypogea)*. Sola *et al.* (2014) showed that *A. indica* contained Azadirachtin in higher concentration (0.2 – 0.6%) in the seeds of the neem more than any other parts of the neem tree. Application of neem products prior to inoculation of crops by pathogens inhibits leaf spots, rusts, mildews, rot diseases and moulds (Mariappan, 1998). Pre inoculation spray of *C. papaya* , *A. indica*, *Z. officinale*, *P. guineense*, *N. tabacum* inhibit mycelial growth of *Colletotricum* species, *A. niger*, *P. expansum* and reduced disease in yam (Gwa and Ekefan 2017, Gwa and Ekefan, 2018, Gwa *et al.*, 2018b). Sopialena *et al.* (2018) used *A. indica* leaves as organic pesticides in controlling Chili pepper (*Capsicum frutescens*) diseases in Indonesia.

Seed and seedling treatment: Seed and seedling treatment is very important because a lot of diseases associated with seed and wherever such seeds are used for sowing, the mycoflora become active and initiate seed rot and seedling mortality. The aim of seed and seedling treatment is to destroy mycoflora which are responsible for causing seed decay and seedling blights, and also to reduce attack of soil-inhabiting microbes. Seed treatment with 5 % leaf extracts of neem, marigold and garlic bulb extract significantly reduced the wilt disease complex (*Rhizoctonia solani, Fusarium oxysporum, Sclerotium rolfsii* (*Corticium rolfsii*) and *Macrophomina phaseolina*) in lentil (Sinha and Sinha, 2004). According to Okigbo *et al.* (2009b), application of *A. sativum* extract recorded high rot reduction of 62.80 % in yam tubers. Similarly, application of *A. indica, Z. officinale, P. guineense* on yam tubers reduced rot development on yam tubers in storage (Gwa and Richard, 2018). Also, treatment of cowpea seeds with leaf extract of *Moringa oleifera* and *Vernonia amygdalina* inhibited mycelial growth of *Aspergillus flavus* and *Botryodiplodia theobromae* causing rot of Cowpea (*Vigna unguiculata* (L.) Walp) seeds (Mamkaa and Gwa, 2018). According Sani and Gwa (2018), application of *A. indica* and *Z. officinale* extracts in the ccontrol of *Fusarium oxysporum* and *Rhizoctonia Solani* on Tomato (*Solanum Lycopersicum*) fruits reduced the growth of the fungi on the fruits. Oil extracted from seeds of *A. indica* (neem oil) and *Annona* sp. (custard apple oil) and treated on rice seedlings was found to significantly reduce the life-span of *Nephotettix virescens* (Dist.) carrying rice tungro virus which consequently reduced virulence of the pathogen (Mariappan, 1998). Findings by Okigbo *et al.* (2009b) reported that aqueous and ethanol extracts of seeds of *Aframomum melegueta* were effective in controlling *Aspergillus niger, Botryodiploidia theobromae, Fusarium solani* and *Penicillium oxalicum* on cassava tubers. Similarly, Amaeze *et al.* (2013) reported that *A. niger, A. flavus, F. solani, F. oxysporum, Geotrichum candidum, Rhizoctonia carotae* were significantly controlled on potato and carrot with aqueous extracts of bulbs of *Allium sativum*. It has also been reported that different parts of *M. oleifera* such as the leaves, roots, fruits, flowers and seeds contain phytochemicals compounds like alkaloids, carotenoids, tannins, anthraquinones, anthocyanins and proanthocyanidns (Padayachee and Baijnath, 2012) that are capable of inhibiting the growth of the pathogens. Hubert *et al.* (2015) also inhibited the growth of *Pyricularia grisea* with bulbs extract of *A. sativum*. Report by Zakari *et al.* (2015) revealed that application of ethanol leaf extracts of *Azadirachta indica, Tridax procumbens* and *Vernonia amygdalina* on pepper fruits inhibited the growth of *Aspergillus niger, Aspergillus flavus, Botrytis cinerea, Colletotrichum capsici* and *Phytophthora capsici.*

Soil application: Soil is known to be a very good habitat of a large number of plant pathogenic fungi. The sources of primary inoculum of these fungi come from the soil, since soil has a lot of decomposed organic matter which contain both the active and dormant stages of pathogens. This means most diseases are initiated from the soil which contained primary source of inoculum. If the soil is therefore treated, disease management will be economically easier and transmission from one soil to another or from plant to another will be less. Report by different authors confirmed that leaves, rhizomes, seeds of botanicals such as *Azadirachta indica, Zingiber officinale, Piper guineense, Moringa oleifera, Vernonia amygdalina, Nicotiana tabacum* and *Carica papaya* can be incorporated into the soil to inactivate the various stages of soil-borne plant pathogenic diseases. Mandal (2001) reported that farmers in Bengal applied leaves of garari (*Cleistanthus collinus*) as manure for pest control in paddy. The leaves of the plant help to control so many pests, and at same time potent against brown spot disease (*Drechslera oryzae*) of rice. These leaves are said to deter almost all pests, but are particularly effective against the brown spot disease (*Drechslera oryzae*) (Mandal, 2001). Results obtained by Ekefan *et al.* (2018) showed that leaves of *A. indica*, rhizomes of *Z. officinale*, seeds of *P. guineense,* leaves of *N. tabacum* and *C. papaya* inhibited the growth of various pathogenic fungi on yam setts in the soil and increased germination of yam.

According to Hundekar *et al.* (1998), application of organic matter such as groundnut, cotton, neem, safflower cakes reduced the build up of inoculum of *Fusarium moniliforme* and *Macrophomina pheseolina* in the soil, thereby reducing the incidence of stalk rot of sorghum. Ahuja and Ahuja, (2008) found that when dhaincha (*Sesbania cannabina*) is planted as green manure prior to planting rice, it reduces incidence of sheath blight and increased disease resistance in rice.

Post harvest application of botanicals

Pathogens easily gained access to harvested crops mostly through wounds created during the processes of harvesting, handling and transportation. If theses wounds are not treated after harvest, rots initiated by pathogens on fruits, vegetables and tubers are common. The harvested produce can therefore be protected from invading pathogens to prevent rots and increase the shelf life of the harvested crops by the application of botanical pesticides. Extracts from plants contain a large number of chemical compounds which are efficacious against post-harvest pathogens of different crops. Results obtained on essential oils, aromatic compounds and volatile substances proved that botanical pesticides are effective in controlling plant pathogens.

Essential oils extracted from fruits of *Xylopia aethiopica* and seeds of *A. indica* were found to be effective in controlling *Colletotrichum destructivum* on cowpea (Obi and Barriuso-Vargas (2013). Similarly, extracts of *Ocimum canum Mentha arvensis* and *Zingiber officinale* were found to be potent against *Penicillium italicum*, which is responsible for causing blue mould rot of lime and orange fruits during storage (Tripathi *et al.*, 2004). In a related development, oil extracted from *Ocimum sactum* leaves was used to inhibit the growth of *Botrytis cenerea*. Aisha (2013) reported the inhibition of the growth of *Colletotrichum gloeosporiorides, C. orbiculare, Cladosporium sp, Curvularia* sp, *Fusarium* sp,

73

Pestalotiosis sp with ethanol extracts from rhizomes of *Emblica officinalis*. Zakari *et al.* (2015) found that ethanol leaf extract of *Vernonia amygdalina* has high fungitoxic activity against *A. niger, A. flavus, C. capsici, B. cinerea, P. capsici* on pepper.

Conclusion

The use of botanical pesticides is of paramount importance and one of the best alternatives in the field of crop protection for the management of plant diseases. In view of these obvious advantages, botanical pesticides are therefore easily integrated with other plant disease management strategies to reduce the risk associated with the application of synthetic pesticides to increase export quality by producing chemical-free crops. Farmers are therefore encouraged to adopt to the use of botanical pesticides as alternative way of managing crop diseases both in the field and in store, since they are cheap, non-hazardous, biodegradable, easily available and environmentally friendly. However, it is recommended that field trials be conducted to access the effectiveness of botanicals rather than *in vitro* and green house experiments. In addition, studies on biosafety of botanical pesticides should be carried out to determine the level of toxicity on humans consuming the crops, animals feeding on them, as well as the quantity of residual chemicals in the harvested crop. It is also recommended that mixtures of two or more botanicals be combined to reduce the resistance that will possibly develop from pathogens as a result of continued use of a particular plant pesticide over a long period of time. This will provide food security for the country at economically low cost without affecting quality and will also prolong postharvest shelf life of the produce.

References

Ahuja, S. C. and Ahuja, Uma. (2008) Learning from farmers– Traditional rice production technology. Asian Agi-History. 12(1): 19-41

Aidah, N., Abdullah, N., Oskoueian, E., Sieo, C. C. and Saad, W. Z. (2014) Membrane-active antibacterial compounds in methanolic extracts of Jatropha curcas and their mode of action against Staphylococcus aureus S1434 and Escherichia coli E216. Int. J. Agric. Biol., 16: 723–730

Aisha, M. H. A. (2013) Antifungal activity of plant extracts against fungal pathogens of Piper nigrum. Intl. J. Trad. Herb. Med., 1 (4):116-123,

Amadioha, A. C. and Obi, V. I. (1998) Fungitoxic activity of extracts from Azadirachta indica on Colletotrichum lindemuthianum in cowpea. J. Herbs, Spices, Medicinal Plants 6(1): 33-40.

Amadioha, A. C and Obi, V. I. (1999) Control of anthracnose disease of cowpea Cymbopogon citratus and Ocimum gratissimum. Acta phytopathol. Entomol. Hungerica. 34(92): 85-89

Amaeze, N. J., Ezeh, P. A. and Dan-kishiya, A. S. (2013) Evaluation of Garlic (Allium Sativum) and uziza (Piper Guineense) on the Control of Tuber Rot Fungi of Potato and carrot. American Journal of Research Communication, 1(9): 167-174

Anukwuorji, C. A., Anuagasi, C. L. and Okigbo, R. N. (2013) Occurrence and Control of Fungal Pathogens of Potato (Ipomoea Batatas L. Lam) With Plant Extracts. PhTechMed 2(3):278-289

Azwanida, N. N. (2015) A Review on the Extraction Methods Use in Medicinal Plants, Principle, Strength and Limitation. Med Aromat Plants 4: 196. doi:10.4172/21670412. 1000196

Baig, D. N., Bukhari, D. A. and Shakoori, A. R. (2010) cry Genes profiling and the toxicity of isolates of Bacillus thuringiensis from soil samples against American bollworm, Helicoverpa armigera. J Appl Microbiol 109, 1967–1978.

Bauer, A. W., Kirby, W. M. M. and Sherris, J. C. (1996) Antibiotic susceptibility testing by a standardizedsingle disk method. Am. J. Clin. Pathol. 45: 493-496.

Bediakao, A., Showemimo, F. A., Asiamo, Y. O. and Amewowor, D. H. A. K (2007) In vitro Analysis of Growth Media and the Control of Yam Minisett-rot. Biotechnology, 6: 1- 4.

Bimakr, M. (2010) Comparison of different extraction methods for the extraction of major bioactive flavonoid compounds from spearmint (Mentha spicata L.) leaves. Food Bioprod Process. 1-6.

Bloomquist, J. R., Boina, D. R., Chow, E., Carlier, P. R., Reina, M. and Gonzalez- Coloma, A. (2008) Mode of action of the plant-derived silphinenes on insect and mammalian GABAA receptor/chloride channel complex. Pestici Biochem & Physiol 91:17–23

Brown, A. E. (2005) Mode of action of insecticides and related pest control chemicals for production agriculture, ornamentals and turf. Pesticide Info Leaflet Nr 43:1–13. http://pesticide.umd.edu

Cares, M. G., Vargas, Y., Gaete, L., Sainz, J. and Alarcon, J. (2009) Ultrasonically assisted extraction of bioactive principles from Quillaja Saponaria Molina. Physics. Procedia. 3: 169-178.

Chengala, L. and Singh N. (2017) Botanical pesticides – a major alternative to chemical pesticides: A review. Int. J. of Life Sciences, 5(4.

Chima, N. (2012) Antifungal Potencies of Leaf Extracts of Carica papaya on Fungi implicated in Soft Rot of Yam. Annals of Food Science and Technology, 3:11-16

Cowan, M. M. (1999) Plant products as antimicrobial agents. Clinical microbiology reviews. 12(4):564-582.

Council on Scientific Affairs (1997) Education and Information Strategies to Reduce Pesticide Risks. American Medical Association

Das, K., Tiwari, R. K. S. and Shrivastava, D. K. (2010) Techniques for evaluation of medicinal plant products as antimicrobial agent: Current methods and future trends. Journal of Medicinal Plants Research. 4(2):104-111

Dimetry, N. Z. (2014) Different Plant Families as Bioresources for Pesticides. In: Advances in Plant Biopesticides, D. Singh (Ed.), Springer India. DOI: 10.1007/978-81-322-2006-0_1

Dubey, R. K., Rajesh, K. Jaya, and Dubey, N. K. (2007) Evaluation of Eupatorium cannabinum Linn. Oil in enhancement of shelf life of mango fruits from fungal rotting. World Journal of Microbiology and Biotechnology, 23: 467-473

Ekefan, E. J., Nwankiti, A. O. and Gwa, V. I. (2018) Comparative Assessment of Antimicrobial Potency of Some Selected Plant Extracts Against Seed Borne Pathogens of Germinating Yam Setts. J Plant Pathol Microbiol 9 (7):1-8 doi: 10.4172/2157-7471.1000444

Food and Agriculture Organization of the United Nations (FAO) (2002) International Code of Conduct on the Distribution and Use of Pesticides. Retrieved on 2017 – 07 – 11.

Grdisa, M. and Grsic, K. (2013) Botanical Insecticides in Plant Protection. Agriculturae Conspectus Scientificus, 78 (2): 85 – 93.

Gomez, K. A. and Gomez, A. A. (1984) Statistical Procedures for Agricultural Research 2nd Edition John Wiley and sons. 680pp.

Gwa, V. I. and Akombo, R. A. (2016) Studies on the Antimicrobial Potency of Five Crude Plant   Extracts and Chemical Fungicide in in vitro Control of Aspergillus flavus, Causal Agent of White Yam (Dioscorea rotundata) Tuber Rot. Journal of Plant Sciences and Agricultural Research, 1(1): 1-8.

Gwa, V. I. and Ekefan, E. J. (2017) Fungal Organisms Isolated from Rotted White Yam (Dioscorea rotundata) Tubers and Antagonistic Potential of Trichoderma harzianum against Colletotrichum Species. Agri Res & Tech: Open Access J. 10(3): 555787. DOI:10. 19080/ARTOAJ.2017.10.555787

Gwa, V. I. and Nwankiti, A. O. (2017a) Efficacy of some plant extracts in in-vitro control of Colletotrichum species, causal agent of yam (Dioscorea rotundata Poir.) tuber rot. Asian Journal of Plant Science and Research, 7(2):8-16

Gwa, V. I. and Nwankiti, A. O. (2017b) In Vitro Antagonistic Potential of Trichoderma harzianum for Biological Control of Fusarium moniliforme isolated from Dioscorea rotundata Tubers. Virol-mycol 6(2):1-8. 166. doi:10.4172/2161-0517.1000166

Gwa, V. I., Ekefan, E. J. and Nwankiti, A. O. (2017) Antifungal Potency of Some Plant Extracts in the Control of White Yam (Dioscorea rotundata poir) Tuber Rot. Adv Biotech & Micro, 7(1): 555703. DOI: 10.19080/AIBM.2017.07.555703

Gwa, V. I. (2018) Fungitoxic Activities of Plant Extracts on Mycelial Growth inhibition of Fusarium Oxysporum Causal Agent Yam Tuber Rot in Zaki-Biam, Benue State,

Nigeria. Journal of Pesticides and Biofertilizer 1(1):1-6 DOI: http;//doi.org/03.2018/ 1.100015

Gwa, V. I. and Ekefan, E. J. (2018) Fungicidal Effect of Some Plant Extracts against Tuber Dry Rot of White Yam (Dioscorea rotundata Poir) Caused by Aspergillus Niger. Int. J. Hort. Agric, 3(3): 1-7. DOI: http://dx.doi.org/10.15226/2572-3154/3:3/00123

Gwa, V. I. and Nwankiti, A. O. (2018) In vitro and In vivo antimicrobial potency of selected plant extracts in the control of postharvest rot-causing pathogens of yam tubers in storage. Global Journal of Pests, Diseases and Crop Protection. 6(1):276-287

Gwa, V. I. and Richard, I. B. (2018) Susceptibility of White Yam (Dioscorea rotundata Poir) Tuber to Rot Fungi and Control with Extracts of Zingiber officinale Rosc. Azadirachta indica A. Juss. and Piper guineense Schumach. J Plant Pathol Microbiol 9: 452. doi: 10.4172/2157-7471.1000452

Gwa, V. I., Nwankiti, A. O. and Hamzat, O. T. II. (2018a) Antimicrobial activity of five plant extracts and synthetic fungicide in the management of postharvest pathogens of yam (Dioscorea rotundata Poir) in storage. Acad. J. Agric. Res. 6(6): 165-175.

Gwa, V. I., Nwankiti, A. O. and Ekefan, E. J. (2018b) "Antifungal Effect of Five Aqueous Plant Extracts on Mycelial Growth of Penicillium expansum Isolated from Rotted Yam Tubers in Storage". Acta Scientific Agriculture 2(6): 65-70.

Hadian, S. (2012) Antifungal activity of some plant extracts against some plant pathogenic fungi in Iran. Asian Journal of Experimental Biological Sciences, 3(4):714-718.

Handa, S. S., Khanuja, S. P. S., Longo, G. and Rakesh, D. D. (2008) Extraction Technologies for Medicinal and Aromatic Plants, (1stedn), no. 66. Italy: United Nations Industrial Development Organization and the International Centre for Science and High Technology Trieste, 21-25.

Hubert, J., Mabagala, R. B. and Mamiro, D. P. (2015) Efficacy of Selected Plant Extracts against Pyricularia grisea, Causal Agent of Rice Blast Disease. American Journal of Plant Sciences, 6: 602-611

Hundekar, A. R., Anahosur, K. H., Patil, M. S., Kalapannavar, I. K. and Chattannavar, S. N. (1998) In vitro evaluation of organic amendments against stalk rot of sorghum. J. Mycol. Pl. Pathol. 28 (1):26-30

Ijato J. Y., Oyeyemi, S. D., Ijadunola, J. A., Ademuyiwa, J. A. (2010) Allelopathic effect of leaf extract of Azardirachta indica and Chromolaena odorata against post harvest and transit rot of tomato (Lycopersicum esculentum L). Journal of American Science. 6(12):1595-1599.

Ijato, J. Y. (2011) Antimycotic effects of aqueous and ethanol plant extracts on yam rot pathogens in Ado-Ekiti, Nigeria. Academia Arena, 3(1):115-119

Ijato, J. Y., Otoide, J. E., Ijadunola, J. A. and Aladejimokun, A. O. (2011) Efficacy of antimicrobial effect of Vernonia amygdalina and Tridax procumbens in in vitro control of tomato (Lycopersicum esculentum) post harvest fruit rot. Report and Opinion 3(1):120-123]. (ISSN: 1553-9873). http://www.sciencepub.net

Jehan Bakht, A., and Mohammad, S. (2012) Antimicrobial activity of Nicotiana tabacum using different solvents extracts Pak. J. Bot., 44(1): 459-463.

Jeyalakshmi, C. and Seetharaman, K. (1998) Biological control of fruit rot and die-back of chilli with plant products and antagonistic microorganisms. Plant Disease Research 13: 46 -48.

John, M. E. and Gideon, I. O. (2018) Phytochemical Content and In1 Vitro Antimycelial Efficacy of Colocasia esculenta (L), Manihot esculenta (Crantz) and Dioscorea rotundata (Poir) Leaf Extracts on Aspergillus niger and Botryodiplodia Theobromae. Journal of Horticulture and Plant Research. 1:9-18

Korpraditskul, V., Rattanakreetakul, C., Korpraditskul, R. and Pasabutra, T. (1999) Development of plant active substances from sweetflag to control fruit rot of mango for export. Proceedings of Kasetsart University Annual Conference. Bangkok, pp. 34.

Korsten, L. and De Jager, E. S. (1995) Mode of action of Bacillus subtilis for control of avocado post harvest pathogens. S. Afr. Avocado Growers Assoc. Yearb, 18: 124-130

Kumar, A.S., Reddy, N. P. E., Reddy, K. H and Devi, M. C. (2007) Evaluation of Fungicidal Resistance Among Colletotrichum gloeosporioides isolates Causing Mango Anthracnose in Agric. Export Zone of Andhra Pradesh, India. Plant Path. Bull., 16: 157–160

Maji, M. D., Chattopadhyay, S., Kumar, P. and Saratchandra, B. (2005) In vitro screening of some plant extracts against fungal pathogens of mulberry (Morus spp.). Archives Phytopath. Pl. Protec. 38(3):157 – 164.

Malkhan, S. G., Shahid, A., Masood, A and Kangabam, S. S. (2012) Efficacy of plant extracts in plant disease management. J. Earth. Envir. Sci., 3: 425–433

Mamkaa, D. P. and Gwa, V. I. (2018) Effect of Moringa oleifera and Vernonia amygdalina Leaf Extracts against Aspergillus flavus and Botryodiplodia theobromae Causing Rot of Cowpea (Vigna unguiculata (L.) Walp) Seeds. Appl Sci Res Rev. 5 (1,2): 1-7 DOI: 10.21767/2394-9988.100067

Mandal, T. K. (2001) Pest control of paddy. Honey Bee. 12(3): 9.

Mariappan, V. (1998) Neem for the management of crop diseases. Associated Publishing Ccompany, New Delhi. Pp.220

Nagarajan, M.K., Avon, L., William, R., Masler, III, Hinckley (1990) Effect of plant extracts and oils on rice Yellow dwarf infection. Madras Agric. J. 77: 197-201.

Nahunnaro, H. (2008) Effects of Different Plant Extracts in the Control of Yam Rot Induced by Rhizopus stolonifer on Stored Yam (Dioscorea spp.) in Yola, Adamawa State, Nigeria. Medwell. Journal of Agricultural Science, 3(5):382- 387

Nair, R. and Chanda, S. (2005) Anticandidal activity of Punica granatum exhibited in different solvents. Pharm. Biol. 43: 21-25.

Nene, Z. H and Thapilyal, (2002) Management of mushroom pathogens through botanicals. Ind. Phytopathol. 58:189-193

Nwankiti, A. O. and Gwa, V. I. (2018) Evaluation of Antagonistic Effect of Trichoderma Harzianum against Fusarium oxysporum causal Agent of White Yam (Dioscorearotundata poir) Tuber Rot. Trends Tech Sci Res. 1(1): 555554

Nweke, F. U. (2015) Effect of Some Plant Leaf Extracts on Mycelia Growth and Spore Germination of Botryodiplodia theobromae Causal Organism of Yam Tuber Rot Journal of Biology, Agriculture and Healthcare, 5: 8

Obi, V. I. and Barriuso-Vargas, J. J. (2013) Effect of some botanicals on Colletotrichum destructivum O`Gara of cowpea. African Journal of Microbiology Research.7(37):4576-4581

Ogwulumba, S. I., Ugwuoke, K. I. and Iloba, C. (2008) Prophylactic Effect of Pawpaw Leaf and Bitter Leaf Extracts on the Incidence of Foliar Mycopathogens of groundnut (Arachis hypogaea L.) in Ishiagu, Nigeria. African Journal of Biotechnology, 7(16):2878-2880

Okigbo, R. N. and Nmeka, I. A. (2005) Control of Yam Tuber with Leaf Extracts of Xylopia aethiopica and Zingiber officinale. African Journal Biotechnology, 4 (8):804-807.

Okigbo, R. N., Putheti, R. and Achusi, C. T. (2009a) Post-harvest Deterioration of Cassava and its Control using Extracts of Azadirachta indica and Afromonium meleguata. E-Journal of Chemistry, 6(4):1274-1280.

Okigbo, R. N., Anuagasi, C. L. and Amadi, J. E. (2009b) Advances in Selected Medicinal and Aromatic Plants Indigenous to Africa. Journal of Medicinal Plant Research, 3(2):86-95.

Okigbo, R.N. and Emeka, A.N. (2010) Biological Control of Rot-inducing Fungi of Water Yam (Dioscorea alata) with Trichoderma harzianum, Pseudomonas syringe and Pseudomonas chlororaphis. Journal of stored product Research, 1(2): 18-23

Okigbo, N. R., Enweremadu, C.E., Agu, C.K., Irondi, R.C., Okeke, B. C., Awah, S. N., Anaukwu, C. G., Okafor, I.O., Ezenwa, C. U. and Iloanusi, A.C. (2015) Control of White Yam (Dioscorea rotundata) Rot Pathogen using Peel Extract of Water Yam (Dioscorea alata). Advances in Applied Science Research, 6(10):7-13

Padayachee, B. and Baijnath, H. (2012) An overview of the medicinal importance of Moringaceae. J. Med. Plants Res, 6(48), 5831-5839.

Pandey, A. and Tripathi, S. (2014) Concept of standardization, extraction and pre phytochemical screening strategies for herbal drug. Journal of Pharmacognosy and Phytochemistry 2014; 2 (5): 115-119.

Raghuvansi, N. S., Dubey, K. S. and Kumar, B. (2006) Effect of soil extracts amended with some organic materials on conidial germination of Colletotrichum falcatum. Indian J. Pl. Pathol. 24(1-2): 117-118.

Romanazzi, G., Lichter, A., Gabler, F. M. and Smilanick, J. L. (2012) Recent advances on the use of natural and safe alternatives to conventional methods to control postharvest gray mold of table grapes. Postharvest Biology and Technology, 63:141–147.

Salie, F., Eagles, P. F. K. and Leng, H. M. J. (1996) Preliminary antimicrobial screening of four South African Asteraceae species. J. Ethonopharmacol. 52: 27-33.

Satish, S., Raveesha, K. A. and Janardhana, G. R. (1999) Antibacterial activity of plant extracts on phytopathogenic Xanthomonas campestris pathovars. Letters in Applied Microbiology. 28:145- 147.

Sani, S. and Gwa V. I. (2018) "Fungicidal Effect of Azadiracta Indica and Zingiber Officinale Extracts in the Control of Fusarium oxysporum and Rhizoctonia Solani on Tomato (Solanum Lycopersicum) Fruits". Innovative Techniques in Agriculture 2(4): 439-448

Shafique, S. and Abdul, M. R. (2012) Cymbopogon citrates: a remedy to control selected Alternaria species. J Med Plants Res 6 (18): 79-85

Sharma, N. and Tripathi, A. (2006) Fungitoxicity of the essential oil of Citrus sinensis on post-harvest pathogens. World J. Microbiol. Biotech. 22 (6): 587-593.

Shiva Rani, S. K., Neeti, S. and Udaysree (2013) Antimicrobial Activity of Black Pepper (Piper nigrum L.). Global Journal of Pharmacology, 7 (1): 87-90

Sinha, R. K. P. and Sinha, B. B. P. (2004) Effect of potash, botanicals and fungicides against wilt disease complex of lentil. Annls. Pl. Protec. Sci. 12(2): 454-455

Sola, P., Mvumi, M., Ogendo, J. O., Mponda, O., Kamanula, J. F., Nyirenda, S. P., Belmain, S. R. and Stevenson, P. C. (2014) Botanical pesticide production, trade and regulatory mechanisms in sub – Saharan Africa: making a case for plant based pesticidal products. Food Sec. DOI: 10.1007/s12571-014-0343-7

Sopialena, Sila, S., Rosfiansyah, Nurdiana, J. (2018) The role of mimba (neem) leaves as organic pesticides in chili pepper (Capsicum frutescens). Nusantara Bioscience, 10(4), 246-250. DOI: 10.13057/nusbiosci/n100408

Subramanian, G., Tewari, B. B. and Gomathinayagm, R. (2014) Antimicrobial Properties of Carica papaya (Papaya) Different Leaf Extract against E. coli, S. aureus and C. albicans. AJPP1(1):025-039

Suleiman, M. N. (2010) Fungitoxic Activity of Neem and Pawpaw Leaves Extracts on Alternaria Solani, Causal Organism of Yam Rots: Advances in Environmental Biology, 4(2): 159-161

Sutar, N., Garai, R., Sharma, U. S. and Sharma, U. K. (2010) Anthelmintic activity of Platycladus orientalis leaves extract. International Journal of Parasitology Research. 2(2):1-3

Taiga, A. (2011) Comparative Studies of the Efficacy of some Selected Fungicidal Aqueous Plant Extracts on Yam Tuber Dry Rot Disease. Annals of Biological Research, 2 (2): 332-336

Tijjani, A., Adebitan, S.A., Gurama, A.U, Aliyu, M., Dawakiji, A.Y, Haruna, S.G. and Muhammmed, N.A. (2013) Efficacy of Some Botanicals for the Control of Wet Rot Disease on Mechanically Injured Sweet Potato Caused by Rhizopus Stolonifer in Bauchi State. International Journal of Scientific and Research Publications, 3 (6):1-10

Tripathi, R. K. R. and Tripathi, R. N. (1982) Reduction in bean common mosaic virus (BCMV) infectivity visà-vis crude leaf extracts of some higher plants. Experimentia. 38(3): 349.

Tripathi, P., Dubey, N. K., Banerji, R. and Chansouria, J. P. N. (2004) Evaluation of some essential oils as botanical fungitoxicants in management of post-harvest rotting of citrus fruits. World J. Microbiol. Biotech. 20 (3): 317-321.

Vadashree, S., Sateesh, M. K., Lakshmeesha, T. R., Sofi, M. S. and Vedamurthy, A. B. (2013) Screening and assay of extracellular enzymes in Phomopsis azadirachtae causing die-back disease of neem. J Agricultural Technol, 9, 915-927

Yadav, V.K. and Thirumuty, V.S. (2006) Fungitoxicity of some medicinal plant extracts against Sarocladium oryzae causing sheath rot in rice. Indian J. Pl. Pathol. 24(1-2):93-96.

Zakari, B. G., Chimbekujwo, I. B., Channya, K. F. and Bristone B. (2015) In vitro antifungal activity of selected plant diffusates against post harvest fruit rot of pepper (capsicum spp. l.) in Yola. Global Journal of Biology, Agriculture and Health science. 4 (1):142-148.

CHAPTER 4

The Use of Botanicals in the Control of Insect Pests in Agriculture

Olanrewaju Folusho Olotuah
Department of Plant Science and Biotechnology
Adekunle Ajasin University,
Akungba-Akoko, Ondo State, Nigeria.
Email: olanrewaju.olotuah@aaua.edu.ng, lanrose2002@yahoo.com

Abstract

Considerable efforts have been made world-wide to find safer biodegradable substitutes for synthetic insecticides, as biopesticides have been found to be safer, cheaper and easier to produce for use in field and storage pest management. The use of botanicals in pest control has gained more attention in recent years, and many of the promising plants have been adopted as biopesticides in controlling insect pests. Awareness of botanical pesticides in agriculture has gained much attention over the years, with biological efficacies of some had been determined in form of powders, dusts, synergists, oils and fumigants. Several parts of these plants have been found effective against many insect pests, and such have been adopted for use in pest management. It is observed that the utilization of these plant materials in integrated pest control has been promising, but most investigations are only *in vivo* in laboratories. Formulation of botanical insecticidal materials for economy, efficiency, safety and enhanced effectiveness is an option in increasing the adoption potential of those that have been found effective. This paper gives extensive review of the use and adoption of some botanicals in pest management.
Keywords: Plant extracts, Pesticide, Efficacy, Pest Management, Biopesticides.

Introduction

Adequate protection of stored food commodities from insect infestation and damage is of paramount importance in achieving and maintaining food sufficiency in the world. Insect pests have caused so much damages to crops, farmers and to the world at large. That is why researchers have used so many ways and methods to control it. The primary approach had centred on the use of synthetic insecticides, but these were observed to be associated with several problems. Consequently, the synthetic insecticides are generally perceived not to be cost-effective and environmentally friendly. The option of adoption of botanicals is a panacea to the attendant problems of use of synthetics. Similarly, the perception of many farmers is that botanicals are cheap, available and thought to be safer than conventional pesticides. A large number of plant species from a wide range of families have been eval-uated. Current research efforts on product development are being focused more on ecolog-ically tolerable control measures, including the use of inert materials, plants' powders, oils and extracts. There is also an increasing awareness that plants possess chemicals which naturally protect them from pests and pathogens. The tropical region is well endowed with a wide array of these floristic species with defensive chemicals, and quite a number of them have been used traditionally in protecting insect pests. Jacobson (1989) suggested that the most promising botanicals were to be found in the families *Meliaceae*, *Rutaceae*, *Asteraceae*, *Annonaceae*, *Lamiaceae* and *Canellaceae*.

The plant species that have been investigated are frequently those used locally, within in-dividual countries, as spices or in traditional medicine. According to Ofuya (2003), syn-thetic insecticides involves risks for human health and the environment, especially when improperly used which may be common among uneducated rural farmers in Africa. Since the last decade, Plant-derived insecticide have been vigorously investigated worldwide, as a possible replacement for synthetic insecticide in stored products protection (Lale, 2001). The most successful use of plant products as an insecticide is that of the pyrethroids. The insecticidal properties of several *Chrysanthemum* species were known for centuries in Asia. Even today, powders of the dry flowers of these plants are sold as insecticides.

After elucidation of the chemical structures of the six terpenoid esters responsible for the insecticidal activities of these plants; many synthetic analogues have been patented and marketed. Synthetic pyrethronoids have better photo-stability and are generally more ac-tive than their natural counterparts. Nicotine and non-nicotine components of several mem-bers of the genus *Nicotiana* have been used commercially as insecticides. *N. rustica* is the chief commercial source. Synthetic variations of nicotine such as 5-methl non- nicotine have been demonstrated to be effective as insecticide. Ryanodine an alkaloid from the tropical shrub *Ryana speciosa* has been used as commercial insecticide against European corn borer. Rotenone is a flavonoid derivative that strongly inhibits mitochondria respira-tion. No other phenolic compound has been used commercially as insecticides, although the content of certain phenolic compounds in plant tissues have been correlated with the host plants resistance to insects, and many have been demonstrated to be strong hormone mimics have been found to effective sterilize insects. Plants contain a myriad of com-pounds with potential for commercial development in insect-control (Zhu *et al.,* 2001b).

Attributes of Botanicals

Desirable characteristics of botanicals for use in pest control would probably be that the plant is perennial, easy to grow and not expensive to produce. Plants should also show no potential to become weed or host for plant pathogens themselves, and should if possible offer complementary economic uses. Botanical insecticides tend to have broad spectrum activity. They are safe and relatively specific in their mode of action, easy to produce and use.

Adoption of Botanicals

The conventional control methods earlier adopted in the control of insect pests had gained attention in the last three decades and several plant parts have been adopted. Recent research efforts have been turning more towards selective bio-rational pesticides, that are safer, cheaper and easier to produce than synthetic insecticides. It has been reported that Essential oils from plants have been proved to possess good potential for use as fumigants against stored product insects, including storage bruchids (Papachristos and Stamopoulos, 2002 and Tapondjou *et al.*, 2002). Raja *et al.* (2000) reported that when jute bags treated with different plant leaves extract including *Azadirachta indica*, *Vitex negundo*, *Cleistanthus collinus* and *Jatropha Curas* and then used for cowpea seeds storage, the egg laying rates by the *C. maculatus* adult emergence and seed damage were reduced.

Kim *et al.* (2003) showed the potent insecticidal activity of extracts of cinnamomum cassia bark and oil, horseradish (*Cocholeria aroracia*) oil and mustard (*Brassica juneea*) oil against *C. chinensis*, within one day after *Eucalyptus* seed powder treatment caused the death of emerging adult of *C. chinensis*. Plant products such as vegetable oils essential oils, volatile oils, crude extracts and powders have been tested against *Calosobruchus. maculatus* (Lale, 2001, Boeke *et al.*, 2002). Also, dry powder made from *A. indica* seed, *A. juss*, buds of clove tree, *Eugenia aromatica*, baill, fruits of West Africa brown pepper, piper guineense, seed of "pepper fruit" tree, *Dennetia tripetala* baker and root bark of the "tooth ache plant", *Zanthozylum zanthozyloides* (Lam) waterm, applied at 2% of the weight of seed beetle in storage (Lale, 2001 Ogunwolu *et al.,* 2001; Adedire and Lajide, 2001, Ofuya and Salami, 2002). Tapondjou *et al.* (2002) showed that dry ground leaves of *Chenopodium ambrosioides* inhibited F. progeny production and adult emergence of the *C. chinensis* and *C. maculatus*. The use of plant materials, extracts, oils, serve as repellant against several insect such as weevils, flour beetles, bean- seed beetles, and potatoes moths etc. plant materials, extracts and oils also help to reduce the amount of synthetic pesticides needed thereby, decreasing the pesticides load in food grains.

The extracts and oils of plant materials have been found to be alternatives to conventional synthetic insecticides for the control of stored product insect pests (Lale, 2001 and Boeke *et al.*, 2002). This is due to adverse effects of chemical fumigants used in stored products, for protection, in respect of ozone depletion, high mammalian toxicity, insect resistance and health hazard. It is now established that vegetable oil, essential oil extract from plants are very effective in controlling certain species of bruchids by their effect, such as melon seed oil (Okunola, 2003). In an experiment on the comparative use of botanical oil extracts in pest management, the efficacy of three botanicals, *Cymbopogon citratus*, *Hyptis suaveolens* and *Eucalyptus globulus* were tested on *Callosobruchus maculatus* (F), a pest of cowpea under ambient laboratory conditions. Insect mortality was observed at the different concentration levels for a few minutes and repeated three times for three weeks. In a screen house experiment, 1.0% concentration was observed to be the average and most effective concentration and subsequently selected as a basis for comparison in all the selected botanicals. The application of *Hyptis suaveolens* at 1.0 % was observed to be the most effective and significantly different from other botanicals, while *Eucalyptus globulus* was the least effective. This is as shown in Tables 1, 2 and 3 below.

Table 1: Cumulative mortality rate of insect pests determined at different times at the first week of the experiment

	2 MINUTES	4 MINUTES	6 MINUTES
Hyptis suaveolens	40.0±0.0a (100%)	0±0.0c (0%)	0±0.0b (0%)
Cymbopogon citratus	37.5±0.5b (93.75%)	2.5±0.5b (6.25 %)	0±0.0b (0%)
Eucalyptus globulus	27.0±1.0c (67.5%)	7.5±1.5a (18.75%)	5.5±0.5a (13.75%)
Control	0±0.0d (0%)	0±0.0c (0%)	0±0.0b (0%)

Means in each column bearing the same letter are not significantly different at the 5 % level of probability using Tukey's Test

Table 2: Cumulative mortality rate of insect pests determined at different times at the second week of the experiment

	2 MINUTES	4 MINUTES	6 MINUTES
Hyptis suaveolens	40.0±0.0a (100%)	0±0.0c (0%)	0±0.0b (0%)
Cymbopogon citratus	35.5±0.5b (88.75%)	4.5±0.5b (11.25%)	0±0.0b (0%)
Eucalyptus globulus	26.0±0.5c (65%)	8.5±0.5a (21.25%)	5.5±0.5a (13.75%)
Control	0.0±0.0d (0%)	0.0±0.0c (0%)	0.0±0.0b (0%)

Means in each column bearing the same letter are not significantly different at the 5 % level of probability using Tukey's Test

Table 3: Cumulative mortality rate of insect pests determined at different times at the third week of the experiment

	2 MINUTES	4 MINUTES	6 MINUTES
Hyptis suaveolens	40.0±0.0a (100%)	0±0.0c (0%)	0±0.0b (0%)
Cymbopogon citratus	38.0±0.5b (95%)	2.0±0.0b (5%)	0±0.0b (0%)
Eucalyptus globulus	28.0±0.5c (70%)	8.0±0.5a (20%)	4.0±0.5a (10%)
Control	0±0.0d (0%)	0±0.0c (0%)	0±0.0b (0%)

Means in each column bearing the same letter are not significantly different at the 5 % level of probability using Tukey's Test

The insecticidal properties possessed by some essential oils are due to their monoterpenoid contents, this include fatty acids, phenolics, alkaloids and terpenes, especially monoter-penes which are the bioactive consistuent of plant products (Lale, 2001). In this study, the biological control of *Sitophilus oryzae*, *Sitophilus zeamais*, *Callosobruchus maculatus* and *Tribolium castaneum* (red flour beetle) using oil extract of *Eugenia aromatica* was as-sessed.

Table 4 shows the comparison of mean mortality count with 100% concentration

Insect Pests	Mean Mortality count (min)			
	10mins	15mins	20mins	30mins
S. zeamais	4.67	3.33	0	0
S. oryzae	6.33	0	5	4.33
C. maculatus	6.33	4.33	4.67	0
T. castaneum	0	5	0	6

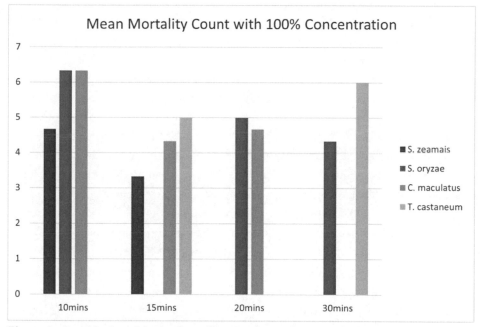

Figure 1: Graphical representation of comparison of mean mortality count with 100% con-centration oil extract of *Eugenia aromatica*.

This research work has confirmed the efficacy and rate of lethality of *H. suaveolens* in pest control. Despite the high rate of consumption of cowpea, statistics show that less than 25% of farmers engage in cowpea production. Consequently, the minority that deals in cowpea production are beset with series of challenges ranging from soil requirement and environmental to pest infestation. Although, there are efforts to combat these attendant problems but the desired goals had not been met. Infestation is perceived to be the most significant of all the challenges. Raja *et al.* (2007) reported that *Callosobruchus maculatus* is one of the major pests of stored cowpea in the tropics corroborating the report of Adugna, 2006 that several insect species attack cereals and pulses store and cause a loss of 10-15% with a germination loss ranging from 50-90%. Meanwhile, Adugna *et al.* (2003) and Doharey (1990) had also given the statistics of germination loss due to the attack of storage pests on cereals and pulse grains to range from 3 - 37% and 4 - 88% respectively. In the control of the insect pests of cowpea, several methods have been adopted. The use of powdery forms of plant extract in pest control has also been reported. Raja *et al.* (2000) reported that when jute bags treated with different plant leaf extracts of *Azadirachta indica, Vitex negmado* and then used for cowpea seed storage, the egg lying rates by the *C. maculatus,* adult emergence and seed damage were reduced, while Lale (2001) also suggested the use of powdery extract of plants in controlling storage insect, and that there are more plant powder extracts discovered that are effective in controlling *C. maculatus.*

Peasant farmers in many parts of Africa frequently mix plant materials with stored grains to prevent insect pest damage (Ofuya, 2003). Ethanol extracts of *Eugenia aromatica* is effective as contact biorational against *S. zeamais, C. maculatus, S. oryzae* and *T. castaneum.* However, the effect of essential oil of *E. aromatica* on these insect pests is dependent on the levels of concentration of oil administered. The difference in response by the different insect pest species could be attributed to the morphological and behavioural differences between the insect (Tanpondju *et al.*, 2002). *E. aromatica* is known to have pungent smell and contains eugenol, sesquiterpene and caryophyline. The action of *E. aromatica* on these beetles could be as a result of stomach poisoning through picking lethal dose of the plant extract by the beetles while feeding on whole or fragmented grains. Therefore, the high toxic effect of *E. aromatica oil on S. zeamais* which is known to have thick exo-skeleton that should give them some level of resistance could probably be due to the feeding habits of these pests during which lethal dose of plant material must be taken up. The comparatively lower susceptibility of *C. maculatus, S. oryzae and T. castaneum* might be due to their feeding habit also. Extracts from *E. aromatica* significantly reduce the population of storage pests. Ethanolic extract of essential oil *of E. aromatica* has been effective as contact biorational against *S. zeamais, C. maculatus, S. oryzae and T. castanuem.* However, the effect of essential oil of *E. aromatica* on these insect pests is dependent on the levels of concentration of oil administered.

Several measures have been adopted to curtail the problems of insect infestation. (Okunola, 2003) had also suggested the use of oils extracted from plants for crop protection, since synthetic insecticide tends to be hazardous to man and its environment. The adoptive use of oil of *E. aromatica* could also be intensified as a probable panacea to crop infestation. Lale (2001) and Ofuya (2003) opined that a few plants in the Nigerian flora with confirmed biological efficacies against species of stored products insects were sufficiently insecticidal to merit scientific formulation.

The use of conventional synthetic insecticides to control maize weevils and other stored product pests has a lot of attendant problems notably high mammalian toxicity, high level of persistence in the environment, workers' safety, insect resistance and health hazards (Sighamony et al., 1986, Adedire and Ajayi, 1996). Some interesting developments have also been made in the use of oils and plant extracts. Oil of groundnut, *Arachis hypogea* has been reported to effectively control *S. zeamais* infestation (Ivbijaro, 1984a). Seed powder and oil of black pepper, *Piper nigrum, P. umbellatum* and *Capsicum frutescens* adversely affect the biology of the maize weevil and also cause high adult mortality (Su and Sondegam, 1980), lvbijaro and Agbaje (1986) reported that pulverized seeds of *Uvaria afzelli, P. umbellatum, Eugenia aromatica* and the bark of *Erythrophleum guineense* were highly toxic to maize weevil when used to surface treat maize grains and subsequently infested with the weevils. According to Akou-Edi (1984), neem seed powder and oil are effective in repelling and killing maize weevil. The powder and oily extracts of pepper fruit, *Dennettia tripetala* have been reported as being effective against *S. zeamais* (Agbakwuru et al., 1978; Lale, 1992). Alligator pepper, *Aframomum melegueta* seed powder and oily extract have been found effective in controlling adult *S. zeamais* (Lale, 1992), while Odeyemi (1993) reported effective control of adult maize with *Lippia adoensis* oil. Huang and Ho (1998) demonstrated that a methylene chloride extract of the spice, *Cinnamomum aromaticum*, had strong contact fumigant insecticidal activity against the maize weevil. Most insecticidal plant powders have generally been recommended for use as undiluted active materials. This has been the case with neem seed powder (Ivbijaro, 1983), seed powder of *Piper guineense* (Ofuya and Dawodu, 2002a), root bark powder of *Zanthoxylum zanthoxyloides* (Ogunwolu and Odunlami, 1996), dry bud powder of *Eugenia aromatica* (Adedire and Lajide, 2001) and fruit powder of *Dennetia tripetala* (Ofuya and Salami, 2002).

Attempts to dilute active insecticidal plant powders with inert materials in formulations have also been carried or documented. Many emulsifiable concentrates have been formulated based on azadirachtin, the active ingredient in neem oil. Opportunities certainly exist in formulating botanical insecticidal materials for economy, efficiency, safety and enhanced effectiveness. Studies on the mixture of powders of dry flower buds of *E. aromatica* and seeds of *P. guineense* with organic flours: cassava, yam and plantain flours, to give insecticidal dusts that are toxic to the notorious storage have also been documented. Consequently, the use of Cashew Nut Shell Liquid (CNSL) has been gaining more attention due to its possession of the active phenolic compounds, Anacardic acid and Cardol, which also have corrosive abrasive properties, Flatedas Nigeria Limited, Unpublished. It has been demonstrated that low concentration of CNSL could be effective in the control of *Callosobruchus maculatus* (Echendu, 1991). Considerable efforts have been made worldwide to find safer biodegradable substitutes for synthetic insecticides (Crombie, 1990). Among these efforts is the use of Cashew Nut Shell Liquid (CNSL), whose efficacy as seed protectant against *C. maculatus* L. on cowpea seeds in storage was reported (Echendu, 1991; Ofuya and Fayape, 1999). Similar work on the protection of Okra, *Abelmoschus esculentus* L. in the control of Okra field insect pests, *Podagrica uniforma* and *P. sjostedti*. *Podagrica* beetles was also documented (Olotuah, 2003). CNSL is a viscous liquid, although it has been confirmed to contain the phenolic compound, which is active in pest control and has been tested and found successful as bactericide (Unpublished). Since the use of plant products in pest control had been documented and had been found to be reliable, the use and economic of use of CNSL is as well desirable. Moreover, the hazards often associated with the use of some synthetic insecticides have not been reported from the use of CNSL. Due to the viscosity of CNSL, it became pertinent to standardize this liquid using a suitable solvent and also to determine its optimal level of protective capability in the screen house to obtain a concentration that will be non-toxic to the plant but will effectively protect it against insect pests.

It must be stressed that the more effective the insecticide, the more its protective capacity and consequently the greater the yield. Keita *et al.* (2001) reported that seeds treated with botanical extract oils did not lose their viability and they also established that powder made from essential oil of different basils provided complete protection against *C. maculatus* and did not show any significant effect on the seed germination rate.

In a similar experiment on pest control focused on the adoptive use of extract, Tapondjou *et al.* (2002) showed that the dry ground leaf of *Chenopodium ambrosioides* inhibited F1 progeny production and adult emergence of the *Callosobruchus chinensis* and *C. maculatus*.

Essential oils are used in perfumery, aromatherapy, cosmetics, incense, medicine, household cleaning products and for flavouring food and drink. They are responsible for the aroma and flavour associated with herbs, spices and perfumes. They are also called volatile oils because they easily diffuse into the air where they are then detectable by our olfactory senses. Essential oils are usually terpenoids another large class of secondary chemicals. Their presence in certain plant parts probably reflects their functions. It has also been reported that some organic pest control product such as Orange Guard use a citrus fruit peel base, such as from lemons and oranges. Citrus oils kill many flying and crawling insects on contact by destroying the waxy coating of the insect's respiratory system. Other organic pesticides use natural extracts to repel rather than kill pests. Some products use garlic or hot peppers and essential oils of herbs such as cloves to repel insects and other pest.

In addition to terpenoids, phenolic compounds are responsible for the aroma and flavour of some spices. For example eugenol is a phenolic compound found in both cinnamon (*Cinnamonum spp*) and cloves from *Syzygium aromaticum*. The functions of essential oils were once considered waste products. However, the biosynthetic pathways that yield essential oils are specialized and imply an expenditure of energy by the plant of their production.

Another debated role for essential oils is to inhibit competing plants or allelopathy. For example, essential oils in sheds leaves can leach into the soil where they may inhibits the germination or growth of other plants competing for the same resources (e.g. light, minerals nutrient, water). As with many other secondary compounds, they are now believed to also defer herbivory and prevent infections by pathogens i.e. bacteria and fungi. Nature had provided plant extracts that make very effective pesticides and insect repellants, which seems to be cheaper, safer and more easily produced than the synthetic insecticides.

On this basis, the use of botanical oil extracts should be seen as an effective protective measure in pest control and adoption of use extended to other crops of economic value.

Plant derived oils and powders have been evaluated and shown to be effective against a number of insect pests (Butler and Henneberry, 1990). The use of several botanicals was adopted and found to be effective. The powdered dry fruits of *Piper guineense* (Schum and Thonn) were reported to give protection to cowpea seeds against the adults of the cowpea bruchid, (Mbata *et al.*, 1995; Onu and Aliyu, 1995). Similarly, many vegetable oils such as corn, cotton seed, peanut olive, soybean oil have also been found to be useful as surface protectant for peas and beans against bruchids (Smith *et al.,* 1995). Many chemicals derived from the plant parts have been tested as promising insect control.

Stored products are attacked by many pests that cause serious damages to the products. Amongst these are storage pests which are often considered to be field-to-store pests. Rawsley 1969, reported that *S. zeamais* has been identified as the most primary pest of stored maize in Ghana where it causes about 35% grain loss within or less than eight weeks of storage in the Euro-bran of most small scale farmers while Gallo *et al.*, 2002 reported that losses in warehouses reach figures of 10%; in Brazil and the losses are around 20% since the storage conditions in the countryside are poor. The maize weevil, *S. zeamais* is described as one of the most destructive pests in tropical and sub-tropical regions, (Caswell, 1976). This pest causes direct damage to grains, causing loss of viability and contamination of produce with excrements, lowering the quality and price of produce (Agboola 1982), (Ivbijaro *et al.*, 1985) (Lale, 1992). Dick 1988, also reported the mode of infestation of *Sitophilus* weevils where they infest grains only after the moisture content had dropped to 50%. Mason 2006, reported *S. zeamais* as a major pest of maize and can be found in numerous tropical areas around the world, including the United States. The maize weevil has been observed to infest other types of stored, processed cereal products such as pasta, cassava and various coarse, milled grains and also known to attack fruit while in storage, such as apples, (Meikle *et al.*, 1999).

Lale (1995, 2002) demonstrated repellency and oviposition deterrence of powdered dry chilli pepper fruits to adults of *C. maculatus*. Ofuya, 2001, also reported that powders made from *Nicotiana tabaccum* L., *Erythrophleum suaveolens* Brenan and *Ocimum gratissinum*

L. significantly reduced oviposition and egg hatch in *C. maculatus*. Consequently, pulverized plant parts have also been found effective in the control of storage pests of cowpea. Aku *et al.* (1998) reported that annonacin obtained from the root bark of *Annona senegalensis* L. and applied as powder, significantly reduced oviposition and adult emergence in *C. maculatus,* and seed weight loss caused by the pest.

Several other research works had also consolidated the assertion of efficiency of different plant plants in various forms in the control of both field and storage pests. For more efficiency, the conventional approach to insect pest control had focused on the extraction and formulation of the active ingredients. Although, most insecticidal plant powders have generally been recommended for use as undiluted active materials, but there have been considerable efforts to formulate active ingredients in inert materials as synergists. This approach had been reported to be highly effective in terms of activity and cost. Researches have further shown that insecticides age over time with a concomitant reduction in efficacy. Several cases had been reported on the efficacy of active ingredients of plant parts in pest management. Such documentations which include the activity of geraniol against the larvae of the caterpillar *Lymantria dispar*, devastator of oak forests, the insecticidal activity (insect repellent and fungicide) of citronellol, citronellyl formiate and citronellyl acetate, nematicidal activities of geraniol and citronellol against *Caenorrhabditis elegans* and *Pratylenchus penetrans* respectively Tsao *et al.* (2000), Reeves and Miller (2010) and antioxidant activities of isomenthone and acaricidal activity of citronellol against *Psoroptes cuniculi* which was observed using linalool and citronellol (Keszei *et al.* (2008), Perruci *et al.* (1997) are all in agreement with the efficacy of eugenol, the active ingredient of *E. aromatica*, in the control of storage pest of cowpea. Conversely, the shelf life of most of these active ingredients has not been documented. More importantly, it was observed that *E. aromatica* powder was still effective in the control of *C. maculatus* and *S. zeamais* five years after the dry flower buds were pulverized. This efficacy shows that with proper storage for five years, Eugenol, the active ingredient of *E. aromatica* could still be effective in the control of storage pests. Thus, this study had made a significant contribution in unraveling the shelf life of *E. aromatica* powders which hitherto appeared to either not to have been earlier investigated nor documented. Fumigation by application of synthetic chemical fumigants such as methyl bromide and aluminium phosphate is perhaps the most effective method of stored products protection against insect depredation (Lale, 2002). However, use of chemical fumigants in stored products protection is being phased out worldwide because of their adverse effects to the environment which includes ozone depletion (WMO, 1995) and the development of insect pest resistance (Zettler *et al.*, 1989). Thus, there was an urgent need to develop new fumigants for post-harvest pest control that were safe, of low cost, convenient to use and environment friendly (Zettler *et al.*, 1997; Papachristos and Stamopolos, 2002). Essential or volatile oils extracted from plants have been shown to possess good potential for use as fumigants against stored product insects including storage bruchids (Don-Pedro, 1996; Shaaya *et al.*, 1997; Papachristos and Stamopolos, 2002; Tapondjou *et al.*, 2002). Less refined plant materials such as crushed plant materials, have been found to be effective and more affordable by poor farmers in developing countries. Bulbs of *A. cepa* and *A. sativum* exhibit insect controlling properties. Ofuya (1986) reported that crushed *A. cepa* bulb scale leaves reduced cowpea seed damage by *C. maculatus* through inhibition of oviposition and adult emergence, which is largely consistent with findings in this study. *A. sativum* was more effective in the control of *C. maculatus* than *A. cepa*. Many chemical pesticide components which have been unravelled among *Allium* species include allicin, hydrocyanic acid, oxalic acid, pyrogallol, quercitin and saponin (Dales, 1996). Some of the volatile chemical ingredients present in volatiles of *A. sativum* are allicin, thioacrolein, ajoene, 2-propene sulfenic acid, 2-propene thiol and propylene (Jain and Apitz-Castro, 1993; Gurusubramanian and Krishna, 1996). Since there is no contact between the crushed bulbs and the insect pests, the mortality and subsequent reduction in percentage adult emergence could be presumably due to the diffusion of the volatile chemicals contained in the bulbs which might have affected vital physiological and biochemical processes associated with embryonic development, which were consequently disrupted. Gurusubramanian and Krishna (1996) have also reported that exposure of freshly laid eggs (< 24 h old) of *Earis vitella* Fabricius and *Dysdercus koenigii* (Fabricius) to volatiles from bulbs of *A. sativum* significantly reduced their hatchability. Research efforts on products development are been focused on more ecological tolerable control measures including the use of wood ash, dried chilli pepper fruits and onion scale leaves.(Ofuya, 2002). There is also increasing awareness that plants, especially plants of medicinal importance, possess chemicals, which naturally protect stored grains from pest and pathogens. Dried and Powder tissues of *Dicoma sessiliflora* and *Neorautaunenia mitis* have been revealed to have significant activity against adults and mortality of the development stage (Boeke *et al.,* 2001). Okonkwo and Okoye, 2006 suggested that alternative strategies should include search for new types of insecticides and the re-evaluation and use of safe traditional botanical pest control agents which are already a part of our diet. This now informed researches on the suitability of readily available food crops in pest control, and thus the biological control of *Sitophilus oryzae* using Palm kernel, Coconut and essential oil of *Eucalyptus camaldulensis* were evaluated in a laboratory experiments. Oils of Palm kernel, Coconut and Eucalyptus were effective as contact treatments against *Sitophilus oryzae*, (Table 5-7). In this experiment, the use of some plant oils, Palm kernel, Coconut and Eucalyptus, was evaluated against rice weevils *Sitophilus oryzae* in a local variety of rice "Ofada". The trial involved exposing adult rice weevils *Sitophilus oryzae* to various levels of concentration of oils. All the studies were taken under normal laboratory temperature of 27^0C. The oils were incorporated into the Petri dishes containing fifty (50) whole grains of rice and ten (10) pairs of insect pests (five males and five females) at different levels of concentrations. The *in vitro* experiment was conducted with the oils diluted with

0.1% of ethanol at 100, 50, 25, 10, 5, 4, 3, 2 and 1 percentages level of concentrations. The ability of the plant oils to protect rice grains was assessed in terms of mortality rates after 5, 10 and 15 minutes of post treatments. Each treatment and level of concentrations was replicated three (3) times. In handling of the extracts, there was no phytochemical effect on human. Essential oil of *Eucalyptus camaldulensis* was the most effective, with the fastest insecticidal effect. In this study, it was affirmed that the use of plant oils can serve as alternative biopesticide to synthetic insecticides.

Table 5: Mean mortality rate of *Sitophilus oryzae* after Treatment with Eucalyptus oil

Concentration %	5mins	10mins	15mins
100	8.3±0.33d	1.3±0.33a	0.3±0.33a
50	7.3±0.33d	1.3±0.33a	1.0±0.00b
25	6.6±0.33c	1.3±0.66b	2.0±0.57c
10	5.6±0.33b	2.0±0.57b	2.3±0.33c
5	5.0±0.57b	2.3±0.33b	2.6±0.88c
4	4.0±0.57b	3.0±0.00c	3.0±0.88c
3	3.6±0.33b	4.0±0.00d	3.0±0.57c
2	1.6±0.33a	4.6±0.66d	3.6±0.33c
1	1.0±0.57a	5.0±0.57d	4.0±0.00d

In a column, values with the same alphabets are not significantly different at 5% level of probability using "Turkey" Honestly Significant Test.

Table 6: Mean mortality rate of *Sitophilus oryzae* after Treatment with Coconut oil

Concentration %	5mins	10mins	15mins
100	7.3±0.33e	2.0±0.00a	0.6±0.33a
50	6.6±0.33e	1.6±0.66a	1.6±0.33a
25	5.6±0.33d	2.3±0.33a	1.6±0.33a
10	5.0±0.00c	2.3±0.33a	2.6±0.33b
5	4.0±0.57b	2.0±0.00a	4.0±0.57c
4	3.0±0.57b	2.3+0.33a	4.6±0.33c
3	2.6±0.33b	2.3±0.33a	5.0±0.00d
2	1.3±0.88a	3.3±0.33b	5.3±0.66d
1	0.3±0.33a	3.0±0.00b	6.6±0.33e

In a column, values with the same alphabets are not significantly different at 5% level of probability using "Turkey" Honestly Significant Test.

Table 7: Mean mortality rate of *Sitophilus oryzae* after Treatment with Palm kernel oil

Concentration %	5mins	10mins	15mins
100	7.3±0.33f	2.6±0.33b	0.0±0.00a
50	6.3±0.33e	2.6±0.66b	1.0±0.57b
20	6.3±0.33e	2.0±0.57b	1.6±0.33b
10	5.3±0.33d	1.0±0.57b	3.6±0.33c
5	5.0±0.57d	0.6±0.33b	4.3±0.88c
4	3.6±0.66c	1.3±0.33b	5.0±1.00d
3	2.0±0.57b	0.6±0.33a	7.3±0.88e
2	1.3±0.66b	0.6±0.66a	8.0±0.00e
1	0.3±0.33a	1.6±0.33a	8.0±0.57e

In a column, values with the same alphabets are not significantly different at 5% level of probability using "Turkey" Honestly Significant Test.

Palm kernel oil is edible plant oil derived from the kernel of oil palm *Elaies guineensis*. It is commonly referred to as 'adin dudu' in Yoruba language. It is one of the few highly saturated vegetable fats and does not contain cholesterol found in unrefined animal fats although saturated fats intake increases cholesterol. Palm kernel oil is a common cooking ingredient; its increasing use in the commercial food industry throughout the world is buoyed by its lower cost, the high oxidative stability (saturation) of the refined product when used for frying, and its lack of cholesterol and trans-fatty acids, both viewed as being heart-healthy attributes. Resembling coconut, palm kernel oil is packed with myristic and lauric fatty acid and therefore suitable for manufacture of soaps washing powders and personal care products (Musa, 2009). Coconut oil is edible oil extracted from the kernel of matured coconut harvested from the coconut farm *Cocos nucifera*. It is popularly known as 'adin agbon' in Yoruba language. Coconut oil is very heat stable which makes it suited to method of cooking at high temperature like frying. Because of its stability, it is slow to oxidize and thus, resistance to rancidity, lasting up to 2 years due to high saturated fat content (Fife, 2005). Coconut oil has varied application from cooking in many south Asian countries (Clark, 2011) to biodiesel fuel applications in diesel engines. It has been tested for use as an engine lubricant. As regards personal application, coconut oil can be used as a personal moisturizer helping with dry skin and reduces protein loss when used in hair (Rele, 2003). Its local application as body moisturizer in Yoruba cultures is well known. Eucalyptus oil is the generic oil for distilled oil from the leaf of a Eucalyptus, a genus of the plant family Myrtaceae native of Australia and cultivated worldwide. Eucalyptus oil has a history of wide application of a pharmaceutical antiseptic, repellent, flavouring, fragrance and industrial uses. The leaves of selected eucalyptus species are steam-distilled to extract eucalyptus oil. Eucalyptus oil has antibacterial effects on pathogenic bacteria in respiratory tract (Salari *et al.*, 2006). It is also used in personal hygiene products for anti-microbial properties in dental care (Nagata, 2008) and soaps. Inhaled eucalyptus oil vapour is a decongestant and effective treatment for bronchitis (Lu, 2004). It can also be used as an insect repellent and biopesticide.

Plants are repositories to arrays of compounds with diverse functions of which large percentages are yet untapped. Palm kernel and Coconut oil are edible, the oils and extracts are non-toxic to human and livestock, this quality is an advantage over other bioactive compounds from non-edible plants source. Some plant extracts have also been tested on cereals and grains legumes for absorptive capacity and confirmed efficient as reported by Adedire and Akinkurolere, 2005, Akinkurolere *et al.,* 2006). On the other hand, essential oil of *Eucalyptus camaldulensis* that contains Allo-aromadendrene, Alpha-pinene, Borneol, Citronellic acid and Tannins was the most effective. It is also an aromatic and astringent plant with a pungent smell. The efficacy of various parts of *Jatropha curcas* in the control of termites had been considered in the laboratory where the leaves, seeds, bark and the roots were pulverized and diluted in non-toxic polar (ethanol) and non-polar (n-hexane) solvents to extract the active ingredient. Jatropha (*Jatropha curcas L.)* or Physic nut is a drought resistant monoecious large shrub or small tree (5–8) m tall. It belongs to the genus *Jatropha* which consists of over 175 species, and a member of the Euphorbiaceae family. It is a drought-resistant perennial, growing well in poor soil. *Jatropha curcas* plant produces seeds with an oil content of 37%. It has been recorded that *J. curcas* has been useful medicinally. A good crop can be obtained with little effort. Depending on soil quality and rainfall, oil can be extracted from *Jatropha* nuts after two to five years. The annual nut yield ranges from 0.5 to 12 tons. The kernels consist of oil to about 60 percent; this can be transformed into biodiesel fuel through esterification. Insect pests such as termites are one of the major limiting factors in increasing productivity of many crops. *Microcerotermes spp.* is economically the most important termite because of the damage it causes on agricultural and domestic products. Records show that parts of *Jatropha curcas* had been used in traditional medicine and as raw material for pharmaceutical and cosmetic industries and when adopted as a component of IPM programme, it could considerably reduce pest infestation. This adoption could as well decrease the use of conventional pesticides, besides maintaining high crop yields. N-hexane bark extract of *Jatropha curcas* tested against termites *Microcerotermes beesoni* possesses high and reliable termicidal properties that could be used in the pest management. Observations on the repellent and toxic effects of *Jatropha curcas* on termites has proven that this plant could be adopted in termite control as well as the control of other evasive insect pest. In the same vein, every part of the plant had be confirmed to be suitable for pest control while the extractions could be done by using any non-toxic and readily available solvent including water. The propagation of the crop is however without stress as the plant could germinate under minimal agricultural practice. Therefore, *J. curcas* is a reliable plant that could be employed in the control of termites in order to eradicate the destructive tendencies of the insect and as such alleviate the attendant challenges as often experienced at homes and on the field. This plant should therefore be encouraged to be propagated as edges at home and on the field. Laboratory experiments had been conducted on the insecticidal properties of n-hexane and ethanoic extracts of the leaves and roots of *Chrysopogon zizanioides* in the control of cockroaches and sugar ants where dried and fresh parts of the plant were used for the experiment. Vetiver grass, *Chrysopogon zizanioides (L)* Roberty, commonly known as vetiver is a perennial bunchgrass of the Poaceae family, native to India. It was formerly known as *Vetiveria zizanioides*. Vetiver is mostly related to Sorghum but shares many morphological characteristics with other fragrant grasses such as Lemongrass (*Cymbopogon citratus*), Citronella (*Cymbopogon nardus*) and Palmarosa (*Cymbopogon martini*). Vetiver grass extracts has potent insecticidal properties against the activities of cockroaches and sugar ants which are the commonest household pests (Odebiyi and Sofowora 1998). According to the objectives of this research work, vetiver grass extract has posed to be a healthy and natural method of curbing household insect pests. Also, the use of vetiver plant extracts is an environmentally-friendly solution that can be adopted by both homes and industries in the reduction/eradication of household pests' activities most especially cockroaches and sugar ants, (Olotuah, 2016).

Keital (2001) reported that seeds treated with botanical oil extracts did not lose their viability and also established the powder made from essential oil at different basics provided complete protection against *M. beesoni* and did not show significant effect on the potent on seed germination rate. This also substantiates the use of the shoot and root of *Cymbopogon citratus* and the observed efficacy in termite control. This is an assertion of the use of morphological parts plants as being not restricted to any specific part of the plant. The efficacy of lemon grass is consequently in accordance with Zue, *et al.* (2001) on the use of lemongrass plant extracts as anti-termitic and repellent ingredient against termites where also a better performance was obtained in comparison to a synthetic termicide. The use of lemongrass extract has provided safer alternative against the use of chemical termicides because of its low toxicity to non-target organisms, especially humans and beneficial insects.

Conclusion

According to Ofuya (2003), many synthetic insecticides formulated as dusts such as pirimiphos-methyl are very effective for stored products protection, but their use has several drawbacks such as increasing costs, inconsistent supplies and hazards to man and the environment. Formulation, which is the science of diluting the toxicant (active material or a.m.) with inert materials that impart desirable biological and physical properties to the insecticides (Lale, 2002) was considered as an option in increasing the adoption potential of the insecticidal plant materials and found effective. Olotuah (2016) equally reported that formulation could facilitate effectiveness of active material through better distribution over product and contact with pest insects. Consequently, formulation could also make active materials more economical to use; the concept which corroborates with the report of Pedigo and Rice (2006) that inert materials added to insecticidal materials to form dusts, could include organic flours and inert clays. In the principle of formulation, the botanicals are often pulverized to particular particle sizes for efficiency, but this could vary with the plant products. Organic flours such as yam, cassava and plantain flours are produced in large quantities in Nigeria. It is being hypothesized that some proven insecticidal plants powders in Nigeria can be formulated into effective and more economical dusts using these readily available organic flours as carriers.

References

Adedire, C.O. and Ajayi, T.S. (1996) Assessment of the insecticidal properties of some plant extracts as grain protectants against the maize weevil, *Sitophilus zeamais* Motschulsky. *Nigeria Journal of Entomology*, 13: 93 - 101.

Adedire, C.O. and Akinkurolere, R.O. (2005) Bioactivity of some plant extracts on coleopterous pests of stored cereals and grain legumes in Nigeria. *Journal of Zoology Research*. 26, pp 243-249.

Adedire, C.O. and Lajide, L. (2001) Efficacy of powder of some tropical plant in the controls of the pulse beetle. (*Callosobrucus maculatus* (F) (Coleoptera Bruchidae). *Journal of Applied Tropical Agriculture*, 6, 11-25.

Adugna, H., Dangnew, G., Biniam, Z., Biniam, A. (2003) On Farm Storage Studies in Eritrea. Dry Land Coordination Group/OCG Report NO, 28, 2003.

Agbakwuru, E.O.P., Osisiogu, L.U.W., Ugochukwu, E.N. (1978) Insecticides of Nigerian Vegetable Origin. II. Some Nitro-alkanes as protectants of stored cowpeas and maize against insect pests. *Nigerian Journal of Science*, 12: 493 - 504.

Agboola, S.D. (1982) Research for effective food storage in Nigeria. NSPRI Occasional Paper Series; 4:21.

Akinkurolere, R.O., Adedire, C.O. and Odeyemi, O.O. (2006) Laboratory evaluation of the toxic properties of forest Anchomanes, *Anchomanes difformis* against pulse beetles *Callosobruchus maculatus* (*Coleoptera: Bruchidae*). *Insect Science* 13, pp 25-29.

Akou-Edi, D. (1984) Effects of neem seed powder and oil on *Tribolium congusum* and *Sitophilus zeamais*. In: Schmutterer, H., Ascher, K.R.S. (eds.) Natural Pesticides from the neem tree (*Azadirachta indica* A Juss.) and other tropical plants. Proceedings of the Second International *Neem* Conference, Federal Republic of Germany, 25 - 28 May, 1983. Eschborn, Germany Federal Republic, pp. 445 - 452.

Aku, A.A., Ogunwolu, E.O., Attah, J.A. (1998) *Annona senegalensis* L. (*Annonaceae*): Performance as a botanical insecticide for controlling cowpea seed bruchid, *Callosobruchus maculatus* (F.) (*Coleoptera*: *Bruchidae*) in Nigeria. Journal of plant diseases and protection. 105:513-519.

Boeke, S.J., Van Loon, J.J.A., Van Huss, A., Kossou, D.K. and Dike, M. (2001) The use of plant products to protect stored leguminous seeds against seed beetles: A review: *Backhuys Publishers*, Netherlands pp.108.

Boeke, S.J., Vanloon, J.J.A., Van Huss, A., Kossou, D.K., Dicke, M. (2002) The use of plant products to protect stored leguminous seeds against seed beetles: A review (M). The Netherlands: *Backhuys Publishers*, 108.

Butler, G.D., Henneberry, T.J. (1990) Pest Control on Vegetables and Cotton with Household Cooking Oils and Liquid Detergents. Southwest. *Entomol.* 15, 123-131.

Caswell, G.H. (1976) The storage of grain legumes. In: A. Youdeowei (ed.), Entomology and the Nigerian Economy. Entomological Society of Nigeria.;1 - 142. Chemosphere, 50, 775-780.

Clark, M. (2011) Once a Villain, Coconut oil charms the health food world. The New York Times. Retrieved 2011-03-02

Crombie, L. (Ed.) (1990) Recent Advances in the Chemistry of Insect Control II. Special Publication No. 79, *Roy. Soc. Chem.*, Cambridge, U.K. pp: 296.

Dales, M.J. (1996) A review of plant materials used for controlling insect pests of stored products. NRI Bulletin N° 65. 84 p. Natural Resources Institute (NRI), Chatham, Kent, UK.

Dick, K. (1988) A review of Insect Infestation of Maize in Farm Storage in Africa with special reference to the ecology and control of *Prostephanus truncatus*. Overseas Development of Natural Resources Institute Bulletin U.K.; 18: 420.

Doharey, R. B. (1990) Eco-toxicological Studies on Pulse Beetles Infesting Green Grain. Comparative efficacy of some edible oils for the production of green grain (*Vigna radiata* (L.) Wilezek against pulse beetles *Callosobruchus chinensis* (L.) and *Callosobruchus maculatus* (F.). *Bulletin of Grain Technology* V. 28(2), 116-119.

Don-Pedro, K.N. (1996) Fumigant toxicity of citrus peel oils against adult and immature stages of storage insect pests. Pesticide Science 47:213-223.

Echendu, T.N.C. (1991) Ginger, Cashew and Neem as surface Protectants of Cowpea against Infestation and Damage by *Callosobruchus maculatus* (Fab.) *Tropical Science* 31: 209 – 211.

Fife, B. (2005) Coconut cures, Piccadilly Books Limited. pp 184-185. ISBN 978-0-941599-60-3

Gallo, D., Nakano, O. Silveira Neto, S., Carvalho, R. P. L.; Baptista, G. C. de, Berti Filho, E., Parra, J. R. P., Zucchi, R. A., Alves, S. B., Vendramin, J. D. (2002) *Entomologia Agrícola* (*Agricultural Entomology*). Piracicaba, Fealq. 920 p.

Gurusubramanian, G., and S.S. Krishna (1996) The effects of exposing eggs of four cotton insect pests to volatiles of *Allium sativum* (Liliaceae). Bulletin of Entomological Research 86:29-31.

Huang, Y, Ho, S.H. (1998) Toxicity and antifeedant activities of Cannamaldehyde against the grain storage insects, *Tribolium castaneum* (Herbst) and *Sitophilus zeamais* Motsch. *J. Stored Prod. Res.* 34: 11 - 17.

Ivbijaro, M.F. (1983) Preservation of cowpea, *Vigna unguiculata* (L.) Walp. with neem seed, *Azadirachta indica* A. Juss. *Protection Ecology* 5: 177 - 182.

Ivbijaro, M.F. (1984a) Toxic effects of groundnut oil on the rice weevil, *Sitophilus oryzae* (L.). *Insect Science and its Application* 5: 251 - 252.

Ivbijaro, M.F. and Agbaje, M. (1986) Insecticidal activities of *Piper guineense* and *Capsicum species* on the cowpea bruchid, *Callosobruchus maculatus. Insect Science and its Application* 7: 521 - 524.

Ivbijaro, M.F., Ligan, C., Youdeowei, A. (1985) Control of rice weevils, *Sitophilus oryzae* (L.) in stored maize with vegetable oils. Agricultural Ecosystems and Environment.;14:237 – 242.

Jacobson, M. (1989) *Focus on phytochemical pesticides. Volume1.* The Neem Tree. Boca Raton, Florida, *USA CRC Press Book, Inc* 178p.

Jain, M.K., and R. Apitz-Castro (1993) Garlic: A product of spilled ambrosia. Current Science 65:148-156.

Keital, S.W. (2001)"Efficacy of essential oil of ocimum basilicum L. Insecticidal fumigant L. applied as an insecticidal fumigant and powder to control callosobruchus maculatus [Coleoptera; Bruchidea]," *Journal of Stored Products Research,* vol. 37, pp. 339-349, 2001. *View at Google Scholar | View at Publisher*

Keszei, A., Burbakel, C.L., Foley, W.J. (2008) A molecular perspective on terpene variation in Australian Myritaceae. Aust. J. Bot.; 56:197-213.

Kim, S. I. (2003) "Insecticidal activities of aromatic plant extracts and essential oils against *Sitophilus oryzea* and *Callosobruchus chinensis,*" *Journal of Stored Products Research,* vol. 39, pp. 293-303.

Lale, N.E.S. (1995) An overview of the use of plant products in the management of stored product *Coleoptera* in the tropics. Post-Harvest News and Information. 6:69N-75N.

Lale, N.E.S. (1992) Oviposition deterrent and repellent effects of products from dry chilli pepper fruit, *Capsicum sp* on *C. maculatus.* Post-harvest Biology and Technology.1:343 - 348.

Lale, N.E.S. (2001) The impact of storage insect pests on post-harvest losses and their Management in the Nigerian agricultural system. *Nigerian Journal of Experimental and Applied Biology* 2: 231 - 239.

Lale, N.E.S. (2002) Stored-product entomology and acarology in tropical Africa. 204 p. Mole Publications, Maiduguri, Nigeria.

Lu, X.Q., (2004) Effect of *Eucalyptus globulus* oil on lipopolysaccharide – induced chronic bronchitis and mucin hypersecretion in rats. *Journal of Chinese Materia Medica*. 168-171 (2)

Mason, L.J. (2006) Grain Insect Fact Sheet, E-237- W: Rice, Granary and Maize Weevils *Sitophilus oryzae* (L.), *S. granarius* (L.), and *S. zeamais* (Motsch). Purdue University; 2003.

Mbata, G.N; Oji, O.A. and Nwana, I.E. (1995) Insecticide action on Preparations from the Brown Pepper, *Piper guineense* (Schum) Seeds to *Callosobruchus maculatus* (Fabricius). *Discovery and Innovation* 7 139-142.

Meikle, W.G., Holst, N., Markham, R.H. (1999) Population simulation model of *Sitophilus zeamais* (*Coleoptera*: *Curculionidae*) in grains stores in West Africa. Environmental Entomology. 28(5):836-844.

Musa, J.J., (2009) Evaluation of lubricating properties of Palm kernel oil. Leonardo Electronic, *Journal of Practices and Technologies* (14): pp 107-114. ISSN 1583-1078

Nagata, H., (2008) Effect of eucalyptus extract chewing gum on periodontal health. *Journal of Periodontology*. Vol. 79, No 8 pp 1378-1385(7).

Odebiyi, O. and Sofowora, E. A. (1998). Phytochemical screening of Nigerian Medicinal plants II; Lloyd, 41,234 – 236

Odeyemi, O.O. (1993) Insecticidal properties of certain indigenous plant oils against *Sitophilus zeamais* Mots. *Appl. Ent, Phytopath*. 60: 19 - 27

Ofuya T.I. (2002) Use of wood ash, dry chilli pepper fruits and onion scale leaves for reducing *C. maculatus* (F) in stored cowpea (*Vigna unguiculata)* seeds. *Journal of Agricultural Science;* Cambridge 107, pp 467-468

Ofuya, T.I. (1986) Use of wood ash, dry chilli pepper fruits and onion scale leaves in reducing *Callosobruchus maculatus* (Fabricius) damage in cowpea seeds during storage. Journal of Agricultural Science (Cambridge) 107:467-468.

Ofuya, T.I. (2001) Biology, ecology and control of insect pests of stored food legumes in Nigeria. p. 23- 58. *In* Ofuya, T.I., and N.E.S. Lale (eds.) Pests of stored cereals and pulses in Nigeria. Dave Collins Publications, Nigeria.

Ofuya, T.I. (2003) Beans, insects and man. Inaugural Lecture Series 35. 45 p. The Federal University of Technology, Akure, Nigeria.

Ofuya, T.I., Dawodu, E.O. (2002a) Aspects of insecticidal action powder from *Dennetia tripetala* Bak. against *Callosobruchus maculatus* (F.) (Coleoptera: Bruchidae). *Nigerian Journal of Experimental and Applied Biology* 3 (2): 303 - 308.

Ofuya, T.I. and O.E. Fayape (1999) Evaluation of Cashew Nut Shell Liquid and Powder for the Control of *Callosobruchus maculatus* (Fabricus) (Coleoptera: Bruchidae) Infesting Stored Seeds of Cowpea, *Vigna unguiculata* (L.) Walpers. *Appl. Trop. Agric*. Vol. 4: 72 -77.

Ofuya, T.I., Salami, A. (2002) Laboratory evaluation of different powders from *Dennetia tripetala Bak.*, as protectant against damage to stored seeds of cowpea by *Callosobruchus maculatus* (F). *J. Sust. Agric and the environ* 4 (1): 36-41.

Ogunwolu, E.O., Odunlami, A.T. (1996) *Suppression of seed bruchid (Callosobruchus maculatus F.)* development and damage on cowpea (*Vigna unguiculata* (L.) Walp.) with *Zanthoxylum zanthoxyloides* (Lam.) Waterm. (Rutaceae) root bark powder when compared with neem seed powder and pirimiphos methyl. *Crop Protection* 15: 603 - 607.

Ogunwolu, E.O., Igoli, J.O., Longs, N.N. (2001) Reduction in reproductive fitness of *Callosbruchus maculatus* F. exposed to *Zanthoxylum zanthozyloides* (Lam) Watern. *J. Her. Spi and Med. Plants* 6: 19-27.

Okonkwo, E.U. and Okoye, W.I. (2006) The efficacy of four seed powders and the essential oil as protectants of cowpea and maize grains against infestation by *Callosobruchus maculatus* (F) (Coleoptera: bruchidae*)* and *Sitophilus zeamais* (M) (Coleoptera: curculionidae*)* in Nigeria. *Int.*

Okunola, C.O. (2003) Use of Melon seed oil for the control of bruchid damage in cowpea, *African crop science proceedings*, vol. 6 Nairobi, Kenya.

Olotuah, O.F. (2016) The Use of *Chrysopogon zizanioides* Extracts in the Control of Cockroaches (*Periplaneta americana*) and Sugar Ants (*Camponotus consobrinus*). *International* Journal of Applied Research and Technology Vol. 5, No. 3, 37 – 42.

Olotuah, O.F. (2003) Evaluation of cashew nut shell liquid (CNSL) for the control of *podagrica* beetles infesting okra, *Abelmoschus esculentus* (L) Moench. Applied Trop. Agric. 8: 8-10.

Onu, I. and Aliyu, M. (1995) Evaluation of Powdered Fruits of Four Peppers (*Capsicum spp*) for the Control of *Callosobruchus maculatus* (F) on Stored Cowpea Seeds. *International Journal of Pest Management* 41, 143-145.

Papachristos, D.P., and D.C. Stamopolos (2002) Toxicity of vapours of three essential oils to immature stages of *Acanthoscelides obtectus* (Say) (Coleoptera: Bruchidae). Journal of Stored Product Research 38:365-373.

Pedigo, L.P., Rice, M.E. (2006) Entomology and Pest Management. Pearson Prentice Hall, New Jersey, USA.: 749.

Perruci, S., Cioni, P.L., Cascella, A., Macchioni, F. (1997) Therapeutic efficacy of linalool for the topic treatment of parasitic otitis caused by *Psoroptes cuniculi* in the rabbit and in the goat. Med. Vet. Entomol. 11:300-302.

Raja, N., S. Albert, A. Babu, S. Ignacimuthu and S. Dorn (2000) Role of botanical protectants and larval parasitoid, *Dinarmus vagabundus* (Timberlake) (Hymenoptera: Pteromalidae) against *Callosobruchus maculatus* Fab. (Coleoptera: Bruchidae) infesting cowpea seeds. Mala. *Applied Biol.*, 29: 55-60.

Raja, M., John William S. and M. Jayakumar (2007) Repellent Activity of Plant Extracts against Pulse Beetle *Callosobruchus maculatus* (Fab.) (Coleoptera: Bruchidae). *Hexapoda* 14 (2), 142-145.

Rawsley, J (1969) Crop Storage Food Research and Development Units, Accra, Ghana. Technical Report. (1) FAO PL, SF/GHA. 7 : 8.

Reeves, W.K., Miller, M.M. (2010) Aqueous 2% Geraniol as a Mosquito Repellent Failed against *Aedes aegyption* Ponies. J. Am. Mosquito Control Assoc. 26(3):340-341.

Rele, A. (2003) Effect of mineral oil, sunflower oil and coconut oil on prevention of hair damage (PDF). *Journal of Cosmetic Science,* 54(2). pp 175-192

Salari, M.H., Amine, G., Shirazi, M.H., Hafezi, R., Mohammadypour, M. (2006) Antibacterial effect of *Eucalyptus globulus* leaf extract on pathogenic bacteria isolated from specimens of patients with respiratory tract disorder. *Clinical Microbiology: European Society of Clinical Microbiology and Infect. Diseases.* 12(2):194-196.

Shaaya, E. and M. Kostyukovysky (2006) "Essential oils: Potency against stored product insects and mode of action" *J. Entomol. Acarol. Res., (43),* pp. *245-257.*

Shaaya, E., M. Kostjukovski, J. Eilberg, and C. Sukprakarn (1997) Plant oils as fumigants and contact, insecticides for the control of stored-product insects. Journal of Stored Products Research 33:7-15.

Sighamony, S, Anees, I, Chandrakala, T, Osmani, Z. (1986) Efficiency of certain indigenous plant products as grain protectants against *Sitophilus oryzae* (L.) and *Rhizopertha dominica* (F.). *Journal of Stored Products Research* 22: 21 - 23.

Smith, D.N., King, W.J., Topper, C.P., Borna, F. and Cooper, J.F. (1995) Alternative Techniques for Sulphur Dust for Cashew Trees for the Control of Powdery Mildew Caused by the Fungus *Oidium anacardii* in Tanzania. *Crop. Prot.* Vol 14, No 7 pp555-560.

Su, H.C.F., Sondengam, B.L. (1980) Laboratory evaluation of toxicity of two alkaloidal annides of *Piper guineense* to four species of stored products insect. *Journal of Entomological Society* 15: 47 - 52.

Tapondjou, L.A., C. Adler, H. Bouda, and D.A. Fontem (2002) Efficacy of powder and essential oil from *Chenopodium ambrosioides* leaves as post-harvest grain protectants against six-stored product beetles. Journal of Stored Products Research 38:395-402.

Tsao, R., Yu, Q. (2000) Nematicidal activity of monoterpenoid compounds against economically important nematodes in agriculture. J. Essent. Oil Res.: 12:350-354.

WMO (1995) Scientific assessment of ozone depletion: 1994. World Meteorological Organization, Global Ozone Research and Monitoring Project. Report N° 37. World Meteorological Organization (WMO), Geneva, Switzerland.

Zettler, J.L., J.G. Leesch, R.F. Gill, and B.E. Mackey (1997). Toxicity of carbonyl sulfide to stored product insects. Journal of Economical Entomology 90:832- 836.

Zettler, J.L., W.R. Halliday, and F.H. Arthur (1989) Phosphine resistance in insects infesting stored peanuts in the Southeastern United States. Journal of Economical Entomology 82:1508-1511.

Zhu, B.C., Henderson, G. Chen, F., Fei, H. and Laine R.A. (2001). Evaluation of vetiver oil and seven insect active essential oils against the Formosan subterranean termite. *J. Chem.* 27(8): 1617-25.

Zue, D.Z, Hueng, Cen, Y.J., Fred S. and Hu, X.G. (2001). Anti-feedant activities of secondary metabolites from *Ajuga nipponensis* against adult of striped flea beetle, *Phyllotreta striolata. J. Pest Sci.* 82(2): 195-202.

CHAPTER 5

Strategies in the Use of Botanicals and Microorganisms in Weed Control

by
Olatunde Philip Ayodele
Department of Agronomy,
Faculty of Agriculture, Adekunle Ajasin University,
Akungba-Akoko, Ondo State, Nigeria.
Email*: olatunde.ayodele@aaua.edu.ng*

Abstract

Weed control deters losses from weeds. However, some control methods may adversely affect human health and the environment. Hence, the use of botanicals and microorganisms to exert biocontrol on weeds has gained relevance due to its environmental friendliness and absence of chemical residues in agricultural products. Target specificity and self-perpetuating populations are other benefits associated with the use of botanicals and microorganisms for weed control. This chapter focuses on the attributes of good biocontrol agents and approaches that can be deployed in exerting biocontrol on weeds using botanicals and microorganisms. Also, methods of optimizing the weed control efficiency of these biocontrol agents are discussed.

Keywords: Allelopathic weeding; biocontrol; bioherbicide; microbial weeding; weed control

Introduction

Weeds compete with crops for soil nutrients, light, water, carbon dioxide and space (Zhang and Chen, 2017). Consequently, crop yields are reduced by weed interference. Annual global loss caused by weeds has been estimated to be well above $100 billion U.S. dollars (Swanton *et al.,* 2015). Effective weed control is capable of mitigating the negative effect of weeds on crop yield (DiTommaso *et al.,* 2016). Hence, the need for weed control in agriculture cannot be overemphasised.

Weed control can be implemented through biological, chemical, cultural, physical and prevention methods. These methods have individual limitations and adverse effects. Hence, weed control methods may engender complications. For instance, chemical weed control has ecotoxicological effect on biodiversity (Oladele and Ayodele, 2017), causes crop injury (Singh *et al.,* 2019) and predisposes human health to dangers (Del Pino *et al.,* 2017;

Kwiatkowska *et al.*, 2014). Similarly, physical weed control through hand weeding has health implications; it has been reported to cause back injuries on field workers (Minkoff-Zern, 2017). Also, cultural methods such as the use of tillage for weed control leads to soil degradation through erosion (Reicosky, 2003). Hence, the trend in weed management is to promote safe and eco-friendly weed control methods.

Biological control method is being given attention due to its comparative advantage over chemical and physical control methods. Biological weed control is the deliberate use of living organisms in suppressing weed abundance below level of economic importance (Crump *et al.*, 1999). This definition of biological control has been disputed on the premise that organisms that do not act on weeds in-situ and their products are not captured. For instance, while it is acceptable that a living allelopathic plant that negatively affects the growth of the neighbouring weeds exercised biological control, the use of allelopathic plant residue as mulch in controlling weeds is classified as cultural method.

Agronomic practices such as plant spacing, planting of resistant varieties, intercropping and crop rotation are classified as 'cultural weed control methods' on the basis that they serve other purposes primarily other than weed control. Cultural weed control method comprises agronomic practices that have the ability of putting weeds in check, in addition to their primary objectives. The classification of some agronomic practices involving living organism as 'cultural methods' has introduced ambiguity into the classification of weed control methods. The definition of biological weed control stated earlier and the classification of some agronomic practices involving living organisms as culture control has limited biological weed control to the use of herbivores, antagonistic microorganisms and allelopathic control. However, it has been suggested that cultural control should be regarded as a mean to biological control where organisms are involved (Gabriel and Cook, 1990). Hence, biological weed control discussed in this chapter is addressed from this perspective. More so, Harris (1991) interpreted biological weed control to be the use of organisms and their products in controlling weeds. The term 'biological control' has been abbreviated as 'biocontrol' in many literatures, hence they are used interchangeably. Generally, biocontrol has the advantage of reduced tendency of harm to nontarget organisms, relatively low cost of control compared to other methods, no toxic residue and biodegradable control agents. Establishment of self-perpetuating populations is another advantage of biocontrol where organisms are the biocontrol agents (Culliney, 2005). Among the biocontrol agents use to exert control on weeds, the use of botanicals and microorganisms are discussed in the following sections.

Attributes for selecting botanicals and microorganisms as biocontrol agents

Biocontrol agents of weeds comprise higher plants, herbivores, plant pathogens and natural products (Ghosheh, 2005). The use of appropriate organisms or natural materials as biocontrol agents is germane to the success of biological weed control programme. Hence, the characteristics that a good biocontrol agent should possess are presented in Figure 1 and discussed below:

Host Specificity

This is an important characteristic of a good biocontrol agent. It guarantees that nontarget organisms are not negatively affected when biocontrol agents are applied or introduced for weed control. Day and Riding (2019) subjected *Puccinia spegazzinii,* a rust fungus that exert biocontrol on *Mikania micrantha* to host specificity test. The results showed that this biocontrol agent only infested *M. micrantha* among the 19 plant species of Asteraceae family tested. It was on this basis that permission for the release of this biocontrol agent was forwarded to Austrialia authority. The implication of this is that *P. spegazzinii* may not affect the other 18 species tested when released for the biocontrol of *M. micrantha.*

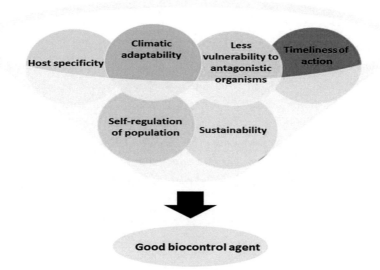

Figure 1: Attributes of a good biocontrol agent

i. Climatic adaptability

This an essential trait that a good biocontrol agent should possess, especially when it is to be introduced to a new environment. Sun *et al.* (2017) reported that optimal climatic match between home and release locations results in good control efficiency through successful establishment of the biocontrol agents. On the contrary, climatic match may not be necessary in some situations, particularly that some biocontrol agents can thrive outside their normal climatic range. However, extreme temperature and humidity may affect the establishment of biological agents in the absence of climatic adaptability (McFadyen, 1998).

ii. Less vulnerability to antagonistic organisms

The antagonistic organisms of biocontrol agents reduce their presence and make them in-effective for weed control. For instance, the presence of termites may reduce plant's mulch used for weed control. Also, predation on organisms that exert biocontrol may affect their establishment for weed control (Pratt *et al.*, 2003; Paynter *et al.*, 2019). Hence, a good biocontrol agent should be less prone to antagonistic organisms capable of drastically re-ducing its population where it is applied.

iii. Timeliness of action

For a crop plant to benefit from the biocontrol exerted on weeds, suppression of weed population below the economic injury level must be achieved at the critical weed control period of the crop that the weed control programme seeks to protect (Korres and Norswor-thy, 2015). Therefore, a good biocontrol agent should suppress weed population within the critical weed control period of the crop that is to be protected.

iv. Self-regulation of population

Botanicals and microorganisms intended for biocontrol of weeds should be able to regulate their presence in terms of population or concentration of natural product released in re-sponse to changes in the population of the target weeds. Thus, this prevents weed abun-dance without further human intervention. Generally, allelopathic plants have the ability of altering the concentration of phytochemicals released to the environment in response to competition from weeds (Sun *et al.*, 2012). Also, microorganisms used for biocontrol in-crease in population with increasing population of the target weeds.

v. Sustainability

It is ideal for a good biocontrol agent to allow some weed population to thrive and be maintained at low equilibrium level that is below economic threshold. This guarantees food supply for the biological agents and prevents environmental degradation as erosion that may occur through elimination of weeds.

Methods of deploying biocontrol agents for weed control in crop-weed environment

Botanicals and antagonistic microorganisms are introduced to crop-weed environments in various ways to exert biocontrol on weeds. This is usually done to the advantage of the crop plant. The use of botanicals for weed control in a crop-weed environment may depend on the bioherbistatic or bioherbicidal nature of the crop plants, biocontrol plants or plant products. Antagonistic microorganisms may be applied as microbial preparations in a way similar to that of herbicide.   The method of introducing antagonistic microorganisms for weed control is different from that of crop plants and biocontrol plants used for weed control. Hence, common methods of deploying microorganisms and botanicals as biocontrol agents are described in Figure 2 and discussed in the following section.

Methods for deploying antagonistic microorganisms as biocontrol agents for weed control in a crop-weed environment.

Fungi, bacteria and viruses are the common antagonistic microorganisms used to exert biocontrol on weeds (Hershenhorn *et al.*, 2016). The use of these organisms for weed control is referred to as 'microbial weed control'. This is a subset of biological control that deploys microorganisms through classical and inundative biological control approaches to suppress the population of weeds.

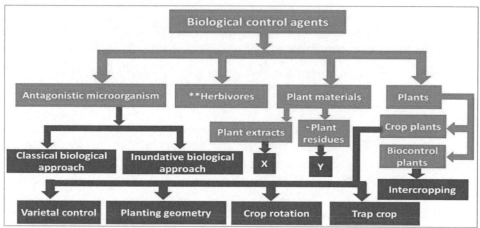

Figure 2: Biological control agents and methods of deployment
= biological control agents, = application methods, X = foliar and soil application,
Y = mulching and soil incorporation, ** = phytophagous insects and grazing animals are
not discussed here.

Classical biological control approach

This involves the use of exotic microorganisms as pathogens on invasive weed species in a new environment (Chutia *et al.*, 2007). This is also referred to as 'inoculative biological control' in many literatures. However, Eilenberg *et al.*, 2001 differentiated classical biological control from inoculative biological control on the basis of additional release of the exotic microorganism that is required in inoculative biological control unlike classical biological control.

Classical biological control through the use of microorganisms has been deployed to control some weed species in some countries (Table 1). In South Africa, an exotic fungus (Uromycladium tepperianum) was introduced to control an invasive tree (Acacia saligna). Remarkably, the tree density reduced by 90 -95% (Wood, 2012).

Table 1: Microbial weed control programmes with classical approach of application

Name of the weed species	Types of infection	Name of the pathogen	Country
Chondrilla junces L.	Weed rust	*Puccinia chondrillina* Bud and Syd	Australia
	Black berry	*Phragmidium violaceum* (Schultz)	
Rubus constrictus L.	rust	Winter	Chile
Acacia saligna (Labill) Wendl	Acacia gall rust	*Uromycladium teperianum* (sac) Mc. Alp	South Africa
Carduus thoermeri Weinm	Carduus rust	*Puccinia carduorum* Jacky	North America
Centaurea diffusa Lam	Centaurea rust	*Puccinia jaceae* Otth	Romania
Ageratina riparia (Regd) K and R	Rust	*Cercosporella* sp.	Hawaii, Jamaica
Lantana camara L.	Rust	*Septoria passiflorae* Syd	Hawaii

Source: Chutia *et al.* (2007)

Table 2: Registered Microbial herbicides since 1964

Name of microbes	Trade Names	Target weed species	Registration detail
Acremonium diospyri	-	Persimmon trees in rangelands	USA, 1960
Collectrotrichum gloeosporioides f. sp cascutae	Libao	*Cuscuta* sp.	China, 1963
Phytopthora palmivora	DeVineTM	*Morrenia odorata*	USA, 1981
Collectrotrichum gloeosporioides f. sp. aeschynomene	Collego	*Aeschynomene virginica*	USA, 1983
Puccina cnaliculata	Dr. BioSedge	*Cyperus esculantus*	USA. 1987
Collectrotrichum gloeosporioides f. sp. Malvae	BioMalTM	*Malva pusilla*	Canada, 1992
Cylindrobasidium leave	StumpoutTM	*Acacia* sp.	South Africa, 1997
Chondrostereum purpureum	BiochonTM	*Prunus serotina*	Netherlands, 1997
Xanthromonas campestries pv. poae	CampericoTM	*Poa annua*	Japan, 1997
Collectrotrichum acutatum	Hakatak	*Hakea gummosis* and *H. sericea*	South Africa, 1999
Puccinia thlaspeos	Woad Warrior	*Isastis tinctoria*	USA,2002
Chondrostereum purpureum	ChontrolTM EcolearTM	Alders and other hardwoods in rights of way and forests	Canada, 2004
Chondrostereum purpureum	Myco_TechTM	Deciduous tree species in rights of way and forests	Canada, 2004
Alterneria destruens	Smolder	Dodder species	USA, 2005

Source: Chutia *et al.* (2007)

Inundative biological control approach

This is the application of concentrated phytopathogenic agents of natural enemy present in an environment to the susceptible weeds therein for growth suppression (Hershenhorn *et al.*, 2016). Inundative approach of biological control can easily be adopted through the application of microbial herbicide preparations of indigenous weed pathogens (Chutia *et al.*, 2007). Details of some microbial herbicide preparations produced commercially are presented in Table 2. The term 'bioherbicide' has been used interchangeably for 'microbial herbicide preparations' in literatures due to fact that the latter is a subset of the former. However, this term is deliberately avoided in this section since it encompasses metabolite of higher plants (Radhakrishnan *et al.*, 2018).

Methods of deploying crop plants as biocontrol agents for weed control in a crop-weed environment

Crop plants can be deployed for weed control based on their characteristics. Some agronomic practices harness these characteristics to control weeds. Varietal control, planting geometry, crop rotation and use of trap crops are examples of such.

i. Varietal control

Some crop varieties display traits that negatively affect weed competing with them. Among such traits is the morphological architecture of the plant which may debar weed growth. For instance, branching cassava varieties control weed population better than the non-branching type (Reshma *et al.*, 2016). Allelopathy is another associated trait of some crop varieties. Allelopathic cultivars have the ability to retard the growth of the neighbouring weeds through the release of phytochemicals (Sun *et al.*, 2012). Hence, the cultivation of allelopathic varieties over non allelopathic varieties has the benefit of exerting biocontrol on weeds.

ii. Planting geometry

Crop plant spacing can successfully be used to manage weed in a crop-weed environment. Narrow spacing is effective for weed control (Ashraf *et al.*, 2016). It facilitates early closing of crop canopies and deter weed growth. Wide spacing permits weed emergence due to light penetration (Laurie *et al.*, 2015; Peerzada *et al.*, 2017). It is important to ensure that spacing does not result into competition among the crop plants when used for weed control.

iii. Crop rotation

Weed associates of crops can be controlled by cultivating different crops in temporal sequence. The use of crop rotation breaks crop -weed specificity commonly found in parasitic weeds and crop mimics. Weeds that are crop mimics depend on human for

111

propagation and survival since their seedbank exists mainly in the granaries of their asso-ciate crop with few in the soil. Sowing a crop different from the one sown in the previous farming season on a farmland exposes weeds that are crop mimics of the previously culti-vated crop, making them vulnerable to control. Crop rotation change selection pressure by presenting different natural enemies and light intensity under the changing crop canopy such that no weed species is at advantage (Murphy, *et al.*, 2006). Mhlanga *et al.* (2015) reported that maize-velvet beans rotation decreased weed density by 92% (from 357 to 30 plant m-2) over time.

iv. Trap crop

The integration of plants that are capable of promoting suicidal germination of parasitic weed into the farming system is effective in reducing the seedbanks of parasitic weeds. For instance, wheat (*Triticum aestivum*) is a false host of *Orobanche minor* which parasitize red clover (*Trifolium pratense*). Lins *et al.* (2006) reported that parasitism on red clover by *Orobanche minor* reduced when red clover was sown as a follow-up crop to wheat.

Methods of deploying biocontrol plants for weed control in a crop-weed environment. Biocontrol plants are plants that are deliberately added to cropping systems for the purpose of enhancing crop productivity through pest (weed) control (Parolin *et al.*, 2014). Biocon-trol plants are introduced into crop-weed environment by intercropping. The types of in-tercropping with biocontrol plants are:

i. Intercropping system with allelopathic plants

Plants with allelopathic tendency sown together with crop plant may be effective in controlling weeds. Fernández-Aparicio *et al.* (2013) reported that *Orobanche crenata* that is a parasitic weed of some legume crops, can be inhibited by intercropping cereals with susceptible legumes. The study showed that 2-benzoxazolinone, its derivative 6-chloroacetyl-2-benzoxazolinone, and scopoletin present in cereals are capable of inhib-iting germination of *Orobanche crenata* seeds. Kandhro *et al.* (2014) evaluated the weed suppression effect of two allelopathic crops (sorghum and sunflower) inter-cropped with cotton. Sorghum and sunflower in cotton plots supressed weeds by 62.4 and 59.6 % respectively and significantly enhanced growth and yield traits of cotton, particularly seedcotton yield as compared to No weeding.

ii. Intercropping system with cover crop

The ability of cover crops to smother weeds growing between crops has been exploited successfully. Cover crops have the ability to physically hinder emerging weeds. Chikoye *et al.* (2001) explored the use of cover crops in controlling cogongrass (*Imperata cylindrica* L.) in cassava and maize., The biomass of cogongrass was lowered by 52% - 71% in dif-ferent locations, at 12 months after the planting of the cover crops. The study showed that velvet bean (*Mucuna cochinchinensis* (Lour.) A. Chev.) was effective in controlling cogongrass across different locations.

iii. Intercropping system with allelopathic cover crop

Generally, cover crops exert biocontrol on weeds by spreading on soil surface and inter-cepting sunlight. However, some cover crops are allelopathic. The integration of allelopa-thic cover crops into intercropping system leverages on the physical ability of cover crops to debar weed emergence and allelopathy. Amossé *et al.* (2013) studied the weed suppres-sion ability of legumes planted in relay-intercropping with wheat and compared with sole planting of wheat crop. The study showed that the intercrops suppressed weed density in wheat and in the follow-up crop. Red clover that was reported to be the most effective cover crops for weed control in the study has allelopathic tendency (Wyngaarden *et al.,* 2015).

Methods for deploying plant products as biocontrol agents for weed control in a crop-weed environment.

Metabolites and materials from plants such as extracted phytochemicals and plant residues can be deployed for weed control. The form in which these plant products exist influences its application.

i. Application of plant extracts

Phytochemicals are usually extracted in water. Hence, plant extracts are in liquid form and they are applied through the use of knapsack sprayer. Extracts from plants with allelopathic tendency are foliarly applied to weeds to suppress their growth. In some cases, these ex-tracts are applied to the soil to reduce the viability of weed seeds present therein. Sunflower and sorghum extracts applied for weed control can cause about 50% reduction in weed dry weight and increase crop yield (Cheema *et al.,* 1997; Khan *et al.,* 2017). The weed control efficiency of these extracts is influenced by the concentration of the plant extract applied to the weeds and its frequency of application (Khan *et al.,* 2017; Naseem *et al.,* 2009). Weeds such as *Avena fatua* L., *Chenopodium album* L., *Coronopus didymus* L. and *Phala-ris minor* Retz can be suppressed by sunflower plant extract (Naseem *et al.,* 2009).

ii. Mulching

Plant residues are usually in solid forms. They are deliberately applied to the soil surface primarily to conserve soil moisture and for weed control purpose. Plant residues physically debar the emergence of weeds. In some cases, they exert biocontrol through allelopathy. The use of plant residue for weed control may be more effective than chemical weed con-trol in some scenarios. For instance, Baraiya *et al.* (2017) reported that the use of grass mulch at the rate of 5 t/ha in okra plots resulted in density and dry weight of weeds that is less than that of pendimentalin at 20 days after planting.

iii. Incoporation of plant residues into soil

Plant residues can be buried into the soil to exert biocontrol on the seedbank of the weeds. Mohler *et al.* (2018) reported that incorporation of cover crop residue into the soil is a means for reducing the weed seedbanks. The study showed that the persistence of common lamb's quarters (*Chenopodium album* L.) and Powell amaranth (*Amaranthus powellii* S. Watson) decreased with the addition of hairy vetch (*Vicia villosa* Roth.) into the soil. It was concluded that nitrate released during the decomposition of hairy vetch promoted fatal germination of weed seeds.

Optimising the biocontrol efficiency of botanicals and microorganisms for weed control

Microorganisms that are biocontrol agents usually affect weeds negatively by causing diseases on them. Hence, it is important to promote factors that are favourable for disease formation to enhance their weed control efficiency. The conceptual model of disease triangle shows that factors that play role in disease formation are host, pathogen and the environment (Hammerschmidt, 2018). Therefore, strategies that promote weed susceptibility, pathogen virulence and favourable environment for disease formation can optimise microbial weed control. Synergy and genetic enhancement are examples of such strategies, and they are applicable to the use of botanicals for weed control.

i. Synergy

Notably, the use biocontrol methods and other methods has superior weed control efficiency than sole adoption of these methods. This has been the justification for Integrated weed control. The use of botanicals, microorganisms and natural products for weed control is meant to suppress weed growth and not to eliminate weeds. However, the degree of weed suppression resulting from these biocontrol agents can be enhanced through the addition of adjuvants, herbicides in sub-lethal dose and combinations of antagonistic microorganisms.

ii. Synergy with sub-lethal dose of herbicides

Application of reduced concentration of synthetic herbicides with antagonistic fungi (myco-herbicides) to weeds has synergetic effect with respect to weed control. Sharon *et al.* (1992) reported from a study that sub-lethal dose of glyphosate enhances the virulence of *Alternaria cassiae*. Jurair and Khan by suppressing the biosynthesis of phytoalexin in the shikimate pathway. The study further showed that the synergetic effect of herbicide does not alter the host specificity of the antagonistic fungi, rather it enhances the biocontrol efficiency on a susceptible host.

Similarly, application of reduced concentration of synthetic herbicides with allelopathic plant extracts to weeds has synergetic effect with respect to weed control. A sulfonylurea herbicide (Atlantis®) added at 3.6 g a.i / ha to a tank mix of sorghum and sunflower extracts that was applied at 10 L ha-1 had improved weed control efficiency to tank mix of sorghum and sunflower extracts that was applied alone twice (Hussain *et al.*, 2014). The benefit from the use of reduced concentration of herbicides with plant extracts for weed control goes beyond improved weed control efficiency. It is a pointer that the use of herbicide for weed control can be reduced through this synergy.

iii. Synergy with adjuvants

Adjuvants can be used to improve the weed control efficiency of myco-herbicides. Adjuvants are surfactants, humectants and stickers capable of improving the performance of foliar applied sprays. Weaver *et al.* (2009) studied the effect of some commercially produced adjuvants on *Myrothecium verrucaria*, a biocontrol fungus that is effective on some weed species. The study showed that *M. verrucaria* that do not spray well under field condition improved with surfactants. The bioherbicidal activity of this fungus that improved was attributed to the alteration of the plant cuticle and the emergence of infection courts. Hence, it was recommended that compatible surfactants with optimised hydrophilic-lipophilic balance can be used to improve the herbicidal efficacy of myco-herbicides.

The mixture of adjuvant and allelopathic plant extract has more negative effect on plant growth than the sole application of allelopathic plant extract (Kamran *et al.*, 2019). This has been exploited for weed control. In a study, application of sorgaab (sorghum water extract) alone for weed control at 12 L ha-1 reduced weed dry weight by 70%, while addition of phenoxaprop-p-ethyl surfactant to this treatment at 375 g a.i ha-1 reduced weed dry weight by 87% (Cheema *et al.*, 2003).

iv. Synergy with other organisms

The use of secondary pathogens may boost the effectiveness of an antagonistic organism for weed control. A typical example of this is the use *Alterneria cuscutacidae* Rudak for biocontrol of dodder that is less effective. The inundative release of *Cladosporium* sp, *Fusarium* sp, and *Rhizoctonia* sp as secondary pathogen improves its ability to control weeds (Cartwright and Templeton, 1989).

The combined use of various allelopathic materials can improve allelopathic activity against weeds due to the presence of diverse allelochemicals and their synergistic effect against weeds. El-Nagdi *et al.* (2016) found that powdered dry leaves of lantana (*Lantana camara* L.) and guava (*Psidium guajava* L.) either alone or in combination reduce the dry weight of both *Corchorus olitorius* (L.) and *Echinochloa colonum* (L.). The observed reduction in the dry weight of these weeds was greater when lantana and guava were applied in combination.

Genetic Enhancement

The genetic make up of the biocontrol agents can be manipulated to achieve environmental stability and enhance virulence.

i. Environmental stability through genetic engineering

Biocontrol agents introduced in a classical biological control approach has to adapt to the temperature and humidity of the new environment. Genetic technique such as mutation or hybridization creates variability for selection. This may be useful in overcoming extreme environmental conditions. Also, engineering biocontrol microorganisms to overexpress a DNA repair photolyase and heat shock protein (HSP) encoding gene increased their resistance to solar radiation and extreme temperatures (Wang and Wang, 2017).

ii. Enhanced Virulence through genetic engineering

The virulence of biocontrol agents can be improved through the manipulation of their genetic make-up. This promotes their ability to exert biocontrol. For instance, the manipulation of the genetic make-up of some fungi to produce certain amino acids capable of inhibiting plant growth by disrupting plant vital biochemical pathway has been a viable means of enhancing weed control ability of these fungi. Valine is an amino acid that inhibits acetolactate synthase (ALS) enzyme in plants. This is the same mode of action displayed by the sulfonylurea herbicides. The mutant of *Fusarium oxysporum* f. sp. *cannabis* that excrete valine exert better biocontrol on weeds than the wild type (Cai and Gu, 2016).

Conclusion

Botanicals and microorganisms exert biocontrol on weeds through various means and their weed control efficiency can be optimised. The need to reduce reliance on synthetic herbicide and promote environmental safety necessitates that the use of this biocontrol agents for weed control should be embraced more.

References

Amossé, C., Jeuffroy, M., Celette, F. and David, C. (2013) Relay-intercropped forage legumes help to control weeds in organic grain production. *European Journal of Agronomy,* Volume 49, pp. 158-167.

Ashraf, U., Abbas, R., Hussain, S., Mo, Z., Anjum, S., Khan, I., and Tang, X. (2016) Consequences of varied planting geometry and early post emergence herbicides for crop-weed interventions in rice under semi-arid climate. *Planta Daninha,* 34(4), pp. 737-746.

Baraiya, M., Yadav, K., Kumar, S., Lal, N., and Shiurkar, G. (2017) Effect of integrated weed management in okra. *International Journal of Chemical Studies,* 5(4), pp. 1103 - 1106.

Cai, X. and Gu, M. (2016) Bioherbicides in organic horticulture. *Horticulturae,* 2(2), p. 3.

Cartwright, D. and Templeton, G. (1989) Preliminary evaluation of a dodder anthracnose fungus from China as a mycoherbicide for dodder control in the US. *Journal of the Arkansas Academy of Science,* 43(1), pp. 15-18.

Cheema, Z., Khaliq, A. and Farooq, R. (2003) Effect of concentrated sorgaab aloon and in combination with herbicides and a surfactant in wheat. *The Journal of Animal and Plant Science,* Volume 13, pp. 10 -13.

Cheema, Z., Luqman, M. and Khaliq, A. (1997) Use of allelopathic extracts of sorghum and sunflower herbage for weed control in wheat. *Journal of Animal and Plant Sciences,* 7(3-4), pp. 91-93.

Chikoye, D., Ekeleme, F. and Udensi, U. (2001) Cogongrass suppression by intercropping cover crops in corn/cassava systems. *Weed Science,* 49(5), pp. 658-667.

Chutia, M., Mahanta, J., Bhattacheryya, N., Bhuyan, M., Boruah, P. and Sarma, T. (2007) Microbial herbicides for weed management: prospects, progress and constraints. *Plant Pathology Journal,* 6(3), pp. 210-218.

Crump, N., Cother, E. and Ash, G. (1999) Clarifying the nomenclature in microbial weed control. *Biocontrol Science and Technology,* 9(1), pp. 89-97.

Culliney, T. (2005) Benefits of classical biological control for managing invasive plants. *Critical Reviews in Plant Sciences,* 24(2), pp. 131-150.

Day, M. and Riding, N. (2019) Host specificity of Puccinia spegazzinii (Pucciniales: pucciniaceae), a biological control agent for Mikania micrantha (Asteraceae) in Australia. *Biocontrol Science and Technology,* 29(1), pp. 19 - 27.

Del Pino, J., Moyano, P., Díaz, G., Anadon, M., Diaz, M., García, J., Frejo, M. (2017) Primary hippocampal neuronal cell death induction after acute and repeated paraquat exposures mediated by AChE variants alteration and cholinergic and glutamatergic transmission disruption. *Toxicology,* Volume 30, pp. 88-99.

DiTommaso, A., Averill, K. M., Hoffmann, M. P., Fuchsberg, J. R. and Losey, J. E. (2016) Integrating Insect, Resistance, and Floral Resource Management in Weed Control Decision-Making. *Weed Science,* 64(4), pp. 743-756.

Eilenberg, J., Hajek, A. and Lomer, C. (2001) Suggestions for unifying the terminology in biological control. *BioControl,* 46(4), pp. 387-400.

El-Nagdi, W., Youssef, M. and El-Rokiek, K. (2016) Allelopathic effect of dry leaves of lantana and guava for controlling root knot nematode, Meloidogyne incognita on cowpea and some associated weeds. *International Journal of ChemTech Research,* 9(6), pp. 55-62.

Fernández-Aparicio, M., Cimmino, A., Evidente, A. and Rubiales, D. (2013) Inhibition of *Orobanche crenata* seed germination and radicle growth by allelochemicals identified in cereals. *Journal of Agricultural and Food chemistry*, 61(41), pp. 9797-9803.

Gabriel, C. J. and Cook, R. (1990) Biological control-the need for a new scientific framework. *BioScience,* 40(3), pp. 204-208.

Ghosheh, H. (2005) Constraints in implementing biological weed control: a review. *Weed biology and management,* 5(3), pp. 83- 92.

Hammerschmidt, R. (2018) How glyphosate affects plant disease development: it is more than enhanced susceptibility. *Pest Management Science,* 74(5), pp. 1054-1063.

Harris, P. (1991) Invitation paper (CP Alexander fund): classical biocontrol of weeds: its definition, selection of effective agents, and administrative–political problems. *The Canadian Entomologist,* 123(4), pp. 827-849.

Hershenhorn, J., Casella, F. and Vurro, M. (2016) Weed biocontrol with fungi: past, present and future. *Biocontrol science and technology,* 26(10), p. 1313–1328.

Hussain, S., Hassan, F., Rasheed, M., Ali, S. and Ahmed, M. (2014) Effects of allelopathic crop water extracts and their combinations on weeds and yield of rainfed wheat. *Journal of Food, Agriculture & Environment,* 12(3&4), pp. 161-167.

Kamran, M., Cheema, Z., Farooq, M., Ali, Q., Anjum, M.and Raza, A. (2019) Allelopathic influence of sorghum aqueous extracts on growth, physiology and photosynthetic activity of maize (*Zea mays* L.) seedling. *Philippine Agricultural Scientist,* 102(1), pp. 33-41.

Kandhro, M., Tunio, S., Rajpar, I. and Chachar, Q. (2014) Allelopathic impact of sorghum and sunflower intercropping on weed management and yield enhancement in cotton. *Sarhad Journal of Agriculture,* 30(3), pp. 311- 318.

Khan, F., Khalil, S., Rab, A., Khan, I.and Nawaz, H. (2017) Allelopathic potential of sunflower extract on weeds density and wheat yield. *Pakistan Journal of Weed Science Research,* 23(2), pp. 221 -232.

Korres, N. and Norsworthy, J. (2015) Influence of a rye cover crop on the critical period for weed control in cotton. *Weed science,* 63(1), pp. 346-352.

Kwiatkowska, M., Nowacka-Krukowska, H. and Bukowska, B. (2014) The effect of glyphosate, its metabolites and impurities on erythrocyte acetylcholinesterase activity. *Environmental toxicology and pharmacology,* 37(3), pp. 1101-1108.

Laurie, S., Maja, M., Ngobeni, H. and DuPlooy, C. (2015) Effect of different types of mulching and plant spacing on weed control, canopy cover and yield of sweet potato (*Ipomoea batatas* (L.) Lam). *American Journal of Experimental Agriculture,* 5(5), pp. 450 - 458.

Lins, R., Colquhoun, J. and Mallory Smith, C. (2006) Investigation of wheat as a trap crop for control of Orobanche minor. *Weed Research,* 46(4), pp. 313-318.

McFadyen, R. C. (1998) Biological control of weeds. *Annual Review of Entomology,* 43(1), pp. 369-393.

Mhlanga, B., Cheesman, S., Maasdorp, B., Muoni, T., Mabasa, S., Mangosho, E.and Thier-felder, C. (2015) Weed community responses to rotations with cover crops in maize-based conservation agriculture systems of Zimbabwe. *Crop Protection,* Volume 69, pp. 1-8.

Minkoff-Zern, L. A. (2017) The case for taking account of labor in sustainable food sys-tems in the United States. *Renewable Agriculture and Food Systems,* 32(6), pp. 576-578.

Mohler, C., Taylor, A., DiTommaso, A., Hahn, R.and Bellinder, R. (2018) Effects of In-corporated Rye and Hairy Vetch Cover Crop Residue on the Persistence of Weed Seeds in the Soil. *Weed Science,* pp. 1-7.

Murphy, S., Clements, D., Belaoussoff, S., Kevan, P.and Swanton, C. (2006) Promotion of weed species diversity and reduction of weed seedbanks with conservation tillage and crop rotation. *Weed Science,* 54(1), pp. 69 - 77.

Naseem, M., Aslam, M., Ansar, M. and Azhar, M. (2009) Allelopathic effects of sunflower water extract on weed control and wheat productivity. *Pakistan Journal of Weed Sci-ence Research,* 15(1), pp. 107 - 116.

Oladele, S. and Ayodele, O. (2017) Glyphosate, 1, 1'-dimethyl-4, 4'-bipyridinium dichlo-ride and Atrazine induces changes in Soil organic carbon, bacterial and fungal com-munities in a tropical alfisol. *Eurasian Journal of Soil Science,* 6(3), pp. 238 - 248.

Parolin, P., Bresch, C., Poncet, C. and Desneux, N. (2014) Introducing the term 'Biocontrol Plants' for integrated pest management. *Scientia Agricola,* 71(1), pp. 77- 80.

Paynter, Q., Peterson, P., Cranwell, S., Winks, C.and McGrath, Z. (2019) Impact of gen-eralist predation on two weed biocontrol agents in New Zealand. *New Zealand Plant Protection,* Volume 72, pp. 260-264.

Peerzada, A., Ali, H. and Chauhan, B. (2017) Weed management in sorghum [Sorghum bicolor (L.) Moench] using crop competition: a review. *Crop Protection,* Volume 95, pp. 74 - 80.

Pratt, P., Coombs, E. and Croft, B. (2003) Predation by phytoseiid mites on Tetranychus lintearius (Acari: Tetranychidae), an established weed biological control agent of gorse (*Ulex europaeus*). *Biological Control,* 26(1), pp. 40-47.

Radhakrishnan, R., Alqarawi, A. and Abd_Allah, E. (2018) Bioherbicide: Current knowlegde on weed control mechanism. *Ecotoxicology and environmental safety,* Volume 158, pp. 131-138.

Reicosky, D. (2003) Global Environmental benefits of soil carbon management. In: G. L., B. J., A. Martinez-Vilela and Holgado-Cabrera, eds. *Conservation Agriculture.* Dor-drecht: Springer Science+Business Media., pp. 3-12.

Reshma, N., Sindhu, P., Thomas, C. and Menon, M. (2016) Integrated weed management in cassava (Manihot esculenta Crantz). *Journal of Root Crops,* 42(1), pp. 22-27.

Sharon, A., Amsellem, Z. and Gressel, J. (1992) Glyphosate suppression of an elicited defense response: increased susceptibility of Cassia obtusifolia to a mycoherbicide. *Plant Physiology,* 98(2), pp. 654-659.

Singh, V., Masabni, J., Baumann, P., Isakeit, T., Matocha, M., Provin, T., Bagavathiannan, M. (2019) Activated charcoal reduces pasture herbicide injury in vegetable crops. *Crop Protection,* Volume 117, pp. 1- 6.

Sun, B., Kong, C.-H., Wang, P. and Qu, R. (2012) Response and relation of allantoin production in different rice cultivars to competing barnyardgrass. *Plant Ecology,* Volume 213, pp. 1917 - 1926.

Sun, Y., Brönnimann, O., Roderick, G., Poltavsky, A., Lommen, S.and Müller Schärer, H. (2017) Climatic suitability ranking of biological control candidates: a biogeographic approach for ragweed management in Europe. *Ecosphere,* 8(4), p. e01731.

Swanton, C. J., Nkoa, R. and Blackshaw, R. E. (2015) Experimental Methods for Crop–Weed Competition Studies. *Weed Science,* 63(Special Issue), p. 2–11.

Wang, C. and Wang, S. (2017) Insect pathogenic fungi: genomics, molecular interactions, and genetic improvements. *Annual Eeview of Entomology,* Volume 62, pp. 73 - 90.

Weaver, M., Jin, X., Hoagland, R. and Boyette, C. (2009) Improved bioherbicidal efficacy by Myrothecium verrucaria via spray adjuvants or herbicide mixtures. *Biological Control,* 50(2), pp. 150-156.

Wood, A. (2012) Uromycladium tepperianum (a gall-forming rust fugus) causes a sustained epidemic on the weed Acacia saligna in South Africa. *Plant Pathology,* 41(3), pp. 255 - 261.

Wyngaarden, S., Gaudin, A., Deen, W. and Martin, R. (2015) Expanding red clover (*Trifolium pratense*) usage in the corn–soy–wheat rotation. *Sustainability,* 7(11), pp. 15487-15509.

Zhang, X. and Chen, Y. (2017) Soil disturbance and cutting forces of four different sweeps for mechanical weeding. *Soil and Tillage Research,* Volume 168, pp 167-175.

CHAPTER 6

Potential of Wood Chips for Weed Control

by

Sabine Gruber*[1], Jialu Xu[1], Sabine Zikeli[2], Regina G. Belz[3]

1*Institute of Crop Production (340a), University of Hohenheim, Germany*
2*Centre for Organic Farming (309), University of Hohenheim, Germany*
3*Institute of Agricultural Sciences in the Tropics (490f),*
University of Hohenheim, Germany
***Corresponding author**: sabine.gruber@uni-hohenheim.de*

Abstract

Wood chips are recognized as a source of various secondary metabolites that act as putative phytotoxins in extracts or mulches. Thus, they have the potential for direct use in agriculture for natural weed control or for the development of bio-herbicides. Aqueous extracts of wood chips originating from hedgerows and trees on farmland significantly affected the germination of seeds of different crop species (*Brassica napus*, *Lactuca sativa*, *Triticum aestivum*) in the laboratory. The effect depended on the crop species and the nature and treatment of the ligneous material used for extraction. Wood chips applied as mulch on farmland reduced the number of weeds, but, depending on the amount applied, also resulted in a reduction in crop yield. If wood chips are used on farmland for weed or erosion control, their application should be only on selected crops with a high risk of weed infestation or soil erosion, and should not be applied on the same area in subsequent years. The documented inhibitory effects of unidentified compounds released from wood chips, primarily from the bark, indicate allelochemical effects that could be used for the development of bio-herbicides in the future.

Keywords: allelochemicals, germination, ligneous mulch, hedgerow, herbicide

Introduction

Weed control is usually performed chemically in conventional farming, but chemical-synthetic herbicides are not permitted in organic farming, according to organic standards. Even in conventional farming, however, the use of herbicides is under discussion due to both possible environmental hazards and because several weeds have developed resistance to specific herbicidal modes of action (Perotti *et al.*, 2020). Traditionally, another option for direct weed control is mechanical weeding, which is widely used in organic farming. Indirect weed control options such as the cultivation of crop varieties that can successfully outcompete weeds, soil inversion tillage, or sophisticated crop rotation practices are measures that can be used in a cropping system to avoid weed problems. Allelochemicals, i.e., bioactive secondary metabolites released by plants, can also be considered for use as direct or indirect weed control methods. A small number of naturally generated herbicidal products already exist; for example organic acids, essential oils, or crude botanical products (Kelderer *et al.,* 2006, Dayan and Duke, 2010). However, the potential for application of these naturally generated herbicidal products is limited. Ideally, natural herbicidal substances would be generated on-farm, or close to the farm environment, to avoid transportation costs and to provide simple solutions at low cost. Wood and bark are often available on or around farms as cuttings from hedgerows, bushes, trees and forests. In the past, farmland was often structured by hedgerows and bushes as wind breaks or field borders. These elements disappeared in the landscape in many industrial countries due to increased mechanization of agriculture and increased working width of the machinery. However, because of their many ecological benefits, hedgerows and bushes should be maintained or even reintroduced into current agricultural landscapes. To sustain their ecological functions, hedgerows must be pruned periodically. The resulting ligneous residues are often too small for transportation and combustion, and therefore are often chopped and left un-used in heaps to decompose at the site. Wood chips can also originate from forestry residues, when small stems and branches are left behind from timber production. Another source is wood residue from municipal or roadside plantings that are cut periodically to maintain the functions of the plantings and for traffic safety. Since these ligneous residues contain various secondary metabolites as putative phytotoxic allelochemicals (Saha *et al.*, 2018), they also have the potential for use in agriculture for natural weed control via allelopathy or as bioherbicides. This potential is explored through the following two approaches: (i) the effects of wood chip extracts on seed germination (bio-herbicide approach), and (ii) the effect of mulching with wood chips on crops and weeds in the field (residue allelopathy approach).

Allelochemicals. Plants and plant residues can interfere with other plants or micro-organisms by growth-promoting or by phytotoxic ingredients (allelopathy) (Rice, 1984). These allelopathic effects are elicited by the release of organic compounds (i.e., allelochemicals) from specific plant parts via leaching, root exudation, volatilization, or decomposition of plant residues in natural and agricultural eco-systems (Rice, 1984, Iqbal *et al.*, 2019). If sensitive plants are exposed to these allelochemicals, seed germination, plant growth, and plant development can be affected. A great number of annual and perennial plant species, among them radish (*Raphanus sativus* L.), mustard (*Sinapis alba* L.), cereal species (*Triticum* L. spp., *Secale cereale* L.), pine (*Pinus* spp.), blue spruce (*Picea pungens* Engelm.), *Amaranthus* ssp., and eucalypts (*Eucalyptus camaldulensis* Dehnh.) (Khan and Marwat, 2005, Machado, 2007, Prinsloo and Du Plooy, 2018) can have germination and growth inhibiting effects on other plants by their release of allelochemicals. Allelopathic effects on adjacent plants by living woody plants are also documented, for example for eucalypts (*Eucalyptus* spp.), leucaena (*Leucaena leucocephala* (Lam.) De Wit) and walnut (*Juglans nigra* L.) (Chou 1992, 2010, Batish *et al.,* 2008). The use of wood chips from these species, therefore, can have allelopathic effects in agricultural systems in the same way (Rathinasabapathi *et al.,* 2005, Law *et al.,* 2006, Gruber *et al.,* 2006, Ferguson and Rathinasabapathi 2008).

Allelopathic effects most often result from secondary plant metabolites, as, for example, phenolic compounds (Modallal and Al-Charchafchi, 2006) or organic acids such as ferulic acid (in *Leucaena* spp. (Chou, 1992). Allelopathic inhibition, however, often results from a complex mixture of several primary compounds and their metabolites, and can include interactions of various classes of organo-chemical compounds such as phenolic compounds, flavonoids, terpenoids, alkaloids, steroids, tannins, carbonates and others (Williams and Hoagland, 1982, Seigler, 1996; Saha *et al.,* 2018). An exact attribution of allelopathic effects to specific plant compounds is therefore often difficult, and is, to date, unknown for wood chips. In black elder (*Sambucus nigra* L.), for example, 24 aromatic metabolites have been identified as potential allelopathic ingredients (D'Abrosca *et al.,* 2001). Cyanogenic glycosides, at high concentrations, more strongly inhibited germination of dicotyledonous plants than of monocotyledonous plants. Sambunigrine and prunasine had a slight stimulatory effect on germination. Abscisic acid from eucalyptus leaf extracts (*Eucalyptus camaldulensis*) and ephedrine from walnut (*Juglans regia*) leaf extracts reduced seed germination of crown vetch (*Coronilla varia)* (Isfahan and Shariati, 2007).

The total spectrum of allelochemical compounds, therefore, is crucial for understanding the mechanisms and the strength of an allelopathic effect of wood chips. There is little evidence as to whether this spectrum of effective compounds differs between wood chips of different provinces and species or in plant age (Dureya *et al.,* 1999), and whether these differences are relevant in terms of their biological effects on plants or not. If the effects of wood chips on plants are to be used efficiently for weed suppression, differences in efficiency between wood species should be determined.

Water is the ubiquitous solvent in natural environments, and many allelochemicals are water soluble (Blum, 2006). The preparation of aqueous extracts would be a simulation of the natural situation. Aqueous preparations from sorghum (*Sorghum bicolor* L.), sunflower (*Helianthus annuus* L.) and oilseed rape (*Brassica napus* L.) are already in use for weed control (Iqbal *et al.,,* 2009; Jamil *et al.,* 2009) in various crops, and ligneous material such as wood chips may also be a suitable source of allelochemicals for weed control.

Wood chips as source for herbicidal allelochemicals. Wood chips are commonly used as mulch in gardening, for example in municipal areas, to suppress the growth of weeds.

There is also some uses in agriculture and ornamental tree cultivation for the same purpose (own research; Siipilehto 2001, Chiroma *et al.*, 2006, Law *et al.*, 2006, Ferguson and Rathinasabapathi 2008, Cregg *et al.*, 2009, Granatstein *et al.*, 2010). Though, application of wood chips on agricultural or horticultural areas is quite common, scientific research on the mechanistic effects of their inhibitory action is scarce. There is undoubtedly a weed suppressing effect by wood chip application (Wang *et al.*, 2012, Lux und Schmidtke 2012), but the results cannot specifically be attributable to this effect exclusively, to mechanical suppression of emergence (by a thick mulch layer), light exclusion (reduced seed germination and reduced photosynthesis of seedlings covered by mulch), and/or allelopathic effects. However, our own research has clearly shown that allelopathic effects are involved, at least to a certain extent (Banhardt *et al.*, 2008, Gruber *et al.*, 2008, Vogel, 2012). The exact proportional concentration of allelopathy or phytotoxicity to weed suppression by wood chips is not yet known.

The species and traits of the respective ligneous plant have significant effects on the level of inhibition of seed germination by aqueous extracts (Banhardt *et al.*, 2008, Wu *et al.*, 2012) or by mixture into the soil (e.g. for *Salix caprea*; Mudrák and Frouz, 2012). Extracts of willow (purple willow, *Salix purpurea* L. ssp. *purpurea*) resulted in reduced germination of oilseed rape (*Brassica oleracea*) seeds and lettuce (*Lactuca sativa* L.) seeds. In a similar way but to a lesser degree, extracts of bird cherry (*Prunus padus* L. ssp *padus)*, spruce (*Picea abies* (L.) H. Karst.), and black elder (*Sambucus nigra*) suppressed germination of oilseed rape and lettuce seeds. There is no evidence to identify which ingredient in the wood chip extracts resulted in these effects. Nevertheless, identification of potent extracts would enable development of bio-herbicides like those that are already in use outside of Europe, for example in the USA, or New Zealand (Kelderer *et al.*, 2006).

In addition, interactions and concentrations of the released phytotoxins, or the amount of mulch, respectively, may also be relevant for the allelopathic effects (Lisanework und Michelsen, 1993; Pérez 1990; Stephen, 2007). Single compounds and their mixtures could provide a dose-dependent stimulating or inhibiting effect (Belz *et al.*, 2007, 2008). The quality of the expected allelopathic effect of wood chips or plant residues is therefore primarily dependent on the amounts of active substances released. However, the released compounds can also play a role in processes of decomposition, re-structuring, and soil absorption in the environment and thus their effects can change dynamically (Belz *et al.*, 2009).

The efficiency of wood chip extracts, as of any other extracts, is supposed to depend on particle size, pre-treatment and storage conditions of the raw material, on the composition of the raw material and concentration of the allelochemicals, and on the extraction process. Here, both the proportion of solvent water to the solid phase and the extraction time seem to be crucial factors. Allelochemicals also vary in heat stability and volatility (Kohli and Singh, 1991; Iqbal *et al.,* 2019). In terms of wood, core and bark extracts most probably differ in their concentration of allelochemical ingredients as they do in chemical composition in general.

Wood chip extracts: model experiment 1 (bio-herbicide approach)

For testing of allelopathic effects, aqueous extracts from wood chips were prepared and used for germination tests of crop seeds. Wood chips were collected shortly after their production, i.e., the cutting of trees and hedgerows in winter at the Research Station Klein-hohenheim, University of Hohenheim, located in southwestern Germany, and then frozen until use at -18 °C. The wood chips were derived from goat willow (*Salix caprea* L.), bird cherry (*Prunus padus* L. ssp *padus)*, spruce (*Picea abies* (L.) H. Karst.), black elder (*S. nigra* L.), sycamore maple (*Acer pseudoplatanus*), common oak (*Quercus robur*), common hornbeam (*Carpinus betulus*), and common ash (*Fraxinus excelsior*). The wood chips varied in water content (34-55 %), in particle size, and in the proportion of bark, which is thought to contain the highest concentration of secondary metabolites.

Depending on the experimental approach, the raw material was separated into core and bark, chopped or ground into different particle sizes, dried at different temperatures (undried, freeze-dried, oven dried at 60 or 105°C), or treated in combinations of these factors. Wood chips were then incubated in water in a proportion of 1:5 (wood chips (g): deionized water (mL)), a concentration that had been found appropriate in preceding tests. After an extraction period (48–96 hrs, depending on the experimental approach) at room temperature, solutions were filtered and stored at 6°C for about one day until used in germination tests. Germination tests were done in Petri dishes, each with 100 seeds of the tested species (oilseed rape, OSR, *Brassica napus*; winter wheat, *Triticum aestivum*; lettuce, *Lactuca sativa*) in the dark and at 20 °C for 14 or 21 days, depending on the trial.

Significant differences in seed germination were found and depended upon the ligneous species that was used for extraction (Figure 1). Extracts from goat willow, spruce, bird cherry, and common ash strongly reduced seed germination, whereas extracts from sycamore acorn and common oak did not suppress germination of lettuce and oilseed rape seeds. There was also evidence of a species-dependent response to the extracts. Common ash, spruce, hornbeam, and bird cherry extracts affected seed germination of lettuce more strongly than seed germination of oilseed rape. Differences in seed mass and seed morphology could have contributed to the differences in germination inhibition. Extracts from bark of goat willow and bird cherry strongly inhibited germination of wheat and oilseed rape seeds (Figure 2), whereas extracts from the core material was not significantly different from the control. Hence, the active compounds appear to be localized in the bark. In addition, processing of the wood chips before extraction affected the efficiency of germination inhibition (Figure 3). Extracts from fresh material and freeze-dried material of goat willow and bird cherry had highest suppressive effects on both oilseed rape and wheat. Drying the wood chips at temperatures above room temperature mainly resulted in lower efficiency of germination suppression. Hence, the activity of the active substances was interrupted by heat impacts, thus indicating heat instability of the substances.

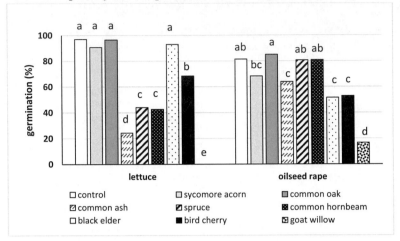

Figure 1: Germination of lettuce (*Lactuca sativa*) and oilseed rape (*Brassica napus*) seeds in Petri dishes with an aqueous extract derived from wood chips of various ligneous species, and a control (water). Lower case letters indicate significant differences at P<0.05. After Banhardt *et al.*, 2008.

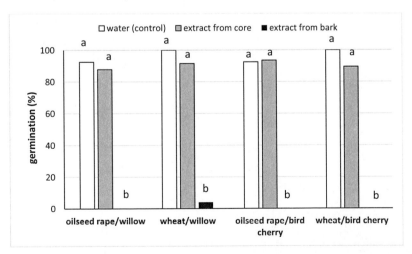

Figure 2: Germination of wheat (*Triticum aestivum*) and oilseed rape (*Brassica napus*) seeds in petri dishes with an aqueous extract derived from wood chips of bark or core of goat willow (*Salix caprea*) and bird cherry (*Prunus padus*), and a control (water). Lower case letters indicate significant differences at P<0.05.

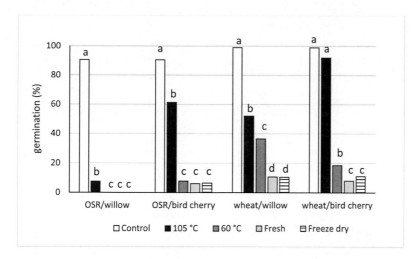

Figure 3: Germination of wheat (*Triticum aestivum*) and oilseed rape (OSR, *Brassica napus*) seeds in petri dishes with an aqueous extract derived from wood chips of goat willow (*Salix caprea*) and of bird cherry (*Prunus padus*) that were dried at different temperatures (freeze dry; 60 °C; 105 °C) or non-dried (fresh) in comparison to a control (water). Lower case letters indicate significant differences at P<0.05.

These findings clearly indicate that the potential of aqueous wood chip extracts for weed suppression is most efficiently achieved by extraction of undried or freeze-dried chips of the bark of eligible species such as goat willow, bird cherry, or common ash. A practical mixture of extracts may be a further option to expand the range of weeds controlled, but exploration of this as well as the identification of the active components remains to be explored.

Wood chip application on fields: model experiment 2 (residue allelopathy approach)
The long-term application of wood chips originating from hedgerow pruning (mixed species) and trees on farmland was shown to have significant effects on weeds and crops (Table 1).

Table 1. Effects of year (Wang *et al.,* 2012: two years; Xu *et al.,* 2018: 16 years), crop (Xu *et al.,* 2018: lentil (*Lens culinaris*)/oats (*Avena sativa*) intercropping in different mixing ratios), and wood chip mulch (WCM) application at 0, 80 or 160 m3 ha-1 on weed infestation and crop yield.

Trait	Effect	Reference, Significance Xu *et al.,* 2018	Reference, Significance Wang *et al.,* 2012
Weed number spring	Year	***	-
	Crop	***	n.s.
	WCM	**	**
Weed biomass harvest	Year	***	**
	Crop	n.s.	**
	WCM	n.s.	*
Soil seed bank	Year	-	-
	Crop	***	-
	WCM	n.s.	-
Crop grain yield	Year	***	**
	Crop	***	**
	WCM	n.s.	n.s.
Crop straw yield	Year	***	-
	Crop	***	-
	WCM	**	-
Crop biomass yield	Year	***	***
	Crop	***	***
	WCM	*	n.s.

When regularly applied in the amounts of 0, 80 or 160 m3 ha-1 in a trial on arable land, weed number in spring was significantly reduced in plots with wood chip application (Xu *et al.,* 2018). The response of weeds to wood chip application was not uniform and varied both from year to year and by the crop grown. The composition of weed species varied and depended on the amount of wood chips used (Gruber *et al.,* 2008). The indicator species *Veronica persica* Poir., for example, increased significantly if wood chips had been applied (Gruber *et al.,* 2008). Generally, weed density in spring was significantly influenced by the growing year, the amount of wood chips applied, and the crop grown (Xu *et al.*, 2018). Weed biomass at harvest and weed seed bank depended only on the respective crop species. However, over the time period of 16 experimental years, the relative crop yield declined under wood chip application when compared to the control (Xu *et al.,* 2018). Though, this yield depression was an undesirable result, it provided evidence of the physical and/or biochemical effect of wood chips on plants in general.

These findings document the potential of wood chip mulch for weed suppression in the field but also point to some practical constraints that should be considered if wood chips are to be used directly and for a long period of time.

Discussion

The observed differences in weed responses to wood chips in field experiments can be attributed to the annual crop rotations and/or on the heterogeneous composition of the wood chips, with changing proportions of ligneous species from year to year. Results from the field experiment correspond with results from the laboratory: wood chips have a spe-cies-specific allelochemical effect. This effect probably interacts with the differences in moisture content of the wood chips, which may vary from year to year, and with the different proportions of bark to core, which also may change by ligneous species, as well as by the age and mass of the wood. The species whose seeds turned out to be sensitive to wood chip extract (*B. napus, T. aestivum, L. sativa*) can be expanded to include field poppy (*Papaver rhoeas*) and blackgrass (*Alopecurus myosuroides*; Gruber et al., 2008). This set of species includes both crops and weeds, as well as seeds with different thousand kernel masses and variations in seed composition (oil, starch). Similar to the findings of Kumar et al. (2006), there were interactions of the crop or weed species with the ligneous species. We presume that the weed suppressing effect of wood chip mulch is mainly attributable to the allelochemicals which are naturally leached. Besides that, effects on plant growth by the comparatively wide C:N ratio of the material (approximately 50; Gruber et al., 2008) are possible as well, just as exclusion of light, mulch cover of seedlings, and/or temperature effects have physical effects on plant growth. Application of wood chips on crops that are susceptible to soil erosion and weed infestation (Lux and Schmidtke, 2012), for example faba beans, peas, or lentil, can be recommended. A frequent and long-term application, however, could lead to yield suppression, probably through the accumulation of allelo-chemicals in the soil.

The treatment of wood chips before application turned out to be crucial for the efficiency of the extracts. Allelochemicals probably underlie degradation or modification processes which can affect their efficiency, as indicated by Julkunen-Tiitto and Sorsa, S. (2001) for purple willow (S. *purpurea* L.).

Selected ligneous species would be suitable for the development of bio-herbicides. These species could grow in hedgerows on farmland and simultaneously increase biodiversity on the fields. Existing hedgerows could gain in value and thus be maintained together with their ecosystem services because of their additional, economical applications. Besides hedgerows, targeted planting of short rotation coppice, for example willow, with a shorter rotation time to gain proportionately more bark material, would provide ligneous material as a basis for the preparation of uniform extracts. Though, these extracts have not yet been systematically tested for their efficiency in direct spraying on weeds, this creates an opportunity for the development of bio-herbicides. Particularly in recent times when weeds have become tolerant to conventional herbicides, new approaches to weed control are needed. The unavoidably return to mechanical weed control instead of the current chemical-synthetic herbicide approach would adversely affect conservation tillage systems. Application of wood chips mulch, and – in the future – of target-specific allelochemicals or their mixtures could be an option for sustainable cropping systems.

References

Banhardt, A., Gruber, S., Claupein, W., 2008. Wirkung von Gehölzhäcksel auf das Unkrautvorkommen im Ökologischen Landbau. *Mitteilungen der Gesellschaft für Pflanzenbauwissenschaften* 20:291-292.

Batish, D.R., Singh, H.P., Kohli, R.K., 2008. Allelopathic tree-crop interactions under agroforestry systems. *In*: Batish, D.R., Kohli, R.K., Jose, S., Singh, H.P. (eds.): Ecological Basis of Agroforestry. CRC Press, Boca Raton FL, pp. 37-50.

Belz, R.G., Reinhardt, C.F., Foxcroft, L., Hurle, K., 2007. Residue allelopathy in *Parthenium hysterophorus* L. – does parthenin play a leading role? *Crop Protection* 26:237-245.

Belz, R.G., Cedergreen, N., Sørensen, H., 2008. Hormesis in mixtures – can it be predicted? *Science of the Total Environment* 44:77-87.

Belz, R.G., Van der Laan, M., Reinhardt, C.F., Hurle, K., 2009. Soil degradation of parthenin – does it contradict the role of allelopathy in *Parthenium hysterophorus* L.? *Journal of Chemical Ecology* 35:1137-1150.

Blum, U. 2006. Allelopathy: a soil system perspective. In: (Reigosa, M.J., Pedrol, N., González, L.; eds.) Allelopathy. A Physiological Process with Ecological Implications. p. 299–340. Springer, Dordrecht.

Chou, C.H., 1992. Allelopathy in relation to agricultural productivity in Taiwan: problems and prospects. *In*: Rizvi, S.J.H., Rizvi, V. (eds.): Allelopathy. Basic and applied aspects. Chapman and Hall, London, UK, pp. 179-204.

Chou, C.H., 2010. Role of allelopathy in sustainable agriculture: use of allelochemicals as naturally occurring bio-agrochemicals. *Allelopathy Journal* 25:3-16.

Chiroma, A.M., Alhassan, A.B., Yakubu, H., 2006. Growth, nutrient composition and straw yield of sorghum as affected by land configuration and wood-chips mulch on a sandy loam soil in Northeast Nigeria. *International Journal of Agriculture and Biology* 8:770-773.

Cregg, B.M., Nzokou, P., Goldy, R., 2009. Growth and physiology of newly planted Fraser Fir (*Abies fraseri*) and Colorado Blue Spruce (*Picea pungens*) Christmas trees in response to mulch and irrigation. *HortScience* 44:660-665.

D'Abrosca, B., DellaGreca, M., Fiorentino, A, Monace, P., Previtera, L., Simonet, A.M., Zarrelli, A., 2001. Potential allelochemicals from *Sambucus nigra*. *Phytochemistry* 58:1073-1081.

Dayan, F.E., and S.O. Duke. 2010. Natural products for weed management in organic farming in the USA. Outlooks Pest Manag. 21(4): 156–160.

Dureya, M.L., English, R.J., Herrmansen, L.A., 1999. A comparison of landscape mulches: chemical, allelopathic, and decomposition properties. *Journal of Arboriculture* 25:88-97.

Ferguson, J., Rathinasabapathi B., 2008. Southern redcedar and southern mangnolia wood chip mulches for weed suppression in containerized woody ornamentals. *HortTechnology* 18:266-270.

Granatstein, D., Wiman, M., Kirby, E., Mullinix K., 2010. Sustainability trade-offs in organic orchard floor management. *Acta Horticulturae* 873:115-122.

Gruber, S., Ruopp. T., Beierl, B., Claupein, W., 2006. Wood chips used for mulching in Organic Farming. *Proc. 9th ESA Congress*, Warschau, Polen, Bibliotheca Fragm. Agron. 11:649-650.

Gruber, S., Acharya, D., Claupein, W., 2008. Wood chips used for weed control in Organic Farming. *Journal of Plant Diseases and Protection*, Special Issue XXI, 395-400.

Iqbal, J., Z.A. Cheema, and M.N. Mushtaq. 2009. Allelopathic crop water extracts reduce the herbicide dose for weed control in cotton (Gossypium hirsutum). Int. J. Agric. Biol. 11(4): 360–366.

Iqbal, A., Shah, F., Hamayun, M., Khan, Z.H., Islam, B., Rehman, G., Khan, Z.U., Shah, S., Hussain, A., Jamal, Y. (2019). Plants are the possible source of allelochemicals that can be useful in promoting sustainable agriculture. Fresenius Experimental Bulletin 28, 140-149.2019,

Isfahan, M.N., Shariati, M. (2007). The effect of some allelochemicals on seed germination of Coronilla varia L. seeds. American-Eurasian J. Agric. & Environ. Sci., 2 (5), 534-538.

Jamil, M., Z.A. Cheema, M.N. Mushtaq, M. Farooq, and M.A. Cheema. 2009. Alternative control of wild oat and canary grass in wheat fields by allelopathic plant water extracts. Agronomy for Systainable Development 29(3), 475–482.

Julkunen-Tiitto, R., Sorsa, S. (2001). Testing the effects of drying methods on willow flavonoids, tannins, and salicylates. Journal of Chemical Ecology 27(4), 779–789.

Kelderer, M., Casera, C., Lardschneider, E. (2006). What can we expect from the commercially available bio-herbicides. In: Boos, M. (ed.) ecofruit - 12th International Conference on Cultivation Technique and Phytopathological Problems in Organic Fruit-Growing, 31st January to 2nd February 2006, Weinsberg/Germany, Conference Proceedings 172-177.

Khan, M.A., Marwat, K.B., 2005. Bioherbicidal effects of tree extracts on seed germination and growth of crops and weeds. *Pakistan Journal of Weed Science Research* 11:179-184.

Kohli, R.K., and D. Singh. 1991. Allelopathic impact of volatile components fromEucalyptus on crop plants. Biol. Plant. 33(6): 475–483.

Kumar, M., Lakiang, J.J., Gopichand, B., 2006. Phytotoxic effects of agroforestry tree crops on germination and radicle growth of some food crops of Mizoram. Lyonia – a journal of ecology and application 11, 83-89.

Lisanework, N., Michelsen, A., 1993. Allelopathy in agroforestry systems: the effects of leaf extracts of *Cupressus luisitania* and three *Eucalyptus* ssp. on four Ethiopian crops. Agroforestry Systems 21:63-74.

Law, D.M., Rowell, A.B., Snyder, J.C., Williams, M.A., 2006. Weed control efficacy of organic mulches in two organically managed bell pepper production systems. HortTechnology 16, 225-232.

Lux, G., Schmidtke, K., 2012. Reduzierung der Verunkrautung durch Einsatz von Grünguthäcksel unter den Bedingungen des ökologischen Landbaus. *Mitteilungen der Gesellschaft für Pflanzenbauwissenschaften* 24:128-129.

Machado, S., 2007. Allelopathic potential of various plant species on downy brome: implications for weed control on wheat production. *Agronomy Journal* 99, 127-132.

Modallal, M.M., Al-Charchafchi, F.M.R. (2006). Allelopathic Effect of Artemisia harba alba on Germination and Seedling Growth of Anabasis setifera. Pakistan Journal of Biological Sciences 9, 1795 – 1798

Mudrák, O., Frouz, J. (2012). Allelopathic effect of Salix caprea litter on late successional plants at different substrates of post-mining sites: pot experiment studies. Botany 90(4): 311–318.

Prinsloo, G., Du Plooy, C.P. (2018). The allelopathic effects of Amaranthus on seed germination, growth and development of vegetables. Biological Agriculture & Horticulture 34(4), 268–279.

Pérez, F.J., 1990. Allelopathic effect of hydroxamic acids from cereals on *Avena sativa* and *A. fatua. Phytochemistry* 29:773-776.

Perotti, V.E., Larran, A.S., Palmieri, V.E., Martinatto A.K., Permingeat, H.R. (2020). Herbicide resistant weeds: A call to integrate conventional agricultural practices, molecular biology knowledge and new technologies. Plant Science 290, in press

Rathinasabapathi, B., Ferguson, J.J., Gal, M., 2005. Evaluation of allelopathic potential of wood chips for weed suppressing in horticultural production systems. Horticultural Science 40,711-713.

Rice, E.L. 1984. Allelopathy. Academic press, Minnesota, USA, 422.

Saha. D., Marble, S.C., Pearson, B.J. (2018). Allelopathic effects of common landscape and nursery mulch materials on weed control. Frontiers in Plant Science 9:733, doi: 10.3389/fpls.2018.00733

Seigler, D.S. 1996. Chemistry and mechanisms of allelopathic interactions. Agronomy Journal. 88(6): 876–885.

Siipilehto, J., 2001. Effect of weed control with fibre mulches and herbicides on the initial development of spruce, birch and aspen seedlings on abandoned farmland. *Silva Fennica* 35:403-414.

Stephen, M., 2007. Allelopathic potential of various plant species on downy brome: implications for weed control on wheat production. *Agronomy Journal* 99:127-132.

Vogel, L., 2012. Entwicklung einer Methode zur Charakterisierung der allelopathischen Wirkung von Extrakten aus Häcksel verschiedener Gehölzarten. Masterthesis, University of Hohenheim.

Wang, L, Gruber S, Claupein W., 2012. Effects of woodchip mulch and barley intercropping on weeds in lentil crops. *Weed Research* 52:161-168.

Williams, R.D., and R.E. Hoagland. 1982. The effects of naturally occurring phenolic compounds on seed germination. Weed Science 30(2)206–212.

Xu, J., Gauder, M., Gruber, S., Claupein, W. (2018). Effects of 16-year woodchip mulching on weeds and yield in organic farming. Agronomy Journal 110 (1), 359-368.

Wu, Y.H., X. Cheng, X., Cai, Q.N., Lin. C.W. (2012). Allelopathic effects of twelve hedgerow plant species on seed germination and seedling growth of wheat (*Triticum astivum* L.). Advanced Materials Research, Vols. 356-360, 2767–2773; https://doi.org/10. 4028/www.scicntific.net/AMR.356-360.2767, page visited 11/2019.

CHAPTER 7

Effective Biological Control Strategies of Plant Diseases Using Microorganisms

by

Timothy Olubisi Adejumo[1] and Ralf Thomas Voegele[2]
[1]Dept. of Microbiology, Adekunle Ajasin University,
P.M.B. 001, Akungba-Akoko,
Ondo State, Nigeria.
[2]University of Hohenheim, Faculty of Agricultural Sciences,
Institute of Phytomedicine, Department of Plant Pathology,
Otto-Sander-Straße 5, 70599 Stuttgart, Germany.
Corresponding Author: *timothy.adejumo@aaua.edu.ng*

Abstract

The effectiveness of any microbial biological control agents to achieve optimum disease control is dependent on a range of modes of action, which may be without direct interaction with the targeted pathogen or through nutrient or spatial competition or other mechanisms like hyperparasitism and antibiosis, that directly interfere and affect the growth conditions of the pathogen. Compounds, enzymes and metabolites are produced *in situ* at low concentrations during the interaction and antagonism. The understanding of the mode of action is also useful to characterize possible risks for humans, the environment and risks for resistance development, as well as for the registration of microbial biological control agents. New approaches that plants may have their own language that allows them to communicate with their leaf and root associated microbiomes, to attract and select specific microbes which can eventually influence plant health and growth is highlighted.
Keywords: biological control, plant diseases, mode of action, antagonists, biopesticides

Introduction

Due to the growing world population, the worldwide attempt to expand food production to answer the need for food has become a serious and one of the most important international issues in recent times (Godfray *et al.*, 2010, Ingram, 2011; Keinan and Clark, 2012). The growing concern for global food security will make food production increase by at least 70% for the coming 40 years due to the increasing human population and consumption. Crop protection plays a significant role in defending crop productivity against competition from pathogens (Oerke and Dehne, 2004). Plant diseases have been found to cause an estimated 40 billion dollars losses worldwide every year (Roberts, 2006), either directly or indirectly, and at least 20–40% of losses in crop yield are caused by pathogen infections (Savary *et al.*, 2012). Weller (1988) mentioned the big challenges that await microbiologists and plant pathologists that are trying to search for and develop environmentally friendly control agents against plant diseases, in order to limit the use of large amounts of chemical pesticides. Alternatively, the use of beneficial microorganisms as biopesticides is one of the most effective methods for safe crop-management practices that works under low to medium disease pressure (McSpadden Gardener and Fravel, 2002). In recent years, interest in this research area has gradually increased, evidenced by a good number of publications, books and reviews on this topic. Researchers have suggested that biological control will continue to grow in significance and play a major role in modern agriculture in the future.

Beneficial biocontrol microbes may be one of the few options that show potential by competing with pathogens or by directly antagonising plant pathogens. The biocontrol microbes modulate plant defense mechanisms, deploy biocontrol actions in plants and offer new strategies to control plant pathogens. Efforts are now being made by researchers to improve, facilitate and maintain long-term plant colonization using new approaches of "plant-optimized microbiomes" (microbiome engineering) (Syed Ab Rahman, 2018).

Plant Disease Management

The management of plant diseases are achieved through the use of resistant varieties, chemicals and biological control methods.

i. Resistant varieties

Breeding for resistant varieties has been found to be one of the most successful and reliable management options for controlling plant diseases (Maloy, 2005), which has been used for many crops, being of high consumer acceptance, relatively inexpensive compared to the use of synthetic pesticides (Reddy, 2017). However, the drawback is that these varieties often take decades to develop, with extremely high regulatory approval cost and breakdown of new varieties' resistance within several years of their release, due to several causes such as mutations of the pathogens toward virulence, sexual and asexual recombination events, issues with variety uniformity in genetics, and decreasing field resistance (Syed Ab Rahman, 2018). However, new promising approaches to disease resistant varieties come from genome editing using CRISPR/Cas9 (Belhaj *et al.,* 2015) and other technologies that may be deregulated and considered *in par* with classical breeding approaches.

ii. The use of chemicals

The use of chemicals is central in the prevention of losses and damages caused by plant diseases, as this has brought improvements to crop quality and agricultural output

(Ragsdale, 1991). However, it has caused undesirable side effects such as food contamination, environmental dispersal and higher costs of food production (Carvalho, 2006). Many chemicals used in agriculture also destroy beneficial endophytic fungi and bacteria (Berg, 2009). Some chemical fungicides are lethal to beneficial insects and fungi inhabiting the soil, and may also enter the food chain (Budi *et al.*, 2000). Reports from FAO-WHO and data provided by the US Food and Drug Administration indicate that persistent organic pollutants (POPs) are present in virtually all types of food, including fruits, vegetables, poultry and dairy products (Schafer and Kegley, 2002). Syed Ab Rahman and co-workers (2018) reported that the use of Oryzalin and Trifluralin has been found to inhibit growth of certain species of mycorrhizal fungi that aid plant roots in nutrient uptake (Kelley and South, 1980). Triclopyr, a common landscape herbicide inhibits beneficial ammonia-oxidizing bacteria (Pell *et al.*, 1998), while Glyphosate, commonly used as weed killer greatly reduces growth and activity of beneficial free-living heterotrophic bacteria that aid in nitrogen fixation (Santos and Flores, 1995). More attention is now concentrated to finding and developing alternative inputs that are less toxic, less disruptive and those that will facilitate plant growth to control pests and pathogens.

iii. Biological control

Biological control of plant diseases is the suppression of populations of plant pathogens by living organisms. This appears to be the best option for the development of low cost, eco-friendly and sustainable management approaches for protecting plants and crops (Heimpel and Mills, 2017, Syed Ab Rahman *et al.*, 2018, Köhl *et al.*, 2019). Biocontrol often results in multiple interactions, such as suppressing the pest organism using other microorganisms or the application of antagonistic microorganisms to suppress disease development. Heydari and Pessarakli (2010) reported that more narrowly, biological control refers to the purposeful utilization of introduced or resident living organisms, other than disease resistant host plants, to suppress the activities and populations of one or more plant pathogens. This may involve the use of microbial inoculants to suppress a single type or class of plant diseases. This may also involve managing soils to promote combined activities of native soil and plant-associated organisms that contribute to general suppression (Cook, 1993). Most narrowly, biological control refers to the suppression of a single pathogen by a single antagonist, in a single cropping system. Most specialists in the field would concur with one of the narrower definitions presented above. Other examples include the application of natural products and chemical compounds extracted from different sources like plant extracts, natural or modified organisms or gene products (McSpadden Gardener and Fravel, 2002; Pal and McSpadden Gardener, 2006).

Biological Control Interactions

The different biological control interactions between the populations include: mutualism, protocooperation, commensalism, neutralism, antagonism, competition, amensalism, predation and parasitism.

i. Mutualism is an association among several species where all of them are benefited from the association without any noticed adverse effect e.g. algae and fungi in lichens. Sometimes, it is an obligatory life-long interaction involving close physical and biological contact, such as those between plants and mycorrhizal fungi. However, they are generally facultative and opportunistic. For instance, *Rhizobium* can reproduce either in the soil or, to a much greater degree, through their mutualistic association with Papilionaceous plants e.g. Cowpea. For example, Rhizobium bacteria reproduce either in the soil or, to a much greater degree through their mutualistic association with legume plants. These types of mutualism can contribute to biological control by providing plants with improved nutrition and/or by stimulating host defense mechanism (Chisholm *et al.*, 2006).

ii. Protocooperation is a form of mutualism, but the organisms involved do not depend exclusively on each other for survival. Many of the microbes isolated and classified as biocontrol agents can be considered facultative mutualists involved in protocooperation, because survival rarely depends on a specific host and disease suppression will vary depending on the prevailing environmental conditions (Pal and McSpadden Gardener, 2006).

iii. Commensalism is a symbiotic interaction between two living organisms, where one organism (commensal) benefits and the other (host) is neither harmed nor benefited (Fitter and Garbaye, 1994). An association in which the commensal is not directly dependent on the host metabolically and can survive if separated from the host e.g. *Nitrobacter* and *Nitrosomonas*; *Cellulomonas* and yeasts. Yeasts produce a number of different vitamins for bacteria to use; aerobes and anaerobes (aerobes deplete oxygen for anaerobes to grow).

iv. Neutralism is a biological interaction in which the population density of one species has absolutely no effect on the other (Berg *et al.*, 2005). An inability to associate the population dynamics of pathogen with that of another organism would indicate neutralism, a form of biological control, while

v. Antagonism between organisms results in a negative outcome for one or both.

vi. Competition is when both populations compete for the same limited resources (space/nutrients) and thus, have a negative effect on each other such that the fittest outcompetes the weaker one (Cook, 1993). Biocontrol can occur when non-pathogens compete with pathogens for nutrients and space in the host plant.

vii. Amensalism: In this type of interaction, one population gains a competitive advantage over the other by producing a growth-inhibiting substance e.g. production of a growth-inhibiting antimicrobial substances by the actinomycete *Pseudonocardia* against parasitic fungi *Escovopsis weberi* (Barke *et al.*, 2011).

viii. Predation usually refers to the hunting and killing of one organism by another for consumption and sustenance. It is applied to the actions of microorganisms such as protists and mesofauna, e.g. fungal feeding nematodes and microarthropods that consume pathogen biomass for sustenance (Cook, 1993). Predation occurs when a population has a negative effect on the size of another population by feeding on it and reducing its numbers e.g. protozoa predate on several bacterial species; *Bdellovibrio, Vampirococcus* and *Daptobacter* also prey on bacterial cells. Fungi (e.g. *Arthrobotrys)* trap protozoa and nematodes by sticky hyphae, constricting rings or invasive conidia.

ix. Parasitism is also a symbiotic relation in which two organisms coexist over a prolonged period of time, one organism, usually the physically smaller (parasite) benefits and the other (host) is harmed e.g. the fungi *Escovopsis weberi* parasitizes on other fungi. There is usually some dependence/co-existence between the host and the parasite unlike the predator. The activities of various hyperparasites, for example those agents that parasitize plant pathogens, can result in biocontrol (Lo *et al.*, 1997). Another interesting contribution to biocontrol is when host infection and parasitism by relatively avirulent pathogens lead to biocontrol of more virulent pathogens through the stimulation of host defense systems.

Mechanisms of Biological Control

i. Direct antagonism results from physical contact and/or a high-degree of selectivity for the pathogen by the mechanism(s) expressed by the biocontrol microorganism. In this interaction, hyperparasitism by obligate parasites of a plant pathogen would be considered the most direct type of mechanism because the activities of no other organism would be required to exert a suppressive effect (Harman et al., 2004). Hyperparasites invade and kill mycelium, spores, and/or resting structures of fungal pathogens and cells of bacterial pathogens (Ghorbanpour et al., 2018). Indirect antagonism results from activities that do not involve targeting a pathogen by a biocontrol microorganism. Lafontaine and Benhamou (1996) and Silva et al. (2004) claimed that improvement and stimulation of plant host defense mechanism by non-pathogenic microorganisms is the most indirect form of antagonism. It was discovered that the most effective biocontrol microorganisms studied appear to antagonize plant pathogens employing several modes of action (Cook, 1993). For example, Pseudomonads known to produce the antibiotic 2,4-diacetylphloroglucinol (DAPG) may also induce host defenses (Lafontaine and Benhamou, 1996 and Silva et al., 2004). Additionally, DAPG producing bacterial antagonists can aggressively colonize roots, a trait that might further contribute to their ability to suppress pathogen activity in the rhizosphere of a plant through competition for organic nutrients. The most important modes of action of biocontrol microorganisms are as follows:

ii. Mycoparasitism: In hyperparasitism, the pathogen is directly attacked by a specific biocontrol agent that kills it or its propagules. Four major groups of hyperparasites have been identified which include viruses, facultative parasites, obligate bacterial pathogens and predators. An example is the virus that infects *Cryphonectria parasitica*, the causal agent of chestnut blight, which causes hypovirulence, a reduction in pathogenicity of the pathogen. This phenomenon has resulted in the control of chestnut blight in many places (Milgroom and Cortesi, 2004).

In addition to viruses, several fungal hyperparasites have also been identified including those that attack sclerotia (*Coniothyrium minitans*) or others that attack fungal hyphae (*Pythium oligandrum*). In some cases, a single fungal pathogen can be attacked by multiple hyperparasites: *Acremonium alternatum*, *Acrodontium crateriforme*, *Ampelomyces quisqualis*, *Cladosporium oxysporum* and *Gliocladium virens* are some of the fungi that have the capacity to parasitize powdery mildew pathogens (Milgroom and Cortesi, 2004).

In contrast to hyperparasitism, microbial predation is more general, non-specific and generally provides less predictable levels of disease control. Some biocontrol agents exhibit predatory behaviour under nutrient-limited conditions. They include *Trichoderma*, a fungal antagonist that produces a range of enzymes that are directed against cell walls of pathogenic fungi. However, when fresh bark is used in composts, *Trichoderma* does not directly attack the plant pathogen *Rhizoctonia solani*. By contrast, in the decomposing bark the concentration of readily available cellulose decreases. This activates the chitinase genes of *Trichoderma* sp., which in turn produces chitinase to parasitize *R. solani* (Benhamou and Chet, 1997).

iii. Antibiosis by Antimicrobial Metabolites: Many microbes produce and secrete one or more compounds with antibiotic activity (Leclère *et al.*, 2005; Shahraki *et al.*, 2009). Antibiotics are microbial toxins that can, at low concentrations, poison or kill other microorganisms. It has been shown that some antibiotics produced by microorganisms are particularly effective against plant pathogens and the diseases they cause (Islam *et al.*, 2005). Antibiotics have been shown to be particularly effective at suppressing growth of target pathogens *in vitro* and/or *in situ*. An effective antibiotic must be produced in sufficient quantities (dose) close to the pathogen. Production of antimicrobial metabolites, mostly with broad-spectrum activity, has been reported for biocontrol bacteria belonging to the genera *Agrobacterium, Bacillus, Pantoea, Pseudomonas, Serratia, Stenotrophomonas, Streptomyces*, and many others. In *Bacillus*, especially lipopeptides such as iturin, surfactin, and fengycin have been investigated (Ongena and Jacques, 2008). In *Pseudomonas* many antibiotic metabolites such as DAPG, pyrrolnitrin and phenazine have been studied (Raaijmakers and Mazzola, 2012).

Many antibiotics are produced only when a microbial population reaches a certain threshold. This quorum-sensing phenomenon is well described for phenazine-producing

140

Pseudomonas species. Genomic information revealed that these genera also have the potential to produce many still unknown secondary metabolites with possible antimicrobial activity. Fungal antagonists can produce antimicrobial compounds as well. For *Trichoderma* and closely related *Clonostachys* (former *Gliocladium*) species, 6-PAP, gliovirin, gliotoxin, viridin and many more compounds with antimicrobial activity have been investigated (Ghorbanpour *et al.*, 2018). Arseneault and Filion (2017) reported half-lifes for antibiotics produced by biocontrol strains in soil ranging between 0.25 and 5 days depending on the biocontrol strain, antibiotic and experimental conditions. Such short life spans can be due to microbial decomposition, but also to chemical and/or physical inactivation. Arseneault and Filion (2017) discussed modulation of gene expression by low concentrations of antibiotic instead of inciting cell death at high concentrations as a mode of action. Antibiotics at low concentrations can be involved in signalling and microbial community interactions, communication with plants, and regulation of biofilm formation (Köhl et al., 2019). Raaijmakers and Mazzola (2012) discussed a range of functions of antimicrobial metabolites at low concentrations: there is evidence that antimicrobials including lipopetides protect bacteria from grazing by bacteriovorus nematodes such as *Caenorhabditis elegans*. Also, volatile antibiotic compounds may play a role in long-distance interactions amongst soil organisms including bacterial predators. Lipopeptides of *Bacillus* and *Pseudomonas* are involved in the surface attachment of bacterial cells and biofilm formation by activating signalling cascades finally resulting in the formation of extracellular matrices which protect microorganisms from adverse environmental stresses. Some antibiotics, especially lipopeptides, support the mobility of bacteria, most likely via changing the viscosity of colonized surfaces (Köhl *et al.*, 2019).

Other groups of antibiotics influence the nutritional status of plants. For example, DAPG-producing *Pseudomonas* upregulate nitrogen fixation by plant growth-promoting *Azospirillum brasilense*, and redox-active antibiotics support mobilization of limiting nutrients such as manganese and iron (Köhl *et al.*, 2019). Such a use of microbial metabolites is outside the scientific definition of biological control which is defined as the use of living beneficial organisms to suppress populations of plant pathogens (Heimpel and Mills, 2017), but in a broader definition, use of metabolites is also considered as biological control (Glare *et al.*, 2012).

Several reports demonstrate variability within pathogen populations in their sensitivity to antimicrobial secondary metabolites. Selected isolates of *Pseudomonas* spp. produce DAPG with antimicrobial activity against several plant pathogens. A high diversity in sensitivity to DAPG between isolates of *Gaeumannomyces graminis* var. *tritici* has been reported by Mazzola *et al.* (1995) and for *B. cinerea* by Schouten *et al.* (2008). Isolates of *B. cinerea* also differ in sensitivity to pyrrolnitrin (Ajouz *et al.*, 2011). Other secondary metabolites with proven antimicrobial activity which are produced by Microbial Biocontrol Agents (MBCAs) in bioreactors and applied as formulated bioactive compounds included in the end product in amounts effective in disease control (Glare *et al.*, 2012) are relevant metabolites which need to be assessed for potential toxicological and eco-toxicological risks.

However, for the majority of MBCAs, antimicrobial metabolites are produced at low concentrations *in situ* in microniches with low nutrient availability. Concentrations are sub-inhibitory if modes of action different from antibiosis are exploited (Raaijmakers and Mazzola, 2012). In other situations, metabolite production may be locally and temporally above a minimal inhibitory concentration resulting in inhibition or killing of the targeted pathogen. Such an antibiosis will be restricted in time because of the short life span of antimicrobial metabolites in the environment. Furthermore, the producing antagonist population will drop after application (Scheepmaker and van de Kassteele, 2011).

iv. Metabolite production: Many biocontrol microorganisms produce other metabolites that can interfere with pathogen growth and activities. Lytic enzymes are among these metabolites that can break down polymeric compounds, including chitin, proteins, cellulose, hemicellulose and DNA (Anderson *et al.*, 2004; Press *et al.*, 2001; Wilhite *et al.*, 2001). Studies have shown that some of these metabolites can sometimes directly result in the suppression of plant pathogens. For example, control of *Sclerotium rolfsii* by *Serratia marcescens* appeared to be mediated by chitinase expression (Ordentlich *et al.*, 1988). The antagonistic activities of these metabolites are indicative of the need to degrade complex polymers in order to obtain carbon nutrition. Examples are Lysobacter and Myxobacteria that produce lytic enzymes and have been shown to be effective against some plant pathogenic fungi (Bull *et al.*, 2002).

The effectiveness of such metabolites against plant pathogens is dependent on the composition and carbon and nitrogen sources of the soil and rhizosphere. For example, in post-harvest disease control, addition of chitosan which is a non-toxic and biodegradable polymer of beta-1,4-glucosamine produced from chitin by alkaline de-acylation stimulated microbial degradation of pathogens (Benhamou, 2004). Amendment of plant growth substratum with chitosan suppressed root rot caused by *Fusarium oxysporum* f. sp. *radicis-lycopersici* in tomato (Lafontaine and Benhamou, 1996).

In addition to the above-mentioned metabolites, other microbial byproducts may also play important roles in plant disease biocontrol (Phillips *et al.*, 2004). For example, Hydrogen Cyanide (HCN) effectively blocks the cytochrome oxidase pathway and is highly toxic to all aerobic microorganisms at picomolar concentrations (Ramette *et al.*, 2003). The production of HCN by certain fluorescent pseudomonads is believed to be effective against plant pathogens. Howell *et al.* (1988) reported that volatile compounds such as ammonia produced by *Enterobacter cloacae* were involved in the suppression of cotton seedling damping-off caused by *Pythium ultimum*.

v. Competition: Nutrient sources in the soil and rhizosphere are frequently not sufficient for microorganisms. For a successful colonization of phytosphere and rhizosphere, a microbe must effectively compete for the available nutrients (Elad and Baker, 1985; Keel *et al.*, 1989; Loper and Buyer, 1991). On plant surfaces, host-supplied nutrients include exudates, leachates, or senescent tissue. In addition to these, nutrients can also be obtained from waste products of other organisms such as insects and the soil. It is a general believe that competition between pathogens and non-pathogens for nutrient resources is an important issue in biocontrol (Elad and Baker, 1985; Keel *et al.*, 1989; Loper and Buyer, 1991). It is also believed that competition for nutrients is more critical for soil borne pathogens, including *Fusarium* and *Pythium* species that infect through mycelial contact than foliar pathogens that germinate directly on plant surfaces and infect through appressoria and infection pegs (Elad and Baker, 1985; Keel *et al.*, 1989; Loper and Buyer, 1991).

Competition for rare but essential micronutrients, such as iron, has also been shown to be important in biological disease control. Iron is extremely limited in the rhizosphere, depending on soil pH. In highly oxidized and aerated soil, iron is present in Ferric form (Kageyama and Nelson, 2003; Shahraki *et al.*, 2009), which is insoluble in water and the concentration may be extremely low. This very low concentration cannot support growth of microorganisms. To survive in such environments, organisms have been found to secrete iron-binding ligands called siderophores having high ability to obtain iron from the environment (Shahraki *et al.*, 2009). Almost all microorganisms produce siderophores, of either the catechol or the hydroxamate type (Kageyama and Nelson, 2003). A direct correlation was established *in vitro* between siderophore synthesis in fluorescent pseudomonads and their capacity to inhibit germination of chlamydospores of *F. oxysporum* (Elad and Baker, 1985). The increased efficiency in iron uptake of the commensal microorganisms is thought to be a critical factor in their root colonization ability which is a major factor in biocontrol performance of bacterial antagonists.

vi. Induction of resistance: Plants actively respond to a variety of environmental stimulating factors, including gravity, light, temperature, physical stress, water and nutrient availability and chemicals produced by soil and plant associated microorganisms (Audenaert *et al.*, 2002; Moyne *et al.*, 2001; Vallad and Goodman, 2004). Plant pathologists have characterized the determinants and pathways of induced resistance stimulated by biological control agents and other non-pathogenic microorganisms (Audenaert *et al.*, 2002; Moyne *et al.*, 2001; Vallad and Goodman, 2004). The first pathway: Systemic Acquired Resistance (SAR) is mediated by Salicylic Acid (SA), a chemical compound which is usually produced after pathogen infection and typically leads to the expression of Pathogenesis-Related (PR) proteins (Vallad and Goodman, 2004). These PR proteins include a variety of enzymes, some of which may act directly to lyse invading cells, reinforce cell wall boundaries to resist infections, or induce localized cell death (Vallad and Goodman, 2004).

The second pathway: Induced Systemic Resistance (ISR) is mediated by Jasmonic Acid (JA) and/or ethylene, which are produced following applications of some non-pathogenic rhizobacteria (Audenaert *et al.*, 2002; De Meyer and Höfte, 1997; Kloepper *et al.*, 1980; Leeman *et al.*, 1995; Moyne *et al.*, 2001; Van Loon *et al.*, 1998; Van Peer and Schippers, 1992; Van Wees *et al.*, 1997). Interestingly, the SA- and JA-dependent defence pathways can be mutually antagonistic and some bacterial pathogens take advantage of this to overcome SAR. For example, pathogenic strains of *Pseudomonas syringae* produce coronatine, which is similar to JA, to overcome the SA-mediated pathway (Vallad and Goodman, 2004). Since the various host-resistance pathways can be activated to variable degrees by different microorganisms and insect feeding, it is possible that multiple stimuli are constantly being received and processed by the plant. Thus, the magnitude and duration of host defense induction will likely vary over time. Only if induction can be controlled, i.e., by overwhelming or synergistically interacting with endogenous signals, will host resistance be increased (Audenaert *et al.*, 2002; De Meyer and Höfte, 1997; Kloepper *et al.*, 1980; Leeman *et al.*, 1995; Moyne *et al.*, 2001).

Some strains of root-colonizing microorganisms have been identified as potential elicitors of plant host defenses. For example, some biocontrol strains of *Pseudomonas* sp. and *Trichoderma* sp. are known to strongly induce plant host defenses (Haas and Defago, 2005; Harman *et al.*, 2004). In other instances, inoculation with Plant Growth Promoting Rhizobacteria (PGPR) have been shown to be effective in controlling multiple diseases caused by different fungal pathogens. A number of chemical elicitors of SAR and ISR such as salicylic acid, siderophore, lipopolysaccharides and 2,3-butanediol may be produced by PGPR strains upon inoculation (Ryu *et al.*, 2004; Van Loon *et al.*, 1998).

A substantial number of microbial products have been reported to elicit host defenses, indicating that host defenses are likely stimulated continually during the plant's lifecycle (Ryu *et al.*, 2004; Van Loon *et al.*, 1998). These inducers include lipopolysaccharides and flagellin from Gram-negative bacteria; cold shock proteins of diverse bacteria; transglutaminase, elicitins and a-glucans in Oomycetes; invertase in yeast; chitin and ergosterol in all fungi; and xylanase in Trichoderma (Ryu *et al.*, 2004). These findings indicate that plants would detect the composition of their plant-associated microbial communities and respond to changes in the quantity, quality and localization of many different signals. The importance of such interactions is indicated by the fact that further induction of host resistance pathways, by chemical and microbiological inducers, is not always effective in improving plant health or productivity in the field (Vallad and Goodman, 2004).

Biocontrol Agents
Most commonly used biocontrol agents are isolated by screening organisms from the rhizosphere or endophyte populations for inhibition of growth of the target pathogen *in vitro*.

144

Those that show inhibition are assessed further, although it should be stressed that *in vitro* inhibition is not always a successful indicator of a successful used biocontrol agent. There are other characteristics required such as the ability for mass production and persistence under field conditions (Elliott *et al.* 2009; Martin *et al.*, 2015). Prominent among those species of rhizosphere and endophytic bacteria that are effective BCAs are the actinobacteria and species from the genera *Pseudomonas* and *Bacillus*. Among the well represented fungi that constitute effective biocontrol agents are species of the genus *Trichoderma*. All of these are capable of synthesizing an array of secondary metabolites.

Mixtures of biocontrol agents
Using mixtures of biocontrol agents has been found to increase the consistency of biocontrol across sites with different conditions. Examples include the use of mixtures of PGPR strains in the biocontrol of postharvest dry rot of potato caused by *Fusarium sambucinum* (Recep *et al.*, 2009). Enhanced biocontrol has also been found effective for control diseases of poplar (Gyenis *et al.*, 2003), chilli (Muthukumar *et al.*, 2011), and cucumber (Raupach and Kloepper 1998). It is also possible that different mixtures may need to be used in different climatic areas. Thus, there is a need to identify a number of potential biocontrol agents. However, mixtures do not always give increased control. In some cases, there may be antagonism between the biocontrol agents that results in reduced control compared to single strains. In evaluating biocontrol agents for control of fire blight in pear, Stockwell *et al.* (2011) found that mixtures of *Pseudomonas fluorescens* A506, *Pantoea vagus* C9–1 and *Pantoea agglomerans* Eh252 were less effective than individual strains. It was found that the *Pantoea* strains exert their effect through the production of peptide antibiotics. These reports highlight the importance of considering possible antagonism between strains when developing a biocontrol formulation.

How effective are biocontrol agents?
The level of disease control achieved by application of biocontrol agents to a crop can be close to or equivalent to that achieved by application of a fungicide. The efficacy of a biocontrol agent can be enhanced by mixing with a fungicide provided the fungicide does not have adverse effects on the BCA. Infection of strawberry by *Botrytis cinerea* was reduced to low levels by application of *Trichoderma atroviridae*, but was eliminated by application of the biocontrol agent together with a fungicide. The fungicide alone was less effective than the biocontrol agents alone (O'Brien, 2017). Nakayama and Sayama (2013) reported a similar enhancement in disease control using a biocontrol agent - fungicide mix to inhibit powdery scab of potato. A number of studies have demonstrated that biocontrol can also be used effectively against postharvest diseases (Spadaro and Droby, 2016; O'Brien, 2017). Some endophytes protect against multiple pathogens. A strain of *Bacillus pumilis* isolated from the endosphere of poplar suppressed the growth of three pathogens

Cytospora chrysosperma, *Phomopsis macrospora* and *Fusicoccum aesculi* in greenhouse tests (Ren *et al.*, 2013). Some examples for BCAs in commercial production are provided in Table 1.

A Disadvantage with BCAs is the lack of consistency in disease suppression. O'Brien (2017) reported that differences occur in response to host genotype. The degree of control of *Phytophthora meadii* infection by *Alcaligenes* sp. differed between two cultivars of *Hevea brasiliensis* (Abraham *et al.*, 2013) and biocontrol of diseases in pepper (Lee *et al.*, 2008) and strawberry (Card *et al.*, 2009).

The specificity effect may be related to the production of plant molecules that activate transcriptional activators of the LuxR family in the bacterium (O'Brien, 2017). Endophytes produce a large array of different types of secondary metabolites many of which have not been detected directly but have been inferred from genomic analysis (Demain and Sanchez, 2009; Brader *et al.*, 2014). There are examples where the synthesis of secondary metabolites stimulates changes in plant metabolite production and vice-versa (Ludwig-Muller, 2015).

Mechanisms Associated with Protection of Plants by Biocontrol Agents

Plants have evolved a broad range of strategies to counter-attack and ward off attackers (Ponce de León and Montesano, 2013, Köhl *et al.*, 2019). Defense mechanisms against various environmental challenges of abiotic and biotic stresses, such as drought, herbivores and potentially pathogenic fungi, oomycetes, nematodes, bacteria and viruses include passive defences; non-host resistance, physical and chemical barriers, rapid active defences and delayed active defences. These mechanisms include i to viii discussed below.

i. Stimulation of plant growth

A common effect of the application of a rhizospheric or endophytic BCA to a plant is accelerated growth of the plant. Many bacterial and fungal BCAs produce analogues of plant growth regulatory hormones and volatile compounds that stimulate growth (Harman *et al.*, 2004; Taghavi *et al.*, 2009). *Trichoderma harzianum* produces a butenolide metabolite called harzianolide that both stimulates growth and induces defense mechanisms (Cai *et al.*, 2013). Analogues of plant hormones produced by endophytic bacteria not only promote growth of the plant but they alleviate other stresses such as drought. For example, abscisic acid and gibberellins produced by the bacterial endophyte *Azospirillum lipoferum* have been shown to be involved in alleviating drought stress symptoms in maize (Cohen *et al.*, 2009).

Table 1: Examples of biological control agents in commercial production

Biocontrol agent	Crop	Supplier	Country
Ampelomyces quisqualis M-10	Powdery mildews	EcoGen	USA
B. subtilis MB1600	Fungal pathogens of cotton, large seeded legumes, soybean	Beker Underwood	USA
B. subtilis MB1600 + Rhizobium	Fungal pathogens of soybeans, peanut	Beker Underwood	USA
Bacillus amyloliquefaciens GB99+ *B. subtilis* GB122	Bedding plants in potting mixes	Gustafson, Inc.	USA
Bacillus lichenformis SB3086	Turf Grass, Sclerotinia	Novozymes Biologicals,	USA
Bacillus pumillus GB34	Soybean fungal diseases	Gustafson, Inc.	USA
Bacillus subtilis GB03	Cotton, legume fungal diseases	Gustafson, Inc.	USA
Bacillus subtilis GB03, other *B. subtilis, B. lichenformis, B. megateriu*	Fungal pathogens of greenhouse and nursery plants.	Growth Products Ltd	
Bacillus subtilis QST 713	Vegetables, Fruits	AgraQuest	USA
Coniothyrium minitans	Root rot	Prophyta Biologischer	Germany
Coniothyrium minitans	Root rot	Bioved	Hungary
Escherichia coli phage	Bacterial pathogens of fruit and vegetables	Intralytix	USA
Fusarium oxysporum non-path	Wilt	S.I.A.P.A. Italy	
Fusarium oxysporum non-path	Wilt	Natural Plant Protection	France
Gliocladium catenulatum	Vegetables, Fruits	ArgaQuest	
Gliocladium catenulatum	Root rot wilt	Verdera	Finland
Listeria monocytogenes phage	Bacterial pathogens of fruit and vegetables	Micreos	The Netherlands
Listeria monocytogenes phage	Bacterial pathogens of fruit and vegetables	Intralytix	USA
P. fluorescens A506, and 1629RS *P syringe* 742RS	Certain fruits, almond, potato, tomato	Frost Technol Corp.	USA
Pseudomonas aureofaciens Tx-1	Fungal opathogens of turf Grass	Turf Science Laboratories	USA
Pseudomonas chlororaphis 63–28	Wilt diseases of ornamentals and vegetables in GH	Turf Science Labs	USA
Pseudomonas syringae	Pome fruit, citrus, cherries, potatoes	JET Harvest Solutions	USA
Pseudozyma flocculosa	powdery mildew	Plant Products	Canada
Pythium oligandrum	Root rot	Bioreparaty	Czech Rep.
Salmonella phage	Bacterial pathogens of fruit and vegetables	Intralytix	USA
Streptomyces griseoviridis	Vegetables, Fruits	ArgaQuest	USA
Streptomyces griseoviridis K61	Field ornamental, vegetable fungal pathogens	AgBio	USA
T. harzianum	Grey mold	Makhteshim Chem. Works	Israel
T. harzianum ATCC20476		Binab	Sweden
Trichoderma polysporum ATCC20475		Binab	Sweden
T. harzianum T-22	Root rot	Bioworks	USA
Trichoderma harzianum	Vegetables, Fruits	ArgaQuest	USA
Trichoderma harzianum	Root rot	Efal Agri	Israel
Trichoderma spp	Root rot wilt	Binab	Sweden
Trichoderma spp.	Root rot wilt	Bioplant	Denmark
Trichoderma spp.	Root rot	Agrimm Technologies	N. Zealand
Trichoderma virens GL-21	Root rot	Certis	USA
Trichoderma viride	Root rot wilt	Ecosense Laboratories	India
Xanthomonas campestris phage *Pseudomonas syrinage* phage	Bacterial pathogens of fruit and vegetables	Omnilytics	USA

Source: O'Brien (2017).

ii. Induction of host defense mechanisms

This occurs as a result of the release of elicitors (proteins, antibiotics and volatiles) by the BCA that induce expression of genes of the salicyclic acid (SAR) or the jasmonic acid/ethylene (ISR) pathway (Nawrocka and Malolepsza, 2013; Pieterse *et al.*, 2014). Induced Systemic Resistance (ISR) is characterised by broad spectrum resistance against pathogens of various types as well as abiotic stresses (An *et al.*, 2010; Shoresh *et al.*, 2010). In addition to volatiles, ISR is induced by siderophores and cyclic lipopeptide antibiotics (Jan *et al.*, 2011).

iii. Secretion of polysaccharide degrading enzymes

Enzymes produced by BCAs are capable of degrading cell walls of fungal (or oomycete) hyphae, chlamydospores, oospores, conidia, sporangia, and zoospores resulting in lysis and thus contribute to the protection of the plant. The oligosaccharides released from degradation of the fungal cell walls act as signalling molecules to induce the host defense mechanisms. However, the production of enzymes capable of degrading the hyphal cell walls of pathogenic fungi *in vitro* does not constitute proof that these enzymes are responsible for biocontrol activity *in planta*.

iv. Production of antibiotics

Many biocontrol bacteria and fungi produce multiple antibiotics (including biosurfactants with antibiotic properties such as lipopeptides) that confer a competitive advantage by eliminating other bacteria and fungi (O'Brien, 2017). In addition to their antibiotic properties, lipopeptides are important signalling molecules and affect processes such as motility, induction of host plant defense mechanisms, and formation of microbial biofilms on the inner and outer surfaces of plants (Jan *et al.*, 2011; O'Brien, 2017). *Trichoderma,* which is widely used as a biocontrol agent, and forms the basis of several commercial products for biocontrol also synthesizes an array of secondary metabolites with antibiotic activity (Druzhinina *et al.*, 2011). Inactivation of antibiotic synthesis genes in various species of *Pseudomonas*, or *Bacillus* has provided strong evidence for the role of antibiotics in biocontrol by these species (Wu *et al.*, 2015). Gene disruption was used to provide evidence for roles for fengycin (Yanez-Mendizabal *et al.*, 2012) and iturin (Zeriouh *et al.*, 2011) in biocontrol of peach and curcubit diseases respectively by strains of *Bacillus subtilis* and of iturin in biocontrol of fruit diseases by *Bacillus amyloliquefaciens* (Arrebola *et al.*, 2010). More recent work suggests that different antibiotics from the same strain interact synergistically to achieve disease suppression. A Pseudomonas strain producing phenazine and two types of cyclic lipopeptide antibiotics (sessilins and orfamides) suppresses infection of Chinese cabbage by *R. solani* AG2–1 (Olorunleke *et al.*, 2015).

v. Biofilms

On plant surfaces bacteria rarely exist as single cells, but form large multicellular assemblages called biofilms (Bogino *et al.,* 2013; Flemming *et al.,* 2016). Biofilms typically contain multiple bacterial, or mixed bacterial and fungal species (Flemming *et al.,* 2016; Frey-Klett *et al.,* 2011). In a biofilm cells are covered by a matrix that protects them from desiccation, UV radiation, predation, and bactericidal compounds such as antibiotics. Biofilms aid in plant protection by preventing access to the surface of the plant by a pathogen, and by the production of antibiotics, many of which are only produced when growing in a biofilm. Just as biofilms may aid the survival and proliferation of biocontrol species on plant surfaces, they may aid the survival and proliferation of pathogenic species (Morris and Monier, 2003). Additionally, cell wall degrading enzymes secreted by a pathogen may bind to the biofilm matrix leading to increased heat tolerance and protection against enzymatic degradation (Flemming *et al.,* 2016).

vi. Competition for nutrients

Competition for nutrients on, or proximal to the plant surface (rhizosphere) is also a mechanism that protects plants from pathogens (O'Brien, 2017) Biocontrol agents compete for sugars on the leaf surfaces or root exudates in the rhizosphere. These food sources are required for initial establishment of the pathogen prior to infection. By utilising these food sources, the BCA prevents establishment of the pathogen (Card *et al.,* 2009; Ellis *et al.,* 1999). They occupy the same niches as the pathogen, utilise the same nutrients, and can occupy entry points to the plant tissue that would be used by the pathogen thereby preventing infection by the pathogen (Sneh, 1998). Biocontrol species are able to sequester iron for their own use by the production of iron binding siderophores. This reduces the availability of iron to other organisms such as pathogens (Santoyo *et al.,* 2012). Because bacterial siderophores have a higher affinity for iron than fungal siderophores, they are effective at depriving fungi of iron (Jan *et al.,* 2011).

vii. Inactivation of pathogen phytotoxins

Many plant pathogens produce phytotoxins that contribute to pathogenicity by disrupting processes in the host plant (Strange, 2007). These toxins either act as enzyme inhibitors (HC toxin of *Helminthosporium carbonum*), interfere with membrane function (syringomycin of *P. syringae*), or prevent induction of host defenses (coronatine of *P. syringae*). BCAs can protect plants from phytotoxins by inactivating them or preventing their production. The potent BCA *Burhholderia heleia* PAK1–2 prevents synthesis of the phytotoxin tropolone by the rice pathogen *Burkholderia plantarii* (Wang *et al.,* 2016). A biocontrol strain of *Bacillus mycoides* inactivates the toxins thaxtomin A(1) and B(2) produced by the potato common scab pathogen *Streptomyces scabei* (King *et al.,* 2000). The rice sheath blight pathogen *R. solani* produces a host specific toxin, RS toxin, that is part of its pathogenicity. Biocontrol strains of *T. viridae* produce an alphaglucosidase that in-activates this specific toxin (Shanmugam *et al.,* 2001). Strains of *Fusarium* and *Tricho-derma* capable of inactivating the toxins Eutypine, 4-hydroxybenzaldehyde, and 3-phenyllactic acid produced by the pathogens causing Eutypa dieback and esca disease, two trunk diseases of grapevine (*Vitis vinifera*) have also been isolated (Christen *et al.,* 2005).

viii. Genetically modified biocontrol agents

Genetic engineering techniques that improve the efficacy of strains of bacteria and fungi can be applied. Examples include the transfer of a chitinase gene from *Serratia* to a *Pseu-domonas* endophyte which created a strain with a greatly increased ability to suppress *R. solani* infection of bean (Downing and Thomson, 2000), or the addition of a glucanase gene to *Trichoderma* resulting in a strain that secretes a mixture of glucanases that shows greatly enhanced protection against the pathogens *Pythium*, *Rhizoctonia*, and *Rhizopus* (Djonovic *et al.,* 2007). Zhou *et al.* (2005) assembled a 2,4-diacetylphloroglucinol (2,4-DAPG) biosynthesis locus (*phlACBDE*) cloned from strain CPF-10 into a mini-Tn5 trans-poson and introduced into the chromosome of the non 2,4-DAPG producing strain *P. flu-orescens* P32. The resultant strain provided significantly better protection of wheat against take-all disease caused by *Gaumannomyces graminis* var. *tritici* and tomato against bacte-rial wilt caused by *Ralstonia solanacearum*. In spite of the results, the newly created BCAs are subject to the regulations that govern the use of organisms that are genetically modified through the use of recombinant DNA. Given the stiff opposition that has faced the use of transgenic plants and even greater difficulties of containment faced with genetically mod-ified microorganisms, it is unlikely that BCAs created by recombinant DNA technology will be approved for general use in the near future in many parts of the world. A more realistic approach would be to use nonrecombinant DNA technology to enhance BCAs. Clermont *et al.* (2011) used genome shuffling to generate improved biocontrol strains of *Streptomyces melanosporofaciens* EF-76. Two rounds of genome shuffling resulted in the isolation of four strains with increased antagonistic activity against the potato pathogens *Streptomyces scabies* and *Phytophthora infestans*. Chemical mutagenesis has been used to enhance biocontrol activity, e.g., nitrosoguanidine mutagenesis of *Pseudomonas au-rantiaca* B-162 resulted in the isolation of a strain with threefold elevated levels of phen-azine production and enhanced biocontrol activity (Feklistova and Maksimova 2008). Marzano *et al.* (2013) isolated strains of *T. harzianum* with greatly enhanced biocontrol activity after UV mutagenesis. Because the genetic techniques used in these studies do not involve recombinant DNA, they simply mimic what happens naturally they do not fall under the regulations governing the use of genetically modified organisms and hence they should be more acceptable to being used for disease control. However, one of the potential problems with such agents is that aside from the desired mutation there may be additional mutations in other genes that can result in undesirable consequences. A more recently developed techniques of genome editing can overcome these limitations. Using tools such as Crispr/Cas, mutations can be introduced with great precision into specific locations in the genome with great efficiency (Barrangou and van Pijkeren, 2016).

Emerging Biocontrol Strategies

Plants may have their own language that allows them to communicate with their associated microbiomes by releasing a broad variety of chemicals through their leaves and roots. Syed Ab Rahman *et al.* (2018) suggested that understanding of how plants communicate will help to fight diseases without the application of chemicals, and that it could be a good approach to develop new biocontrol strategies. The language could be what helps the plant to attract and select specific microbes in the rhizosphere and phyllosphere that can offer specific and needed benefits (Vorholt, 2012), making the microbiome to influence plant health and growth via different mechanisms. The attraction of specific microbes in the rhizosphere is associated with the signalling molecules, hormones (Carvalhais *et al.*, 2013) and specific root exudates (Carvalhais *et al.*, 2015) secreted by the plants to match their needs. The symbiotic interaction of legume-rhizobia is a good example of chemical language, where secretions of specific compounds by the plants attract specific rhizobacteria (Fierer *et al.*, 2007, Cooper, 2007).

The new promising approaches that may lead to improved crop yields and potentially more resilient plants include:

i. Using exudates as an approach to attract beneficial microbes to control various plant diseases.

Evidence has shown that plants provide benefits for the microbes by attracting and maintaining specific microbiomes using chemical exudates. For example, flavonoids released from legumes attract specific nitrogen-fixing rhizobacteria (Cooper, 2007), while some beneficial rhizobacteria were found to activate the plant defense responses that prevent foliar diseases (Ryu *et al.*, 2004). Recent studies have shown that the correlation between hormone-treated plants and defense signalling mutants resulted in different exudate profiles and an attraction of different microbial populations (Carvalhais *et al.*, 2013; 2015).

ii. The use of substrates to maintain beneficial biocontrol microbes near crops

Bai *et al.* (2015) showed that the majority of associated microbes can be cultured by employing systematic bacterial isolation approaches. This can be an addition and maintaining of beneficial microbes for biocontrol of plant diseases by providing the right substrates as media of growth.

151

iii. Phyllosphere biocontrol

Aysan *et al.* (2003) have found out that several bacterial antagonists inhibited the growth of bacterial stem rot caused by *Erwinia chrysanthemi* on tomato plants under greenhouse condition. The biological product Serenade (*B. substilis* QRD137) suppresses floral infection of blueberries and reduces fungal growth in flowers treated with the bacterial strain (Scherm *et al.*, 2004). Plants defend themselves on the leaf surface by producing antimicrobial compounds or by promoting growth of beneficial microbes through the release of nutrients and/or signals (Vorholt, 2012).

iv. Breeding microbe-optimized plants

Efforts should be geared towards breeding plants that are optimized to attract and maintain beneficial biocontrol microbes. Genetic engineering and plant breeding can generate microbe-optimized plants that produce the right exudates, to attract and maintain beneficial microbes at the right time, either at the root or on the leaf (Trivedi *et al.*, 2017).

v. Microbiome engineering, plant-optimized microbes and plant-optimized microbiomes

Breeding of individual microbes or entire microbial consortia that harbor beneficial microbes, and maintain them for crop plants in different soil types to produce plant/soil-optimized microbes, and plant/soil-optimized microbiomes that can be used as inoculum for different crops in different soils. Evidence abound that soil microbiomes adapt to their crops over time leading to improved plant-microbe interactions (Berendsen *et al.*, 2012).

vi. Matching microbe-optimized plant seed with the optimal microbiome and soil amendment practices for each soil type

Research is ongoing to find the right microbes that help specific crops grow better. Seeds laced with the right microbiomes would be the best options compared to other applications like sprays or root soaks when considering the transient nature of the microbiomes. To make sure that beneficial microbes are maintained, some soil amendments may be required. Rice *et al.* (1995) showed a successfully commercialized co-culturing of the phosphate solubilizing fungus *Penicillium bilaii* with Rhizobium as a legume inoculant. Also, Liu and Sinclair (1990) showed co-inoculation of soybean with *Bradyrhizobia* and *Bacillus megaterium* enhanced nodulation of soybean.

Conclusion

O'Brien (2017) reported the inadequateness of confrontation assays, as they do not take into account all mechanisms of antagonism, and do not replicate the environment in which the BCA must function. Pliego *et al.* (2011) also claimed that the use of inappropriate screening methods is a major contributor to the failure of biocontrol strategies, as screening for BCAs must be done with an *in planta* assay or an assay with tissue explants. With the continued application of genomics to identification of genes responsible for maintaining the endophytic state, it is possible to identify effective BCAs based on a genetic profile (Benítez and McSpadden Gardener, 2009). In addition, gene identification opens up possibilities for genetic modification, so that instead of screening for new BCAs, new ones can be made by modification of pre-existing ones. The method of production is crucial as it determines the type of propagules (spores, conidia, vegetative cells) produced and thus the shelf life, as well as persistence of the product in the environment (Bisutti *et al.,* 2015; Hanitzsch *et al.,* 2013; Kakvan *et al.,* 2013; Mocellin and Gessler, 2007). O'Brien (2017) observed that despite the fact that a lot of organisms with biocontrol potential have been identified against a large number of pathogens, there have been relatively few developed commercially. To remedy this and take full advantage of the benefits in biological control, the research focus needs to shift from identification of antagonistic organisms to production, formulation and delivery.

Future Outlook

i. Rhizosphere microbial population can be manipulated by simply spraying plants with signalling chemicals or altering the genotype (plant breeding) to attract beneficial microbes (Carvalhais *et al.*, 2015, Wintermans *et al.*, 2016).

ii. Beneficial biocontrol microbes can be kept by culturing using substrates as medium of growth.

iii. Profiling the phyllosphere microbial (Vorholt, 2012), chemical environment, identifying and making use of important plant-microbe, as well as microbe–microbe interactions on the leaf surface will bring new insights into foliar microbiomes by plants, and may lead to new strategies that will enhance food security.

iv. Plants design their own rhizosphere environment by the secretion of specific exudates to improve nutrient availability and interaction with specific beneficial microbes (Trivedi *et al.*, 2017).

v. Efforts towards microbiome engineering can in the future lead to microbial consortia that are better suited to support plants and improve crop yields to address food security.

vi. Microbe-coating of seeds with promising microbes for the right soil is one of the best options for optimizing plant-microbe interactions.

The integration of microbial biofertilizers, biocontrol microbes, optimized microbiomes, soil amendments and matching microbe-optimized crops for different soil types would be the penultimate goal to benefit most from positive plant-microbe interactions. This holds promise to improve crop yields and address food security in an environmentally-friendly and sustainable manner.

References

Abraham, A., Philip, S., Jacob, C.K., Jayachandran, K. (2013) Novel bacterial endophytes from *Hevea brasiliensis* as biocontrol agent against Phytophthora leaf fall disease. BioControl 58: 675–684.

Ajouz, S., Walker, A.S., Fabre, F., Leroux, P., Nicot, P.C., Bardin, M. (2011) Variability of *Botrytis cinerea* sensitivity to pyrrolnitrin, an antibiotic produced by biological control agents. BioControl 56: 353–363.

An, Y., Kang, S., Kim, K.D., Hwang, B.K., Jeun, Y. (2010) Enhanced defense responses of tomato plants against late blight pathogen *Phytophthora infestans* by pre-inoculation with rhizobacteria. Crop Prot 29: 1406–1412.

Anderson, L.M., Stockwell, V.O., Loper, J.E. (2004) An extracellular protease of *Pseudomonas fluorescens* inactivates antibiotics of *Pantoea agglomerans*. Phytopathology 94: 1228-1234.

Arrebola E., Jacobs R., Korsten L. (2010) Iturin a is the principal inhibitor in the biocontrol activity of *Bacillus amyloliquefaciens* PPCB004 against postharvest fungal pathogens. J Appl Microbiol 108: 386–395.

Arseneault, T., Filion, M. (2017) Biocontrol through antibiosis: exploring the role played by subinhibitory concentrations of antibiotics in soil and their impact on plant pathogens. Can J Plant Pathol 39: 267–274.

Audenaert, K., Pattery, T., Cornelis, P., Höfte, M. (2002) Induction of systemic resistance to *Botrytis cinerea* in tomato by *Pseudomonas aeruginosa* 7NSK2: role of salicylic acid, pyochelin and pyocyanin. Mol Plant-Microbe Interact 15: 1147-1156.

Aysan, Y., Karatas, A., Cinar, O. (2003) Biological control of bacterial stem rot caused by *Erwinia chrysanthemi* on tomato. Crop Prot 22: 807–811.

Bai, Y., Müller, D.B., Srinivas, G., Garrido-Oter, R., Potthoff, E., Rott, M., Dombrowski, N., Münch, P.C., Spaepen, S., Remus-Emsermann, M., Hüttel, B., McHardy, A.C., Vorholt, J.A., Schulze-Lefert, P. (2015) Functional overlap of the *Arabidopsis* leaf and root microbiota. Nature 528: 364.

Barke, J., Seipke, R.F., Yu, D.W., Hutchings, M. I. (2011) A mutualistic microbiome How do fungus-growing ants select their antibiotic-producing bacteria? Commun Integr Biol 4: 41–43.

Barrangou, R., van Pijkeren, J.P. (2016) Exploiting CRISPR-Cas immune systems for genome editing in bacteria. Curr Opin Biotechnol 37: 61–68.

Belhaj, K., Chaparro-Garcia, A., Kamoun, S., Patron, N.J., Nekrasov, V. (2015) Editing plant genomes with CRISPR/Cas9. Curr Opin Biotechnol 32: 76–84.

Benhamou, N. (2004) Potential of the mycoparasite, *Verticillium lecanii*, to protect citrus fruit-against *Penicillium digitatum*, the causal agent of green mold: A comparison with-the effect of chitosan. Phytopathology 94: 693-705.

Benhamou N. and Chet I. (1997) Cellular and molecular mechanisms involved in the interaction between *Trichoderma harzianum* and *Pythium ultimum*. Appl Environ Microbiol 63:2095–2099.

Benítez, M.S., McSpadden Gardener, B.B. (2009) Linking sequence to function in soil bacteria: sequence-directed isolation of novel bacteria contributing to soilborne plant disease suppression. Appl Environ Microbiol 75: 915–924.

Berendsen, R.L., Pieterse, C.M., Bakker, P.A. (2012) The rhizosphere microbiome and plant health. Trends Plant Sci 17: 478–486.

Berg, G., Krechel, A., Ditz, M., Sikora, R.A., Ulrich, A., Hallmann J. (2005) Endophytic and ectophytic potato-associated bacterial communities differ in structure and antagonistic function against plant pathogenic fungi. FEMS Microbiol Ecol 51: 215-229.

Berg, G. (2009) Plant–microbe interactions promoting plant growth and health: perspectives for controlled use of microorganisms in agriculture. Appl Microbiol Biotechnol 84: 11–18.

Bisutti, I.L., Hirt, K., Stephan, D. (2015) Influence of different growth conditions on the survival and the efficacy of freeze-dried *Pseudomonas fluorescens* strain Pf153. Biocontrol Sci Tech 25: 1269–1284.

Bogino, P.C., Oliva, M.D., Sorroche, F.G., Giordano, W. (2013) The role of bacterial biofilms and surface components in plant-bacterial associations. Int J Mol Sci 14: 15838–15859.

Brader, G., Compant, S., Mitter, B., Trognitz, F., Sessitsch, A. (2014) Metabolic potential of endophytic bacteria. Curr Opin Biotechnol 27: 30–37.

Budi, S. W., Van Tuinen, D., Arnould, C., Dumas-Gaudot, E., Gianinazzi-Pearson, V., & Gianinazzi, S. (2000) Hydrolytic enzyme activity of *Paenibacillus* sp. strain B2 and effects of the antagonistic bacterium on cell integrity of two soil-borne pathogenic fungi. Appl Soil Ecol 15: 191–199.

Bull, C.T., Shetty, K.G., Subbarao, K.V. (2002) Interactions between Myxobacteria, plant pathogenic fungi and biocontrol agents. Plant Dis 86: 889-896.

Cai, F., Yu, G., Wang, P., Wei, Z., Fu, L., Shen, Q., Chen, W. (2013) Harzianolide, a novel plant growth regulator and systemic resistance elicitor from *Trichoderma harzianum*. Plant Physiol Biochem 73: 106–113.

Card, S.D., Walter, M., Jaspers, M.V., Sztejnberg, A., Stewart, A. (2009) Targeted selection of antagonistic microorganisms for control of *Botrytis cinerea* of strawberry in New Zealand. Australas Plant Pathol 38: 183–192.

Carvalhais, L.C., Dennis, P.G., Badri, D.V., Tyson, G.W., Vivanco, J.M., Schenk, P.M. (2013) Activation of the jasmonic acid plant defence pathway alters the composition of rhizosphere bacterial communities. PLoS One 8: e56457.

Carvalhais, L.C., Dennis, P.G., Badri, D.V., Kidd, B.N., Vivanco, J.M., Schenk, P.M. (2015) Linking jasmonic acid signaling root exudates, and rhizosphere microbiomes. Mol Plant-Microbe Interact 28: 1049–1058.

Carvalho, F.P. (2006) Agriculture pesticides, food security and food safety. Environ Sci Policy 9: 685–692.

Chisholm, S.T., Coaker, G., Day, B., Staskawicz B.J. (2006) Host-microbe interactions: Shaping the evolution of the plant immune response. Cell 124: 803-814.

Christen, D., Tharin, M., Perrin-Cherioux, S., Abou-Mansour, E., Tabacchi, R., Defago, G. (2005) Transformation of Eutypa dieback and esca disease pathogen toxins by antagonistic fungal strains reveals a second detoxification pathway not present in *Vitis vinifera*. J Agric Food Chem 53: 7043–7051.

Cooper, J. (2007) Early interactions between legumes and rhizobia: disclosing complexity in a molecular dialogue. J Appl Microbiol 103: 1355–1365.

Clermont, N., Lerat, S., Beaulieu, C. (2011) Genome shuffling enhances biocontrol abilities of *Streptomyces* strains against two potato pathogens. J Appl Microbiol 111: 671–682.

Cohen, A.C., Travaglia, C.N., Bottini, R., Piccoli P.N. (2009) Participation of abscisic acid and gibberellins produced by endophytic *Azospirillum* in the alleviation of drought effects in maize. Botany, 87, pp. 455-462.

Cook, R.J. (1993) Making greater use of introduced microorganisms for biological control of plant pathogens. Annu Rev Phytopathol 31: 53-80.

De Meyer, G., Höfte, M. (1997) Salicylic acid produced by the rhizobacterium *Pseudomonas aeruginosa* 7NSK2 induces resistance to leaf infection by *Botrytis cinerea* on bean. Phytopathology 87: 588-593.

Demain, A.L., Sanchez, S. (2009) Microbial drug discovery: 80 years of progress. J Antibiot 62: 5-16.

Djonovic, S., Vittone, G., Mendoza-Herrera, A., Kenerley, C.M. (2007) Enhanced biocontrol activity of *Trichoderma virens* transformants constitutively coexpressing beta-1,3- and beta-1,6-glucanase genes. Mol Plant Pathol 8: 469–480.

Downing, K., Thomson, J.A. (2000) Introduction of the *Serratia marcescens chiA* gene into an endophytic *Pseudomonas fluorescens* for the biocontrol of phytopathogenic fungi. Can J Microbiol 46: 363–369.

Druzhinina, I.S., Seidl-Seiboth, V., Herrera-Estrella, A., Horwitz, B.A., Kenerley, C.M., Monte, E., Mukherjee, P.K., Zeilinger, S., Grigoriev, I.V., Kubicek, C.P. (2011) Trichoderma: the genomics of opportunistic success. Nat Rev Microbiol 9: 749–759.

Elad, Y., Baker, R. (1985) Influence of trace amounts of cations and siderophore-producing pseudomonads on chlamydospore germination of *Fusarium oxysporum*. Phytopathology 75: 1047-1052.

Elliott, M., Shamoun, S.F., Sumampong, G., James, D., Masri, S., Varga, A. (2009) Evaluation of several commercial biocontrol products on European and north American populations of *Phytophthora ramorum*. Biocontrol Sci Tech 19: 1007–1021.

Ellis, R.J., Timms-Wilson, T.M., Beringer, J.E., Rhodes, D., Renwick, A., Stevenson, L., Bailey, M.J. (1999) Ecological basis for biocontrol of damping-off disease by *Pseudomonas fluorescens* 54/96. J Appl Microbiol 87: 454–463.

Feklistova, I.N., Maksimova, N.P. (2008) Obtaining *Pseudomonas aurantiaca* strains capable of overproduction of phenazine antibiotics. Microbiology 77: 176–180.

Fierer, N., Bradford, M.A., Jackson, R.B. (2007) Toward an ecological classification of soil bacteria. Ecology 88: 1354–1364.

Fitter, A.H., Grabaye, J. (1994) Interactions between mycorrhizal fungi and other soil organisms. Plant Soil 159: 123-132.

Flemming, H.C., Wingender, J., Szewzyk, U., Steinberg, P., Rice, S.A., Kjelleberg, S. (2016) Biofilms: an emergent form of bacterial life. Nat Rev Microbiol 14: 563–575.

Frey-Klett, P., Burlinson, P., Deveau, A., Barret, M., Tarkka, M., Sarniguet, A. (2011) Bacterial-fungal interactions: hyphens between agricultural, clinical, environmental, and food microbiologists. Microbiol Mol Biol Rev 75: 583-609.

Ghorbanpour, M., Omidvari, M., Abbaszadeh-Dahaji, P., Omidvar, R., Kariman, K. (2018) Mechanisms underlying the protective effects of beneficial fungi against plant diseases. Biol Control 117: 147–157.

Glare, T., Caradus, J., Gelernter, W., Jackson, T., Keyhani, N., Köhl, J., Marrone, P., Morin, L., Stewart, A. (2012) Have biopesticides come of age? Trends Biotechnol 30: 250–258.

Godfray, H.C.J., Beddington, J.R., Crute, I.R., Haddad, L., Lawrence, D., Muir, J.F., Pretty, J., Robinson, S., Thomas, S.M., Toulmin, C. (2010) Food security: the challenge of feeding 9 billion people. Science 327: 812–818.

Gyenis, L., Anderson, N.A., Ostry, M.E. (2003) Biological control of Septoria leaf spot disease of hybrid poplar in the field. Plant Dis 87: 809–813.

Haas, D., Defago, G. (2005) Biological control of soil-borne pathogens by fluorescent pseudomonads. Nat Rev Microbiol 3: 307-319.

Hanitzsch, M., Przyklenk, M., Pelzer, B., Anant, P. (2013) Development of new formulations for soil pest control. IOBC/WPRS Bulletin 90: 211–215.

Harman, G.E., Howell, C.R., Viterbo, A., Chet, I., Lorito, M. (2004) *Trichoderma* species opportunistic, avirulent plant symbionts. Nat Rev Microbiol 2: 43–56.

Heimpel, G.E., Mills, N. (2017) Biological Control - Ecology and Applications. Cambridge: Cambridge University Press.

Heydari, A., Pessarakli, M. (2010) A review on bological cntrol of fungal plant pathogens using microbial antagonists. J Biol Sci 10: 273-290.

Howell, C.R., Beier, R.C., Stipanovic, R.D. (1988) Production of ammonia by *Enterobacter cloacae* and its possible role in the biological control of Pythium pre-emergence damping-off by the bacterium. Phytopathology 78: 1075-1078.

Ingram, J. (2011) A food systems approach to researching food security and its interactions with global environmental change. Food Secur 3: 417–431.

Islam, M.T., Hashidoko, Y., Deora, A., Ito, T., Tahara, S. (2005) Suppression of damping-off-disease in host plants by the rhizoplane bacterium *Lysobacter* sp. strain SB-K88 is-linked to plant colonization and antibiosis against soilborne peronosporomycetes. Appl Environ Microbiol 71: 3786-3796.

Jan, A.T., Azam, M., Ali, A., Haq, Q.M.R. (2011) Novel approaches of beneficial Pseudomonas in mitigation of plant diseases - an appraisal. J Plant Interact 6: 195–205.

Kageyama, K., Nelson, E.B. (2003) Differential inactiviation of seed exudates stimulation of *Pythium ultimum* sporangium germination by *Enterobacter cloacae* influences biological control efficacy on different plant species. Appl Environ Microbiol 69: 1114-1120.

Kakvan, N., Heydari, A., Zamanizadeh, H.R., Rezaee, S., Naraghi, L. (2013) Development of new bioformulations using *Trichoderma* and *Talaromyces* fungal antagonists for biological control of sugar beet damping-off disease. Crop Prot 53: 80–84.

Keel, C., Voisard, C., Berling, C.H., Kahir, G., Defago, G. (1989) Iron sufficiency, a prerequisit for suppression of tobacco black root rot by *Pseudomonas fluorescnes* strain CHA0 under gnotobiotic contiditions. Phytopathology 79: 584-589.

Keinan, A., Clark, A.G. (2012) Recent explosive human population growth has resulted in an excess of rare genetic variants. Science 336: 740–743.

Kelley, W., South D. (1980) Effects of herbicides on in vitro growth of mycorrhizae of pine (*Pinus* spp.). Weed Sci 28: 599–602.

King, R.R., Lawrence, C.H., Calhoun, L.A. (2000) Microbial glucosylation of thaxtomin, a partial detoxification. J Agric Food Chem 48: 512–514.

Kloepper, J.W., Leong, J., Teintze, M., Schroth, M.N. (1980) Pseudomonas siderophores: A mechanism explaining disease suppression in soils. Curr Microbiol 4: 317-320.

Köhl, J., Kolnaar, R., Ravensberg, W.J. (2019) Mode of action of microbial biological control agents against plant diseases: relevance beyond efficacy. Front Plant Sci 10: 845.

Lafontaine, P.J., Benhamon, N. (1996) Chitosan treatment: An emerging strategy for enhancing resistance of greenhouse tomato plants to infection by *Fusarium oxysporum* f.sp. *radicis-lycopersici*. Biocontrol Sci Technol 6: 111-124.

Lee, K.J., Kamala-Kannan, S., Sub, H.S., Seong, C.K., Lee, G.W. (2008) Biological control of phytophthora blight in red pepper (*Capsicum annuum* L.) using *Bacillus subtilis*. World J Microbiol Biotechnol 24: 1139–1145.

Leeman, M., van Pelt, J.A., den Ouden, F.M., Heinsbroek, M., Bakker, P.A.H.M., Schippers, B. (1995) Induction of systemic resistance by *Pseudomonas fluorescens* in radish cultivars differing in susceptibility to Fusarium wilt, using novel bioassay. Eur J Plant Pathol 101: 655-664.

Leclère, V., Bechet, M., Adam, A., Guez, J.S., Wathelet, B., Ongena, M., Thonart, P., Gancel, F., Chollet-Imbert, M., Jacques, P. (2005) Mycosubtilin overproduction by *Bacillus subtilis* BBG100 enhances the organism's antagonistic and biocontrol activities. Applied Environ Microbiol 71: 4577-4584.

Liu, Z., Sinclair, J. (1990) Enhanced soybean plant growth and nodulation by Bradyrhizobium in the presence of strains of *Bacillus megaterium*. Phytopathology 80: 1024.

Lo, C.T., Nelson, E.B., Harman, G.E. (1997) Biological control of Pythium, Rhizoctonia and Sclerotinia infected diseases of turfgrass with *Trichoderma harzianum*. Phytopathology 84: 1372-1379.

Loper, J.E., Buyer, J.S. (1991) Siderophores in microbial interactions of plant surfaces. Mol Plant-Microbe Interact 4: 5-13.

Ludwig-Müller, J. (2015) Plants and endophytes: equal partners in secondary metabolite production? Biotechnol Lett 37: 1325–1334.

Maloy, O.C. (2005) Plant disease management. Plant Health Instructor 10: DOI: 10.1094/PHI-I-2005-0202-01.

Martin, J.A., Macaya-Sanz, D., Witzell, J., Blumenstein, K., Gil, L. (2015) Strong *in vitro* antagonism by elm xylem endophytes is not accompanied by temporally stable in planta protection against a vascular pathogen under field conditions. Eur J Plant Pathol 142: 185–196.

Marzano, M., Gallo, A., Altomare, C. (2013) Improvement of biocontrol efficacy of *Trichoderma harzianum* vs. *Fusarium oxysporum* f. Sp *lycopersici* through UV- induced tolerance to fusaric acid. Biol Control 67: 397–408.

Mazzola, M., Fujimoto, D.K., Thomashow, L.S., Cook, R.J. (1995) Variation in sensitivity of *Gaeumannomyces graminis* to antibiotics produced by fluorescent *Pseudomonas* spp. and effect on biological control of take-all of wheat. Appl Environ Microbiol 61: 2554–2559.

McSpadden Gardener, B.B., Fravel, D. (2002) Biological control of plant pathogens: research, commercialization, and application in the USA. Plant Health Progress 3: 17.

Milgroom, M.G., Cortesi, P. (2004) Biological control of chestnut blight with hypovirulence: a critical analysis. Annu Rev Phytopathol 42: 311-338.

Mocellin, L., Gessler, C. (2007) Alginate matrix based formulation for storing and release of biocontrol agents. Bulletin OILB/SROP 30: 553–555.

Morris, C.E., Monier, J.M. (2003) The ecological significance of biofilm formation by plant-associated bacteria. Annu Rev Phytopathol 41: 429–453.

Moyne, A.-L., Shelby, R., Cleveland, T.E., Tuzun, S. (2001) Bacillomycin D: An iturin with antifungal activity against *Aspergillus flavus*. J Appl Microbiol 90: 622-629.

Muthukumar, A., Eswaran, A., Sangeetha, G. (2011) Induction of systemic resistance by mixtures of fungal and endophytic bacterial isolates against *Pythium aphanidermatum*. Acta Physiol Plant 33: 1933–1944.

Nakayama, T., Sayama, M. (2013) Suppression of potato powdery scab caused by *Spongospora subterranea* using an antagonistic fungus *Aspergillus versicolor* isolated from potato roots. Proceedings of the Ninth Symposium of the International Working Group on Plant Viruses with Fungal Vectors, Obihiro, Hokkaido, Japan, 19–22 August 2013: 53–54.

Nawrocka J, Malolepsza U (2013) Diversity in plant systemic resistance induced by Trichoderma. Biol Control 67: 149–156.

O'Brien, P.A. (2017) Biological control of plant diseases. Australas Plant Pathol 46: 293-304.

Oerke, E.C., Dehne, H.-W. (2004) Safeguarding production—losses in major crops and the role of crop protection. Crop Prot 23: 275–285.

Olorunleke, F.E., Hua, G.K.H., Kieu, N.P., Ma, Z.W., Hofte, M. (2015) Interplay between orfamides, sessilins and phenazines in the control of Rhizoctonia diseases by *Pseudomonas* sp CMR12a. Environ Microbiol Rep 7: 774–781.

Ongena, M., Jacques, P. (2008) Bacillus lipopeptides: versatile weapons for plant disease biocontrol. Trends Microbiol 16: 115-125.

Ordentlich, A., Elad, Y., Chet, I. (1988) The role of chitinase of *Serratia marcescens* in biocontrol of *Sclerotium rolfsii*. Phytopathology 78: 84-87.

Pal, K.K., McSpadden Gardener, B.M. (2006) Biological control of plant pathogens. Plant Health Instructor 2: 1117–1142. doi:10.1094/PHI-A-2006-1117-02.

Pell, M., Stenberg, B., Torstensson, T. (1998) Potential denitrification and nitrification tests for evaluation of pesticide effects in soil. Ambio 27: 24–28.

Phillips, A.D., Fox, T.C., King, M.D., Bhuvaneswari, T.V., Teuber, L.R. (2004) Microbial products trigger amino acid exudation from plant roots. Plant Physiol 136: 2887-2894

Pieterse, C.M., Zamioudis, C., Berendsen, R.L., Weller, D.M., Van Wees, S.C., Bakker, P.A. (2014) Induced systemic resistance by beneficial microbes. Annu Rev Phytopathol 52: 347–375.

Pliego, C., Ramos, C., de Vicente, A., Cazorla, F.M. (2011) Screening for candidate bacterial biocontrol agents against soilborne fungal plant pathogens. Plant Soil 340: 505–520.

Ponce de León, I., Montesano, M. (2013) Activation of defense mechanisms against pathogens in mosses and flowering plants. Int J Mol Sci 14: 3178–3200.

Press, C.M., Loper, J.E., Kloepper, J.W. (2001) Role of iron in rhizobacteria mediated induced systemic resistance of cucumber. Phytopathology 91: 593-598.

Raaijmakers, J.M., Mazzola, M. (2012) Diversity and natural functions of antibiotics produced by beneficial and plant pathogenic bacteria. Annu Rev Phytopathol 50: 403-424.

Ragsdale, N.N. (1991) Health and environmental factors associated with agricultural use of fungicides. National Agricultural Pesticide Impact Assessment Program (NAPIAP).

Ramette, A., Moenne-Loccoz, Y., Defago, G. (2003) Prevalence of fluorescent-pseudomonads producing antifungal phloroglucinols and/or hydrogen cyanide in soils naturally suppressive or conducive to tobacco root rot. FEMS Microbiol Ecol 44: 35-43.

Raupach, G.S., Kloepper, J.W. (1998) Mixtures of plant growth-promoting rhizobacteria enhance biological control of multiple cucumber pathogens. Phytopathology 88: 1158–1164.

Recep, K., Fikrettin, S., Erkol, D., Cafer, E. (2009) Biological control of the potato dry rot caused by Fusarium species using PGPR strains. Biol Control 50: 194-198.

Reddy, P. P. (2017). Cultivar Mixtures/Multiline Cultivars. In *Agro-ecological Approaches to Pest Management for Sustainable Agriculture* (pp. 259-271). Springer, Singapore.

Ren, J.H., Li, H., Wang, Y.F., Ye, J.R., Yan, A.Q., Wu, X.Q. (2013) Biocontrol potential of an endophytic *Bacillus pumilus* JK-SX001 against poplar canker. Biol Control 67:421–430.

Rice, W., Olsen, P., Leggett, M. (1995) Co-culture of *Rhizobium meliloti* and a phosphorus solubilizing fungus (*Penicillium bilaii*) in sterile peat. Soil Biol Biochem 27: 703–705.

Roberts, M.J. (2006) The value of plant disease early-warning systems: a case study of USDA's soybean rust coordinated framework. United States Department of Agriculture, Economic Research Service No. 18.

Ryu, C.M., Farag, M.A., Hu, C.H., Reddy, M.S., Kloepper, J.W., Pare, P.W. (2004) Bacterial volatiles induce systemic resistance in Arabidopsis. Plant Physiol 134: 1017-1026.

Santos, A., Flores, M. (1995) Effects of glyphosate on nitrogen fixation of free-living heterotrophic bacteria. Lett Appl Microbiol 20: 349–352.

Santoyo, G., Orozco-Mosqueda, M.D., Govindappa, M. (2012) Mechanisms of biocontrol and plant growth-promoting activity in soil bacterial species of Bacillus and Pseudomonas: a review. Biocontrol Sci Tech 22: 855–872.

Savary, S., Ficke, A., Aubertot, J.-N., Hollier, C. (2012) Crop losses due to diseases and their implications for global food production losses and food security, Springer.

Schafer, K.S., Kegley S. (2002) Persistent toxic chemicals in the US food supply. J Epidemiol Community Health 56: 813–817.

Scheepmaker, J.W.A., van de Kassteele, J. (2011) Effects of chemical control agents and microbial biocontrol agents on numbers of non-target microbial soil organisms: a meta-analysis. Biocontrol Sci Technol 21: 1225–1242.

Scherm, H., Ngugi, H., Savelle, A., Edwards, J. (2004) Biological control of infection of blueberry flowers caused by *Monilinia vaccinii-corymbosi*. Biol Control 29: 199–206.

Schouten, A., Maksimova, O., Cuesta-Arenas, Y., Van Den Berg, G., and Raaijmakers, J. M. (2008) Involvement of the ABC transporter BcAtrB and the laccase BcLCC2 in defence of *Botrytis cinerea* against the broad-spectrum antibiotic 2,4-diace-tylphloroglucinol. Environ Microbiol 10: 1145–1157.

Shahraki, M., Heydari, A., Hassanzadeh, N. (2009) Investigation of antibiotic, siderophore and volatile metabolites production by Bacillus and Pseudomonas bacteria. Iran J Biol 22: 71-84.

Shanmugam, V., Sriram, S., Babu, S., Nandakumar, R., Raguchander, T., Balasubrama-nian, P., Samiyappan, R. (2001) Purification and characterization of an extracellular alpha-glucosidase protein from *Trichoderma viride* which degrades a phytotoxin as-sociated with sheath blight disease in rice. J Appl Microbiol 90: 320–329

Shoresh, M., Harman, G.E., Mastouri, F. (2010) Induced systemic resistance and plant responses to fungal biocontrol agents. Ann Rev Phytopathol 48: 21–43.

Silva, H.S.A., de Silva Romeiro, R, Macagnan, D., de Almeida Halfeld-Vieira, B., Pereira, M.C.B., Mounteer A. (2004) Rhizobacterial induction of systemic resistance in tomato plants: Non-specific protection and increase in enzyme activities. Biol Control 29: 288-295.

Sneh, B. (1998) Use of non-pathogenic or hypovirulent fungal strains to protect plants against closely related fungal pathogens. Biotechnol Adv 16: 1–32.

Spadaro, D., and Droby, S. (2016) Development of biocontrol products for postharvest diseases of fruit: the importance of elucidating the mechanisms of action of yeast an-tagonists. Trends Food Sci Technol 47: 39–49.

Stockwell, V.O., Johnson, K.B., Sugar, D., Loper, J.E. (2011) Mechanistically compatible mixtures of bacterial antagonists improve biological control of fire blight of pear. Phy-topathology 101: 113–123.

Strange, R.N. (2007) Phytotoxins produced by microbial plant pathogens. Nat Prod Rep 24: 127–144.

Syed Ab Rahman, S.F., Singh, E., Pieterse, C.M.J., Schenk, P.M. (2018) Emerging micro-bial biocontrol strategies for plant pathogens. Plant Sci 267: 102–111.

Taghavi, S., Garafola, C., Monchy, S., Newman, L., Hoffman, A., Weyens, N., Barac, T., Vangronsveld, J, van der Lelie, D. (2009) Genome survey and characterization of en-dophytic bacteria exhibiting a beneficial effect on growth and development of poplar. Appl Environ Microbiol 75: 748–757.

Trivedi, P., Schenk, P.M., Wallenstein, M.D., Singh, B.K. (2017) Tiny microbes, big yields: enhancing food crop production with biological solutions. Microbiol Biotech 10: 999–1003.

Vallad, G.E and Robert M. Goodman, R.E. (2004) Review & Interpretation Systemic Acquired Resistance and Induced Systemic Resistance in Conventional Agriculture. Crop Sci. 44:1920–1934.

Van Loon, L.C., Bakker, P.A.H.M., Pieterse, C.M.J. (1998) Systemic resistance induced by rhizosphere bacteria. Annu Rev Phytopathol 36: 453-483.

Van Peer, R. and Schippers, B. (1992) Lipopolysaccharides of plant growth-promoting *Pseudomonas* sp. strain WCS417r induce resistance in carnation to fusarium wilt. Neth J Plant Pathol 98:129-39.

Van Wees, S.C., Pieterse, C.M., Trijssenaar, A., van't Westende, Y.A., Hartog, F., van Loon, L.C. (1997) Differential induction of systemic resistance in Arabidopsis by biocontrol bacteria. Mol Plant-Microbe Interact 10: 716-724.

Vorholt, J.A. (2012) Microbial life in the phyllosphere, Nature Rev Microbiol 10: 828–840.

Wang, M.C., Tachibana, S., Murai, Y., Li, L., Lau, S.Y.L., Cao, M.C., Zhu, G.N., Hashimoto, M., Hashidoko, Y. (2016) Indole-3-acetic acid produced by *Burkholderia heleia* acts as a phenylacetic acid antagonist to disrupt tropolone biosynthesis in *Burkholderia plantarii*. Sci Rep 6: 22596–22596.

Weller, D.M. (1988) Biological control of soilborne plant pathogens in the rhizosphere with bacteria. Annu Rev Phytopathol 26: 379–407.

Wilhite, S.E., Lumsden, R.D., Strancy, D.C. (2001) Peptide synthetase gene in *Trichoderma virens*. Appl Environ Microbiol 67: 5055-5062.

Wintermans, P.C., Bakker, P.A., Pieterse, C.M. (2016) Natural genetic variation in *Arabidopsis* for responsiveness to plant growth-promoting rhizobacteria. Plant Mol Biol 90: 623–634.

Wolfe, M. (1985) The current status and prospects of multiline cultivars and variety mixtures for disease resistance. Annu Rev Phytopathol 23: 251–273.

Wu, L.M., Wu, H.J., Qiao, J.Q., Gao, X.W., Borriss, R. (2015) Novel routes for improving biocontrol activity of Bacillus based bioinoculants. FrontMicrobiol 6: 01395.

Yanez-Mendizabal, V., Zeriouh, H., Vinas, I., Torres, R., Usall, J., de Vicente, A., Perez-Garcia, A., Teixido, N. (2012) Biological control of peach brown rot (*Monilinia* spp.) by *Bacillus subtilis* CPA-8 is based on production of fengycin-like lipopeptides. Eur J Plant Pathol 132: 609–619.

Zeriouh, H., Romero, D., Garcia-Gutierrez, L., Cazorla, F.M., de Vicente, A, Perez-Garcia A. (2011) The iturin-like lipopeptides are essential components in the biological control arsenal of *Bacillus subtilis* against bacterial diseases of cucurbits. Mol Plant-Microbe Interact 24: 1540–1552.

Zhou, H.Y., Wei, H.L., Liu, X.L., Wang, Y., Zhang, L.Q., Tang, W.H. (2005) Improving biocontrol activity of *Pseudomonas fluorescens* through chromosomal integration of 2,4-diacetylphloroglucinol biosynthesis genes. Chin Sci Bull 50: 775–781.

CHAPTER 8

The Use of Botanicals in the Management of Human Diseases

by
Festus Abiose Olajubu
Department of Microbiology
Adekunle Ajasin University, Akungba-Akoko, Nigeria
Email*: festus.olajubu@aaua.edu.ng*

Abstract

The use of botanicals for therapeutic purposes was dated back to human origin on earth. Due to lack of scientific backing, most of the then useful plants were discovered through trials and errors, and close observation of animals that equally feed on them. Research into phytomedicine became popular especially in the areas of infectious diseases due to treatment failures often recorded with the use of existing synthetic antimicrobials. The availability of these botanicals usually at little or no cost encouraged their usage among Sub-Saharan Africans. With research, virtually all communicable or non-communicable disease conditions can be treated with plant parts. In modern medicine, the uses of botanicals are regulated in many countries of the world by government agencies, thereby encouraging reliance on their acclaimed efficacies. The use of botanicals in combination, evaluation of their toxicity or otherwise, dosage and preservation have been extensively researched. Many more uses of botanicals may still be discovered, as more intensive researches are carried out on these nature's provisions for solving mankind's health issues.

Keywords: Phytomedicine, infectious diseases, antimicrobials, herbal combinations, toxicity

Introduction

Human diseases can be broadly classified into communicable and non-communicable (or infectious and non-infectious). Other classifications may include deficiency diseases, physiological diseases and hereditary diseases. In all these cases, the use of botanical interventions has been documented.

Plant and plant parts (leaves, barks, roots, flowers, seeds and fruits) had long been used for therapeutic purposes, and this use of herbs for their therapeutic values is called Herbal or Botanical medicine. It is widely known and accepted that herbal medicine is the oldest form of healthcare known to mankind. Herbs had been used by all cultures throughout history. The World Health Organization estimated that about four billion people, 80% of

165

the world population, use herbal medicine for some aspect of primary healthcare. Of the estimated 250,000 to 500,000 species of plants on earth, less than 10% of these or their parts are used for food or medicinal purposes (Sofowora, 1993; WHO, 1996).

Primitive man observed and appreciated the great diversity of plants available to him. He puts the plants into various uses such as the provision of shelter, food, drink and clothing, with little knowledge of their medicinal values which are often discovered through trial and error, and of observation of the instinctive discrimination of plants by animals. Despite this low knowledge of the medicinal value of the plants, much of the pharmacopoeia of scientific medicine was derived from the herbal knowledge of native peoples. Herbal medicine is a major component in all indigenous people's medicine, and it enjoys a wider acceptability among the people of developing countries than does orthodox medicine (Gbile, 2002; Sofowora, 1993).

According to reports, since the advent of antibiotics in the 1950s from fungal and bacterial sources, the use of plant derivatives as antimicrobials has been low, especially in the United States. The few ones being used are without documented laboratory or clinical assessment (Cowan, 1999). Plants used as herbs can be classified into wild-grown and farm-grown herbal plants. A wild–grown herb is the one that grows naturally without human intervention, they are often liable to contamination from exhaust fumes, especially if found by roadways, chemicals and pesticides. The farm-grown herbs are the ones planted by native healers or herb sellers for easy access, and usually not too far from their habitats. The home or farm-grown herbs are often the most reliable sources during emergencies (Sofowora, 1993; Wild, 1994).

Today, through science, the medicinal properties of a large number of plants are now known and their healing components are being extracted and analyzed. It is known that about 25% of the prescription drugs dispensed in the United States contain at least one active ingredient derived from plant material. While some of these drugs are made of plant extracts, others are synthesized to mimic a natural plant compound. Many plant components are now synthesized in large laboratories for use in pharmaceutical preparations. Examples of these are vincristine (an antitumor drug), digitalis (a heart regulator) and ephedrine (a bronchodilator) which were originally discovered through research on plants. The synthesis of active ingredients from plants can create problems of being toxic at relatively low concentrations. This is because other compounds present in plants like the minerals, vitamins, alkaloids and glycosides play supportive role in the herb's medicinal property (Wild, 1994).

Herbal preparation can either be taken as an individual herb or as complex herbal formulations. They could be taken raw or processed. The use of alcohol to extract the active properties of the herb leads to the production of tinctures. The essences of the herb can be leached out with water producing aqueous extracts. Herbs can be formulated into capsules or tablets containing powdered form of the raw herb. Teas and Lozenges could be produced from raw herbs. It has also been reported that herbs can be formulated into ointments, salves and rubs which are applied topically. Many of these formulations are used to repair broken skin and heal wounds. They could also be used to fight infections, heal burns and to relax muscle aches and strains. In most countries of the world, pharmaceutical drugs are seen to be increasingly overprescribed, expensive, and even dangerous, while herbal remedies are seen as less expensive and less toxic (Odugbemi, 2006; Sofowora, 1993).

History of Herbal Medicine

Early humans recognized their dependence on nature in both health and illness. Led by instinct, taste and experience, primitive men and women treated illnesses by using plants, animal parts and minerals that were not part of their usual diet. The use of higher plants to treat infections is an age-old practice in a large part of the world population. Hippocrates, who lived in the late fifth century B.C., documented 300 to 400 medicinal plants then in common use. In the first century A.D., Discorides wrote 'De Materia Medica' a medicinal plant catalog. This becomes the prototype for modern pharmcopoeia. The records of king Hammurabi of Babylon (1800 B.C) include instructions for using medicinal plants. He described the use of mint for digestive disorders. Research has confirmed that peppermint (by mildly anesthetizing the lining of the stomach) does indeed relieve nausea and vomiting which could be an indication of digestive disorder (Lewis and Elvin-Lewis, 1995; Leen *et al.*, 2018).

To regulate the practice of herbal medicine, the National Association of Medical Herbalist (NAMH) which later transformed into National Institute of Medical Herbalist was founded in 1800s A.D. The use of herbal medicine came to limelight again during the World War 1 when drugs were in short supply to the wounded soldiers. In 1900 A.D., the British Herbal Pharmacopoeia was produced by the British Herbal Medical Association, and in 2000 A.D., the European Union Legislation advocated that all herbal medicine should be subjected to compulsory clinical testings, comparable to that undertaken for conventional drugs and thus be licensed (Balkam, 2007).

Studies have confirmed the use of plants in treating burns, dermatophytes, cough, infectious diseases and as anti-inflammatory in many parts of the world. Prior to the discovery and subsequent synthesis of antibiotics, the herbs Echinacea (purple coneflower) was widely prescribed in United States to fight infections. Herbal medicine is intertwined with modern or orthodox medicine. This is because, many drugs listed as conventional medications were originally derived from plants. Salicylic acid, a precursor of aspirin was originally derived from white willow and meadow sweet barks, just as Cinchona bark is the source of malaria drug called quinine (Lewis and Elvin-Lewis, 1995).

Collection and Preservation of Medicinal Plants

Collection of plants and plant parts like leaves, roots, barks, fruits and flowers can be done around the house, along the roadsides, on the farm, thick forest, savanna, water surface and in some cases at herbal gardens, specifically made for those plants that are scarce or threatened by extinction. Seasonal plants are collected in large quantities during their season and stored for subsequent use (Odugbemi,2006).

Preservation of the plant parts include cutting into smaller pieces, and either sun- dried or dried under controlled temperature, so as to protect the chemical components and then ground into powder. Others are hung in the kitchen or where they can be exposed to constant heat. Traditionally, herbal recipe or preparations are stored in clay ports, horns of animals, shells of some animals, bamboo stem and various bottle gourds of different sizes (Sofowora, 1993).

Herbal Combinations.

Believe that there are many causes of a disease and an attempt to treat them all has given rise to combination of herbs. It is believed that each active component of a plant will be strengthened by the presence of another plant that has such active ingredient (synergism) or can aid its effectiveness in the body (Olajubu, 2014). Some plants serve as preservatives to others when mixed together. Quite a number of herbs are believed to help the active ingredient in a recipe to get to the desired target i. e. aiding penetration, examples are *Zanthoxylum rubenscens* root, *Aframomum melegueta* and *Tetrapluera tetraptera*. Some common ailments treatable with herbal preparations range from communicable bacterial infections (Diarrhea, Typhoid fever, Urinary Tract Infections, Pulmonary tuberculosis, gonorrhea, meningitis, dysentery, bacteremia, wounds, boils and skin infection) viral infections (Hepatitis, Yellow fever, HIV, Lassa, chicken pox, small pox) fungal infections (Candidiasis, Dermatitis) parasitic infections (Malaria, Schistosomiasis, Tape/Hook worm infestation, Lice, Leshimaniasis, Trichomoniasis and Amoebic dysentery) to non-communicable diseases (hypertension, diabetics, insomnia, kidney stone, fibroid, haemorrhoid, cancer, acute/chronic liver disease, migraine, low sperm count, erectile dysfunction, coated tongues, pile, infertility, menstrual disorders, rheumatism, fevers of all types, convulsion, eye problems, burns, and ulcers). Herbal treatment has also been employed in cases of snake bites, arrest of bleeding, depression, child delivery/labour, free bowel movement, mental alertness and as insect repellant (Osuntokun and Olajubu, 2015; Chrysant and Chrysant, 2017; Mollik *et al.*, 2019).

Uses of Medicinal Herbs

Chewing sticks

Medicinal herbs are used in many African homes as chewing sticks for clearing of teeth. The root or slim stem is chewed to produce a brush-like end, which is used to brush teeth thoroughly. Though, this practice is common in the early hours of the day, in some part of West Africa, e.g. Senegal, chewing sticks are used during the day. The use of herbs as chewing stick is not peculiar to Africans alone, as many other cultures around the globe do practice this act. These chewing sticks impart varying taste sensations; a tingling, peppery taste and numbness is produced by *Fagara zanthoxyloides* Lam root, *Masularia acuminata* gives a strong bitter taste and frothing the same way *Vernonia amygdalina* does. The root of *Terminalia glaucescens* Planch produces discoloration of the mouth. Freshly cut specimens are always desirable because they are more easily chewed into brush. The commonly used herbs are the ones that have good flavour and texture, and a recognized effect on the teeth and supporting tissues. The chewing sticks remove food particles from between the teeth and the crevices, as does the tooth paste and brush method. In addition, the chewing sticks stimulates the gum and destroy microbes present in the mouth, a feature that is absent in the common toothpastes in the market. This explains why many Africans have stronger teeth than the Caucasians, although Africans eat less sweet than the Caucasians, a condition that encourages dental carries (Oloke *et al.*, 2017; El-Said *et al.*, 1971).

This antimicrobial activity of some herbs commonly used as chewing sticks has been demonstrated. They all showed antimicrobial activity when tested by the agar diffusion method, though to varying degrees. The antimicrobial activity of *Fagara xanthoxyloides* Lamb has been shown to be due to benzoic acid derivatives which are active at pH of about 5 and alkaloids which are active at a pH of 7.5. Floride ions, silicon, tannic acid and sodium bicarbonate are substances that have been found in some common chewing sticks, which might be responsible for reduction of bacterial colonization and plague formation (Akpata and Akinrimisi, 1977).

Diarrhea

Herbs are equally used in the treatment of diarrhea. The volatile oil from *Ocimum gratissimum* leaf which contains up to 75 percent of thymol, might be responsible for its antimicrobial activity. In preparations where *O. gratissimum* is used as a cold infusion, the antimicrobial effect of the extracted thymol is probably sufficient explanation for the anti-diarrhoeal effect of the plant. However, in certain other preparations, *O. gratissimum* is boiled with water to form a decoction which contains little of the steam-volatile thymol. Although, such aqueous decoctions were shown to be devoid of antimicrobial activity, they do relax the guinea pig ileum and rat jejunum *in vitro*. Such decoctions could therefore still be effective in calming an over-active gut, thus curing diarrhea. *Ocimum gratissimum* and *Cymbopogon citratus* are cultivated near houses to repel insects. The essential oils of both plants have been shown in an empirical manner to possess insect–repellant properties. Sugar soaked in either of the two oils was not eaten by ants, whereas in a control experiment, sugar which had not been treated with the oils was readily consumed by red ants (Deji-Agboola and Olajubu, 2006; Odetola *et al.*, 2006; Olajubu *et al.*, 2012b).

Skin diseases

Skin diseases of bacteria and fungi origins are commonly treated with herbal remedies with huge success. Herbal recipes for treatment of skin infections are often made up of many components. Most of the herbal remedies are used together with traditional black soap, which on its own possess antimicrobial property. Others are prepared in powdered form to be dusted onto the rashes after bathing with a decoction and the traditional black soap. Yet, others are made into cream to be applied topically. Researchers examined three remedies commonly used for treatment of skin disease. Though, none of the remedies has less than nine components, only three plants were shown to contain potent antimicrobial agents. These plants are *Combretum micranthum, Dracaena manii* and *Terminalia avicenniodes*. One wonders whether the inactive components of these remedies were added to stabilize the decoctions or merely to impress the patient, since they had no effect on overall activity of the remedies. The antimicrobial activity of the various components appeared to be additive, as no synergism could be demonstrated. The practice of adding 'inactive' components to preparations either as flavours, diluents or preservatives; or the art of using mixtures of active components to preparations which need not be synergistic is not uncommon even in modern pharmaceutical practice. Mixtures of compounds are employed, especially in cheap antiseptic preparation to increase the spectrum of activity of the preparation. Some plants or plant parts are made into teas for easy use and their efficacy have been scientifically proven (Olajubu, 2017; Sofowora,1993).

Another popular remedy for treating skin disease in general in Nigeria is the leaf of *Acalypha wilkesiana* for which scientific evidence in support of the traditional use has been documented. The roots and barks of *Zanha africana* and *Polygala nyikensis* are used in many African countries for the treatment of various skin diseases of fungal origin. This is probably due to the presence of xanthones which is a class of natural products with broad spectrum of biological activities including antifungal, antibacterial and anti-inflammatory properties. Antifungal activities of these plants have been demonstrated against *Trichophyton rubrum, T. mentagrophytes, Aspergillus niger, A. flavus, Candida albicans, Histoplasma capsulatum* and *Coccidoides imminites* (Alade and Irobi, 1993; Olajubu *et al.*, 2012a).

Wounds

Scientific report has it that *Staphylococcus aureus, Enterococcus faecalis, Pseudomonas aeruginosa, Proteus mirabilis, Klebsiella pneumonia* and some *Escherichia coli* that were resistant to more than four antibiotics including Pefloxacin, a fourth generation quinolone, were well inhibited by the ethanol extracts of the stem bark of *Lannea welwitschii,* which is being used for the treatment of wounds (Tajbakhsh *et al.*, 2018; Deji-Agboola and Olajubu, 2006).

Malaria fever

Many herbs or herbal combinations are variedly used in different parts of the world for the treatment of Malaria fever which is endemic in Africa. These plants include *Cymbopogon citratus, Dioscorea dumetorum, Enantia chlorantha, Carica papaya* and *Azardirachta indica.* A decoction of the leaves or combination of leaves and stem bark of *A. indica* is drunk while in some cases, the stem bark is also used to treat fevers by inhalation or hydrotherapy. This plant has been shown to possess a steam-volatile, oily constituent in trace amounts, which showed only one component common to the leaf, stem and root bark when examined chromatographically. Fall in parasitic count has been reported with the use of a decoction of the leaf of *A. indica* on *Plasmodium berghei* infected mice. Inhibitory effect of the extract of *A. indica* on *Plasmodium falciparum* culture and the antimalarial effect of the boiled leaf were found to be approximately half of the therapeutic dose of chloroquine sulphate on dry weight basis (Gbile, 2002).

Mentally disturbed patients

Rauwolfa vomitoria is a component of herbal preparations used throughout Africa to treat mentally disturbed patients. The root is given in the form of a decoction or soaked in locally distilled gin. It is also prescribed in the form of a dried powder mixed with maize pap. The use of this plant has to do with its ability to sedate and calm the patient, thereby making him more amenable to other forms of treatment. The root of *R. vomitoria* is known to contain the alkaloid reserpine which possesses hypotensive and sedative properties. It is also known that at least, seventy two alkaloids classified into nineteen types occur in this plant (Wild, 1994).

Antihelmintics

Hagenia abyssinica, Glinus lotoides, Combretum mucronatum and *Mitragyna stipulosa* are common herbs used as antihelmintics in many parts of the world especially in Africa. While *H. abyssinica* is often used for the treatment of Taenia infestation, *G. lotoides* is used for the treatment of hookworms and whipworms infestation.

Diabetics

The secrecy attached to the release of information on herbs has adversely affected investigations into the claimed efficacy of many of such plants. Many herbs or herbal recipes are used for the treatment of diabetics, but very little information about these herbs is available. *Bridelia ferruginea* is one of such plants originally known as a mouthwash, but later as a potent antidiabetic agent. Report has confirmed the clinical efficacy of *Bridelia ferruginea* therapy on diabetics by observing human volunteers been treated in a herbal home. The subjects were earlier diagnosed as diabetics of about 5 years. The blood sugar level of eight out of the ten subjects treated with *B. ferruginea* came down to less than 120mg/100ml from over 250mg/100ml and remained so after eight weeks of daily treatment. Methanolic and aqueous extracts of *B. ferruginea* leaves significantly lowered the fasting blood sugars of rats, but failed to protect the animals against alloxan induced diabetes. *Curcuma longa* has demonstrated similar efficacy on diabetics (Leen *et al.*, 2018).

HIV/AIDS

Though, there are not enough scientific evidence of herbs been used for the treatment of HIV/AIDS, however, a brief course of *Milk thistle* is often used to improve liver functions without a significant interaction with antiretroviral drugs. Loss of appetite, reaction to antiretroviral drugs in form of stomach upset, nausea, pain has been effectively controlled with the use of medical Marijuana which is only legally allowed in few nations of the world. It is advocated that HIV/AIDS patients should always discuss the use of herbs with their healthcare providers before embarking on such course of treatment (Passmore and Jaspan, 2018; Daniel, 2018; Orisatoki and Oguntibeju, 2010).

Sexually Transmitted Infections (STIs)

About twenty-one (21) different plants or in combination have been documented for the treatment of syphilis and gonorrhea, the most common STIs with proven efficacies. Examples of these are *Amaranthus spinosus, Piper betle, Gucuma longa, Mucuna pruriens, Gloriosa superba* and *Aloe vera* (Mohammadreza *et al.*, 2019).

Some common botanicals that have been researched for common ailments

Plate 1: Picture of *Euphorbia kamerunica*
Source: http://home-and-garden.webshots.com

Plate 2: Picture of *Dialium guineense* Source: Agro Forestry Tree Database

Plate 3: Picture of *Kigelia Africana*
Source: www.tradewindsfruit.com

Major Groups of Bioactive Compounds from Plants

Plants have the ability to synthesize aromatic substances, most of which are phenols or phenol oxygen-substituted derivatives. These are secondary metabolites that often serve as plant defense mechanism against predation by microorganisms, insects and herbivores. Some of the compounds are responsible for the plants odour (e.g. terpenoid), pigment (e.g. quinone and tannin) and flavour. The commonly analyzed groups are (1) Phenolics and polyphenols (2) Quinones (3) Flavoniods (4) Tannins (5) Alkaloids (6) Coumarins (7) Terpenoids (8) Essential oils (9) Lectins (10) and polypeptides (Osuntokun and Olajubu, 2015; Cowan, 1999).

Phenolics and Polyphenols
Caffeic and Cinnamic acids are common of this group. The common herbs terrapin and thyme both contain caffeic acid. This acid is effective against viruses, bacteria and fungi. The mechanism thought to be responsible for phenolic toxicity to microorganisms includes enzyme inhibition by oxidized compounds.

Quinones
These are aromatic rings with two ketone substitutions. They are responsible for the browning reaction in cut or injured fruit and vegetables. Researchers discovered an anthraquinone from *Cassia italica* which was bacteriostatic for *Bacillus anthracis, Corynebacteruim pseudodiphthericum* and *Pseudomonas aeruginosa*, and bacterial for *Pseudomonas pseudomalline*. This compound attacks surface-exposed adhesions, cell wall polypeptides and membrane-bound enzyme. Quinines may also render substrates unavailable to the microorganism.

Flavonoids

Cetechin, a prominent member of this group has been found in teas where it demonstrated antimicrobial activity. This compound also inhibited *Vibro cholerae, Streptococcus mutans, Shigella* sp. and other bacterial and microorganisms. Flavonoid compounds exhibit inhibitory effects against viruses. The effect ranges from reducing intracellular replication to inhibiting infectivity. In bacteria, the effects of flavonoids are targeted at the cell membrane.

Tannins

This is a general name given to a group of polymeric substances capable of tanning leather or precipitating gelatin from solution. They are found in almost every part of a plant, the bark, wood, leaves fruits and root. The mode of antimicrobial action is as described for quinines. A review of the antimicrobial properties of tannins showed that tannins are toxic to filamentous fungi, yeast and bacteria. Tannins were considered responsible for the antibiotic activity of methanolic extracts of the stem bark of *Terminalia alata*. Many other studies have shown tannins to be inhibitory to viral reverse transcriptase.

Alkaloids

The first medically useful example of an alkaloid was morphine isolated in 1805 from *Papaver somniferum*. Diterponoid alkaloids are found to have antimicrobial properties. Berberine, another member to the group is potentially effective against trypanosomes and plasmodia. The antibacterial effect is directed at inhibiting DNA replication.

Coumarins

These are phenolic substance responsible for the characteristic odour of hay. As of 1996, at least 1,300 coumarin types had been identified. They are known for their antithrombotic, anti-inflammatory and vasodilatory activities. Though, data about specific antibiotic properties of coumarins are scarce, general antibiotic activity has been documented. Coumarins have been found to stimulate macrophages, which could have an indirect negative effect on infection. It was found inhibitory to Gram–positive bacteria (Akinpelu and Onakoya, 2006). However, some other compounds in varying concentrations have been discovered, such as quercetin, isoquercitrin, Dammarane-type saponin, terpenoids, steroids, astragalosides, flavonoids and polysaccharides, α-pinene, β-pinene, α-pinene, quercetin, myricetin and luteolin flavonoids, β-pinene, 1,3,8-p-menthatriene, ledene, m-menthane, linalyl acetate and 3-carene β-sitosterol, lupeol, lupeol, sitosterol, spathulenol, β-sitostenone. More are likely to be discovered as researches on phytomedicine advances. With more intensive research into phytomedicine, use of natural endowments such as plants, might be all that is needed to fight various diseases threatening human existence on earth.

Terpenoids

These are large and diverse classes of naturally occurring organic chemicals. They mediate in ecological interactions. Terpenoids are very useful in preventing plants from diseases and herbivores, but attract pollinators. Several classes of terpenoids are biologically active and are exploited in the fight against malaria, infectious diseases, inflammations and cancer.

Essential Oils

Essential oils are volatile chemical compounds naturally occurring in plants that give the plants unique scents. They are often extracted and concentrated for the treatment of acne and wrinkles. They are widely used in aromatherapy.

Lectins

These are groups of proteins commonly found in Legumes and grains. Consumption of lectin in high doses can result in chronic inflammation, auto immune diseases and gout, but these conditions rarely occur because lectins are easily broken down when processed or cooked. Lectins do act as antioxidants and thus protect cells damage caused by free radicals. They also slow down digestion and absorption of carbohydrates which may prevent sharp rises in blood sugar.

Polypeptides

These are chains of amino acids linked covalently by peptide bonds and naturally occurring in plants. They aid collagen production for a healthy and fresh looking skins. They are often included in anti-aging products.

Conclusion

Nature is endowed with enough resources to handle any human health challenges that may come our way. The age long use and wide acceptance of botanicals because of it's relative closeness to nature are clear evidence of this, and gives further assurance of more application of botanicals to enhance human's wellbeing. With technological advancement, scientific innovations, emergence and re-emergence of infectious diseases and more interests from researchers into the use of botanicals, there is a great hope and conviction that most, if not all the ailments will find their solutions in the use of botanicals.

Recommendation

The views presented here is by no means exhaustive, more interest should be shown by researchers in the use of botanicals. It will be a way of demonstrating interest in the existence and wellbeing of their citizens if governments and their agencies, pharmaceutical companies and individual philanthropists can sponsor and finance researches in the use of botanicals, so that human beings can adequately make use of what nature has provided.

Acknowledgement

I am indeed grateful to Dr. O.T. Osuntokun, Dr. Mrs. Deji-Agboola, Mrs. Mabel Oseni and my mentor, Prof. T.O. Adejumo for their contributions and encouragement during the course of preparing this piece.

References

Akinpelu, D.A. and Onakoya, T.M. (2006) Antimicrobial activities of medicinal plants used in folklore remedies in south-western Nigeria. *African Journal of Biotechnology*. **5**(11): 1078-1081

Akpata, E. S. and Akinrimisi, F (1977) Antibacteria activity of extracts from some African chewing sticks. Oral Surg. *Oral Med. Oral Pathol*. 44: 717-722.

Alade, P.I. and Irobi, O.N. (1993) Antimicrobial activity of crude extracts of *Acalypha wilkesinna. Journal of Ethnopharmacology*. 39(3): 171-174.

Balkam, J. (2007) Herbal concepts: History of Herbal Medicine. Bookmart Ltd. 3rd Edition pp 215-237.

Chrysant, S.G. and Chrysant, G. S. (2017) Herbs used for the treatment of hypertension and their mechanism of action. *Curr. Hypertens Rep*. 19(9): 77-81.

Cowan, M.A. (1999) Plant products as antimicrobial agents. *Clinical Microbiology Reviews* 12(14): 563-582.

Daniel, M. (2018) Alternative treatment for HIV and AIDS. *Healthline* March, 29.

Deji-Agboola, A.M and Olajubu, F.A (2006). Antimicrobial activity of *Lannea welwitschii* against wound pathogens. *African Journal of Medicine and Pharmaceutical Science*. **10**: 45-52

El-Said, F., Fadulu, S.O., Kuye, J.O. and Sofowora, E.A. (1971) Native cures in Nigeria; Part II: The antimicrobial properties of the buffer extracts of chewing sticks. *Lloydia*. 34(1): 172

Gbile, Z.O. (2002) Ethnobotany of Nigerian plants In: Book of Abstracts. Plenary Lecture. Workshop on medicinal plants, (Ife, January 1-10, 1986).

Leen, J., Jung, W.,Kim, Y., Kim, B., Kim, K. (2018) Exploring the combination and modular characteristics of herbs for alopecia treatment in traditional Chinese medicine: an association rout mining and network analysis study. *BMC Complement Alternate Medicine* 18(1): 204 – 209.

Lewis, W,H, and Elvin-Lewis, M.P.(1995) Medicinal plants as sources of new therapeutics. *Annals of Modern Botanical Garden* 82: 16-24.

Mohammadreza, N., Saber, A., Mohammad, D., Abdolreza, K., Somayeh, S. and Mona, M. (2019) The most important herbs used in the treatment of Sexually Transmitted Infections in Traditional Medicine. *Sudan Journal of Medical Sciences* 14(2): 41-64.

Mollik, A.H., Islam, T., Khatun, A. and Mohammed, R. (2019) Medicinal plants used against Syphilis and Gonorrhoea by traditional medicinal practitioners of Bangladesh. *Planta Medica* 75(09): PH 40.

Odetola, A,A,, Oluwole, F.S., Adeniyi, B., A,, Olatiregun, A.M., Ikupolowo, O, R., Labode,O., Busari, K.O. and Shorinola, J.A. (2006) Antimicrobial and gastrointestinal protective properties of Parquetina nigerescens. *Journal of Biological Science* 6(4): 701-705.

Odugbemi, T. (2006) Outlines and pictures of medinal plants from Nigeria. 1st Edition. University of Lagos Press. P 283.

Olajubu, F. A. Akpan, I. Ojo, D. A. and Oluwalana, S. A. (2012a). Acute and sub-acute toxicity study of *Euphorbia kamerunica* (Pax) in mice. *Journal of Research in Medical Education and Ethics.* 2(1): 56-62.

Olajubu, F. A. Akpan, I. Ojo, D. A. and Oluwalana, S. A. (2012b). Antimicrobial potential of *Dialium guineense* (Wild) stem bark on some clinical isolates in Nigeria. *International Journal of Applied and Basic Medical Research.* 2: 58-62.

Olajubu, F. A. (2014) Enhancement of therapeutic success through herbal-herbal combination. *Int. J. Current Mic. & Appl. Sci.* 3(8): 418-425.

Olajubu, F.A. (2017) Antimicrobial and antihaemolytic activities of crude extracts of some commonly used Tea and Coffee in Nigeria. *Journal of Tea Science Research.* 7(6): 39-45.

Oloke, J., Odelade, K. and Oladeji, O. (2017) Characterization and antimicrobial analysis of flavonoids in *Vernonia amygdalina*: a common chewing stick in South-western Nigeria. *Bulletin of Pharmaceutical Research* 7(3): 149.

Orisatoki, R,O, and Oguntibeju, O.O. (2010) The role of herbal medicine use in HIV/AIDS treatment. *Archieves of Clinical Microbiology* 1(3): 3-7.

Osuntokun, O.T. and Olajubu, F. A. (2015) Antimicrobial and phytochemical properties of some Nigerian Medicinal plants on *Salmonella typhi* and *Salmonella paratyphi* isolated from human stool in Owo Local Government, Ondo State, Nigeria. *Journal of Scientific Research & Reports.* 4(5):441-449.

Passmore, J.A.S. and Jaspan, H.B. (2018) Vagina microbes, inflammations and HIV risks in African women. *The Lancet Infectious Diseases* 18(5): 474-483.

Sofowora, A. (1993). Medicinal plant and traditional Medicine in Africa, Spectrum Books Ltd., Ibadan, Nigeria. 2 nd Edition. Pp 8-11.

Tajbakhsh, M., Karimi, A., Tohidpour, A. (2018) The antimicrobial potential of a new derivative of cathelicidin from *Bungarus fasciatus* against methicillin-resistant *Staphylococcus aureus*. *Journal of Microbiology* 56(2): 128-137.

Wild, R. (1994) The complete book of Natural and Medicinal curve, Rodales Press Inc, 2nd Edition, pp121-173.

World Health Organization (1996) Traditional Medicine, Fact Sheet. No B4: 25-30.

CHAPTER 9

Efficacy, Properties and Therapeutic Uses of Some Major Medicinal Plants for Human Health

by
Oludare Temitope Osuntokun
Department of Microbiology,
Adekunle Ajasin University, Akungba-Akoko,
Ondo State, Nigeria.
Email: *oludare.osuntokun@aaua.edu.ng*

Abstract

Nigeria is well known for its rich ethno-botanical wealth, particularly regarding medicinal plants which are traditionally used in the treatment of ailments, and could be a good source for discovery of new, safe and biodegradable drugs. As widespread as infectious diseases are in Nigeria, the number of medicinal plant species prescribed traditionally against infectious diseases runs into hundreds. Many of these plants have a prolonged and uneventful use that may serve as an indirect testimony to their efficacy. This write up describes the efficacy, properties and medicinal uses of some noted medicinal plants in Nigeria, which are used by the herb medical practitioners for the cure of different infectious diseases in our daily existence. Some of the medicinal plants that will be discussed during this write up are *Spondias mombin. linn, Nicotiana tabacum, Morinda lucida, Carica papaya Linn, Piper guineense, Cymbopogon citratus, Alchornea laxiflora/cordifolia* and *Anisopus manni*. Each medicinal plant has various degrees of efficacy, therapeutic properties and uses.

Keywords : Efficacy, Properties, Therapeutic index, Medicinal plants, Human diseases

Introduction

Nigeria with her diverse ecological conditions, rich ethnic diversity and strong traditional knowledge base, accounts for about 45,000 plant species, out of which more than 8,000 medicinal plant species are used in some 10,000 drug formulations in human healthcare cultures. Majority of these medicinal flora is also very important in veterinary and agriculture-related applications such as biofertilizers, seed treatment and biopesticides research on plants, with medicinal properties and identification of the chemical components responsible for their activities have justified the ancient traditional healing wisdom, and have proven the enduring healing potential of many plant medicines. Numerous plants and herbs are used all over Nigeria by traditional medicine practitioners. Roots, barks, leaves and stem of various plants are employed in ethno-medicine. Many investigators have demonstrated the antimicrobial activity of the constituents of some higher plants. A good proportion of the world population, particularly those living in developing countries like Nigeria depend mostly on herbal medicines for their health needs (Sofowora, 1982; Sofowora, 1993).

1. *Spondias mombin* Linn

Spondias mombin (Iyeye in Yoruba) is a fructiferous tree having habitat in Nigeria, Brazil and several other tropical forests. This plant is commonly found in South Western Nigeria, and it is commonly used in traditional medicine. It belongs to the family of *Anacardiacae* (Martinez *et al.*, 2008). The fruit approximately 1½- inch long oval yellow plum. It has a leathery skin and a thin layer of fruit pulp with a very exotic taste and most of the time tasted like Vitamin C. It hangs in numerous clusters of more than a dozen or more on the tree, very rich in vitamins B1 and C, the fruit mostly exists as an oval seed (Orwa *et al.,* 2009). The pulp of the fruit is sometimes eaten directly, but is too acidic to be considered attractive; it can also be boiled or dried. It is especially used for local cough syrup, ice cream, drinks and jellies. Juices improve with keeping overnight as the mild astringency of the fresh fruit disappears. Fermented products are also good. About half of the fruit weight is pulp, which is 8% water, 10% sugars, 1-8% fibre, and 0.4% ash. The sugars gives about 40 calories/100 g. The fruit is a good source of vitamins A and C; vitamin C carotenoids and flavonoids are present in reasonable concentrations this may be referred to as the Nutraceuticals. There is great variation in fruit quality from region to region, some being sweet and pleasant and others quite disagreeable in flavor (Moronkola *et al.*, 2003).

Ripe fruits are eaten out-of-hand and stewed with sugar. *S. mombin* extracted juice is used to prepare ice cream, cool beverages and jelly in Costa Rica and Brazil. It is used in Panama and Mexico in fairly large quantities as jams. In Amazon, the ripe fruit is used mainly to produce wine and in Guatemala, the ripe fruit is made into a cider-like drink. Mexicans pickle the green ripe fruits into vinegar and eat them like olives with salt and chili, as they do with unripe purple *S. mombin*. The *S. mombin* tree exudes a gum that is used as glue (Moronkola *et al.*, 2003). The fruit juice of *S. mombin* is drunk as a diuretic and febrifuge. The decoction of the astringent Stem bark serves as an emetic, a remedy for diarrhoea, dysentery, haemorrhoids and a treatment for gonorrhoea and leucorrhoea. *S. mombin* is believed to expel calcifications from the bladder in case there is a bladder infection. The powdered of Stem bark of *S. mombin* is applied on wounds especially during African tradition treatment of infection. A tea made from the flowers and leaves of *S. mombin* extract is taken to relieve stomach ache, biliousness, urethritis, cystitis and eye and throat inflammations (Aregheore and Singh, 2003). *S. mombin* leaves juice and the powder of dried leaves are used as poultices on wounds and inflammations (Ayoka *et al.*,2006). The gum is employed as a cough expectorant and to expel tapeworms (Chaisson and Martinson, 2008; Osuntokun and Olajubu, 2014; Wendakoon *et al.*, 2012; Igwe *et al.*, 2010). Concerning the popular use of this plant, the leaves have been found to be abortifacient (Osuntokun *et al.*, 2018), anti-microbial (Olugbuyiro *et al.*, 2013), anti-viral (Adepoju and Oyewole (2008), Vitamin C (Okwu, 2004), wound-healer (Nworu *et al.*, 2007); but this healing effect could not be confirmed by Osuntokun *et al.* (2018). Hess *et al.* (1995) reported the several uses of the plant based on oral communication, and not on any recorded scientific investigation.

All the parts of *S. mombin* tree are medicinally important in traditional medicine. The fruits decoction is drunk as a diuretic and febrifuge, the decoction of the Stem bark and the leaves as emetic, anti-diarrhoea and used in the treatment of dysentery, haemorrhoids, gonorrhoea and leucorrhoea. The antimicrobial, antibacterial, antifungal, and the antiviral properties of *S. mombin* have been reported (Osuntokun, 2018a; Osuntokun *et al.*, 2018a,b,c; Fred-Jaiyesimi *et al.*, 2009).

S. mombin tea of the flowers and the leaves is taken to relieve stomach ache, various inflammatory conditions and wound healings (Rodrigues and Hasse, 2000). Offiah and Anyanwu (1989) have also reported the abortifacient activity of the aqueous extract. Preliminary reports suggest that the phenolic acid, 6-alkenyl-salicylic acid from *S. mombin* are responsible for the antibacterial and molluscicidal of this plant extract (Coates *et al.*, 1994). In another study, the anacardis acid derivative from the hexane extract of the plant was shown to possess beta lactamase inhibitory properties (Corthout *et al.*, 1991).

S. mombin is one of the best medicinal plants that has a very high medicinal value becomes a viral toots in the production of new and novel antibiotics in the pharmaceutical world, because of its various therapeutic advantages against infectious organisms. More research should be encouraged on *S. mombin*. This will help to totally eradicate recalcitrant infections like HIV, breast cancer, liver infection just to mention a few. Indeed *S. mombin* is a magic bullet against infectious diseases.

2. *Nicotiana tabacum*

Nicotiana tabacum, or cultivated tobacco is a perennial herbaceous plant. Its leaves are commercially grown in many countries to be processed into tobacco. There is an indication that *N. tabacum* can be used to stimulates peristaltic movements and increases digestive fire/bile juice/ enzymatic metabolisms. It is a drug of choice for urinary track disorders and diseases related with urinary bladder. It is bitter and pungent in taste. In proper dosing it can be used in cough, Asthma, itching and antihelminthes. It is very good as analgesic and utilized in dental pain, pain-related with genital organ and pain related with eye. It can control dandruff and hair infections, and can dwindle the poison of scorpion bite and related swelling (Osuntokun and Ajayi, 2014).

Tobacco has been used as an antispasmodic, a diuretic, an emetic, an expectorant, a seda-tive, and a sialagogue, and in homeopathy. *N. tabacum* has a long history of use by medical herbalists as a relaxant, though since it is a highly addictive drug it is seldom employed internally or externally at present. The leaves act as antispasmodics, discutients, diuretics, emetics, expectorants, irritants, sedatives and sialagogues. Wet tobacco leaves are applied externally in the treatment of rheumatic swelling, skin diseases and stings, as the active ingredients can be absorbed through the skin. They are also a certain cure for painful piles. A homeopathic remedy made from the dried leaves is used in the treatment of nausea and travel sickness. Some other activities reported for *N. tabacum* are: analgesic activity, an-aesthetic activity, angiogenesis inhibition, antibacterial activity, anti-convulsant activities, anti-estrogenic effect, antifungal activity, antiglaucomic activity, antioxidant activity, an-tistress effect antiviral activity, aromatase inhibition, arrhythmogenic effect, carcinogenic activity, Nicotine for treatment of Alzheimer disease, Parkinson disease, depression and anxiety, schizophrenia, attention deficit hyperactivity disorder (ADHD), pain, and obesity. Ointments made from crushed leaves are used for baldness, dermatitis, infectious ulcers and pediculicide (Prior, 2007). The juice is applied externally as an insect repellent (Akinyemi *et al.,* 2005), while the leaf is added to the liquid for use as a mild stimulant (Akinside and Olukoya, 1995).

3. *Morinda lucida*

Morinda lucida (Oruwo in Yoruba) is a medicinal plant growing in many African countries and widely used as a medicine in Nigeria. *M. lucida* is one of the four most used plants in the preparation of traditional medicines against fever in many part of Nigeria. The leaves are used as "oral teas", which are usually taken orally for the traditional treatment of ma-laria, and as a generao febrifuge, analgesic, laxative and anti infections (Osuntokun *et al.,* 2014). The leaves have also been reported to possess strong trypanocidal and aortic vaso-relaxant activities (Olawale *et al.,* 2011). The bark or leaf decoction is applied against jaundice and the decoction of the stem bark or leaf is combined with a dressing of pow-dered root bark against itches and ringworm (Abbiw, 1990; Iwu *et al.,* 1999). Adeyemi *et al.* (2004) reported in their work that the bark, root and leaf are bitter. They stated that the infusion or decoction of these parts is used for the treatment of yellow fever and other forms of fever.

4. *Carica papaya* Linn

The papaya fruit is very delicious and nutritious. It provides several vitamins and minerals in significant amounts and low in calories. It has an enzyme that is useful in tenderizing meat and for treatment of indigestion (Chukwujekwu *et al.,* 2005). Beyond the nutritional, commercial, and aesthetic values for people, the trees and fruits also offers ecological val-ues, providing food, habitat and shelter for insects, birds, and other animals (Chukwujekwu *et al.,* 2005). Papaya fruit is a rich source of nutrients such as provitamin A, carotenoids, vitamin C, vitamin B, lycopene, dietary minerals and dietary fibre. The uses of the fruit includes: Laxatives- Ripe papaya fruit is laxative which assures of regular bowel movement. Indigestion- The milky juice which is tapped from the green, mature fruit while still in the tree contains an enzyme known as "papain". Void the heart attack or stroke- The folic acid found in papayas is needed for the conversion of homocysteine into amino acids such as cysteine or methionine. If unconverted, homocysteine can directly damage blood vessel walls, is considered a significant risk factor for a heart attack or stroke (Parle and Gurditta, 2011; Soobitha *et al.,* 2013).

The black seeds of the papaya are edible and have a sharp, spicy taste. They are sometimes ground and used as a substitute for black pepper. Other uses include: Nephro - protective activity- In wistar rats nephroprotective activity was observed in dose. Concentration of urine and creatinine were evaluated.

More potent- The papaya seeds are very pungent and peppery, making them almost unpalatable. However the seeds seem to have more potent medicinal values than the flesh. Papaya seeds have antibacterial properties and are effective against E. coli, Salmonella and Staphylococcus infections, Papaya seeds may protect the kidneys from toxin - induced kidney failure (Soobitha *et al.,* 2013).

The seeds can eliminate intestinal parasites, and help detoxify the liver. Used as a skin irritant to lower fever. Cure for piles and typhoid and anti-helminthic and antiamoebic properties. Dried papaya seeds actually look quite similar to peppercorns and can be used in just the same way. Grinding a couple over a meal, especially protein rich meals, is a simple way to add extra enzymes to your diet and improve your digestive health (Cos *et al.*, 2002).

Papaya leaf has many benefits. In some parts of Nigeria, the young leaves of the papaya are steamed and eaten like spinach. Other medicinal uses include:

Dengue fever- papaya leaf juice helps increase white blood cells and platelets, normalizes clotting, and repairs the liver. Cancer cell growth inhibition- recent research on papaya leaf tea extract has demonstrated cancer cell growth inhibition. It appears to boost the production of key signalling molecules called Th1-type cytokines, which help regulate the immune system. Antimalarial and antiplasmodial activity- Papaya leaves are made into tea as a treatment for malaria. Antimalarial and antiplasmodial activity has been noted in some preparations of the plant, but the mechanism is not understood and not scientifically proven. Facilitate digestion- The leaves of papaya plants contain chemical compounds carpain, Substance which kills microorganisms that often interfere with the digestive function. Additional Benefits of Papaya Leaves: As an acne medicine, Increase appetite, Ease menstrual pain, Meat tenderizer, Relieve nausea.

Papaya peel can also be used in many homes in Nigeria as remedies. The presence of vitamin A helps to restore and rebuild damaged skin. Papaya peel is applied as skin lightening agent. When peel is mixed with honey and applied, it can act as soothe and skin moisturizers. Fight dandruff- The papaya vinegar with lemon juice can be applied to the scalp for 20 minutes prior to shampooing to fight dandruff. Muscle Relaxant- Adding papaya oil and vinegar to bath water, along with essential oils like lavender, orange and rosemary can be nourishing, refreshing and relaxing, and can work as a pain reliever and muscle relaxant.

Juice from papaya roots is used in Nigeria to ease urinary troubles. Papaya leaf when dried and cured like a cigar, is smoked by asthmatic persons. An infusion of fresh papaya leaves is used by person to expel or destroy intestinal worms. Fresh young papaya are also used to remedy colic, a certain stomach disorder or cramp. A decoction formed by boiling the outer part of the roots of the papaya tree in the cure of dyspepsia (Arvind *et al.*, 2013).

The milky sap of an unripe papaya contains papain and chymopapain. It is used as antihelmintic, releaves dyspepsia, cures diarrhoea, bleeding haemorrhoids, stomachic, whooping cough (Arvind *et al.*, 2013).

5. *Piper guineense* (African black pepper)

Guinea pepper (*Piper guineense*) is a vine native Nigeria, West Africa, which is used as a substitute for black pepper (*P. nigrum*). It is a perennial medicinal plant that is characterized by heart shaped (piquantness) leave and oval, petiole, alternate, 12 cm long. *P. guineense* is a spice used to flavoring, seasoning, and imparting aroma to food. *P. guineense* leaves are used as a spice to flavor meat preparation and fresh pepper soup (Dada *et al.*, 2013). They are also processed and consumed as vegetables in meals. The fruits are used as a spice to flavor soup, rice, and stew. The oil distilled from *P. guineense* is used in perfumery and in soup making. Sometimes, it is also grown as an ornamental often indoors in cooler climates. *P. guineense* leaves are considered aperitif, carminative and eupeptic (Jackson, 1989). The leaves are used to treat respiratory infections, rheumatism, and syphilis. In Nigeria, the leaves have been shown to have antibacterial activity. *P. guineense* leaves are aseptic in nature and have the ability to relieve flatulence (Pal and Verma, 2013). The leaves are also used for treating female infertility and low sperm count in male while the Fruits are used as an aphrodisiac. *P. guineense* fruit extract is used in the treatment of epilepsy (Anyanwu and Nwosu, 2014).

In traditional herbal medicine, seeds are put into a variety of uses. In some part of Nigeria, seeds are consumed by women after childbirth to enhance uterine contraction for expulsion of a placenta and other remains from the womb. It is added to food of lactating mothers during postpartum period, as it is claimed that it encourages or stimulates uterine contractions, therefore, aiding in the fast return of uterine muscle to the original shape (Tapsell *et al.,* 2006). It is also used in Eastern part of Nigeria as abortifacient. It is also used as an adjuvant in the treatment of rheumatic pains and as an anti-asthmatic and also for the control of weight and as an aphrodisiac (Okigbo and Igwe, 2007).

P. guineense roots are chewed and juice swallowed as an aphrodisiac (Uhegbu *et al.,* 2015; Nwankwo *et al., 2014*). They are also used as chewing sticks for cleaning the teeth. Ash from the burnt plant is used as a salt substitute for medicinal preparations. Research shows that *P. guineense* also has preservative and antioxidant properties (Omodamiro and Ekeleme, 2013; Juliani *et al.*, 2013). Warm extract of the fruits are used as antivomiting and antihelmintic. Ripe fruits together with the seeds of *Parica biglobosa* and root bark of *Rauwolfia vomitora* are boiled with snail, the soup orally taken to treat rheumatic pains (Sara *et al.*, 2015; Soladoye *et al.,* 2010). Powder from the dried fruits mixed with honey acts as carminative and relieves stomachache (Nwozo et al., 2012). The ground formulation from the fruits of *P. guineense, Dioscorea bulbifera, Aframomum melegueta* and *Capsicum frutescens* is mixed with aqueous extract of *Citrus aurantifolia* (lime) against tonsillitis. The fruits and leaves are used as spice for preparing soup for post-partum women (Ekanem *et al.*, 2010; Okonkwo and Ogu, 2014; Abila *et al.,* 1993; Graw-Hill, 1980).

6. *Cymbopogon citratus*

Figure 1. *Cymbopogon citratus*

Cymbopogon (lemon grass) is a genus of about 55 species of grasses among which is *C. citratus*. Lemon grass is equally versatile in the garden. This tropical grass grows in dense clumps that can grow to 6 ft (1.8 m) in height and about 4 ft (1.2 m) in width, with a short rhizome. This tall perennial grass is native to temperate and tropical regions of the Old World and Oceania. Lemon grass oil is used as a pesticide and preservatives. Research shows that lemongrass oil has antifungal properties. Despite its ability to repel insects, its oil is commonly utilized as a "lure" to attract honeybees. Lemon grass works conveniently

185

as well as the pheromone created by the honeybee's nasonov gland, also known as attract-ant pheromones. Because of this, lemongrass oil can be used as a lure when trapping swarms or tempting to draw the attention of hived bees (Oladele, 2008).

Lemon grass is a bitter, aromatic, cooling herb that increases perspiration and relieves spasms. The essential oil obtained from the plant is an effective antifungal and antibacterial properties (Oladele, 2008). The essential oil contains from *Cymbopogon* is about 70% cit-ral, plus citronellal - both of these are markedly sedative internally, the plant is used prin-cipally as a tea in the treatment of digestive problems, where it relaxes the muscles of the stomach and gut, relieving cramping pains and flatulence (Aiyeloja and Ajewole, 2006).

An essential oil obtained from the *Cymbopogon* plant is used in perfumery, scenting soaps, hair oils, cosmetics (Watt and Berger-Brandwijk, 2002) and as an insect repellent. *Cymbopogon* essential oil consist is mainly of citral and is an important starting material in the perfumery industry. It is also used in the synthesis of vitamin A the essential oil obtained from the leaves has been shown to be an effective fungicide in treating pathogenic fungi on cultivated plant (Aiyeloja and Ajewole, 2006).

7. *Alchornea laxiflora / cordifolia*

Alchornea laxiflora (Ipa in Yoruba) is described as a straggling, laxly branched, evergreen dioecious shrub or small tree up to 8 m tall; young shoots erect, later becoming horizontal, hollow, glabrous. In Nigeria, a decoction of the leaves is taken to treat inflammatory and infectious diseases. It is also a common ingredient in herbal antimalarial preparations. Leaves of *A. cordifolia* are taken in water to treat hernia, and also as packing and preser-vation material for kolanuts in Nigeria. The small branches are used as chewing sticks. The leaves or leafy stems, as an infusion or chewed fresh, are taken for their sedative and anti-spasmodic activities to treat a variety of respiratory problems including sore throat, cough and bronchitis, genital-urinary problems including venereal diseases and female sterility, and intestinal problems including gastric ulcers, diarrhoea, amoebic dysentery and worms (Farombi *et al.*, 2003). As a purgative, they are also taken as an enema; high doses taken orally are emetic. They are also taken as a blood purifier, as a tonic and to treat anaemia and epilepsy. The leaf is used as decoction is taken to treat tachycardia.

Young stem pith is bitter and astringent and is chewed for the same use. The pith may also be rubbed on the chest to treat respiratory problems. The leaves are eaten in West Africa and to facilitate child delivery and abortifacient. A cold infusion of the dried and crushed leaves acts as a diuretic. Leaf and root decoctions are widely used as mouth wash to treat ulcers of the mouth, toothache and caries, and twigs are chewed for the same purposes (Ogundipe *et al.*, 2001). A decoction or paste of leafy twigs is applied as a wash to treat fever, malaria, rheumatic pains and enlarged spleen and as a lotion or poultice to sore feet;

vapour baths can also be taken. Root decoction or maceration is taken to treat amoebic dysentery and diarrhoea and used as eye drops to cure conjunctivitis. In Nigeria a decoction of bruised fruit is taken to prevent miscarriage. The sap of the fruit is applied to cure eye problems and skin diseases. In veterinary medicine a leaf or root infusion is given to livestock to treat trypanosomiasis (David *et al.*, 2014).

A. cordifolia extract has been patented for various other applications: antifouling adjuvant in paints, coatings and polymers, and alchorneic acid was proposed as a raw material for hemi-synthesis of plastic. Production and international trade. The leaves, root bark and fruits of *A. cordifolia* are sold in local markets from November to January. The leaves, roots and stem bark contain terpenoids, steroid glycosides, flavonoids (2–3%), tannins (about 10%), saponins, carbohydrates and the imidazopyrimidine alkaloids alchorneine, alchornidine and several guanidine alkaloids (Adeshina *et al.*, 2010; Osuntokun and Thonda, 2016).

The high tannin content was thought to be responsible for this activity. The ethanol extract of the leaf showed significant activity against castor oil-induced diarrhoea in mice. The presence of tannins and flavonoids may account for the increased colonic water and electrolyte reabsorption. The crude methanol extract of the leaf of *A. cordifolia* has a moderate relaxing effect on smooth muscles in vitro, which is attributed to the flavonoid quercetin and its derivatives (Ogundipe, *et al.,* 2001). The ethanol extract of the root significantly delayed the effect of histamine-induced broncho-constriction characterized by shortness of breath in guinea pig. The crude methanol extract of the leaves of *A. cordifolia* and several fractions of it have shown anti-inflammatory activity in the croton oil-induced ear oedema test in mice and in the egg albumen-induced hind paw oedema test in rats (David *et al.*, 2004).

The cytotoxicity of the crude extract was very low. Alcohol extracts from root bark, stem bark, leaves, fruits and seeds disrupted mitotic cell division in onion (*Allium cepa* L.). A methanol extract of the seed has shown inhibition of vascularization in chicken embryos. The approximate nutrient composition of leaf meal for use in chicken feed was per 100 g dry matter: energy 1930 kJ, crude protein 18.7 g and crude fibre 16.4 g. While the production of leaves is high, their palatability to cattle, goats and sheep is rather low (Fennell *et al.*, 2004).

8. *Anisopus mannii*

Anisopus mannii is a species of flowering plant in the family Asclepiadaceae. It is native to the tropical Americas, including the west indies in which has been identified in the Northern part of Nigeria. It is rarely cultivated, this plant is readily common around the northern part of Nigeria (Hausa) and are commonly used in folk medicine. *A. mannii* is also known as 'Sakayau' or 'Kashezaki' (meaning sweet killer) among the Hausas of the Northern Nigeria, where a cold decoction of the stem is traditionally used as remedy for hyperglycemia. It is a familiar herb in the traditional medicinal preparations in Northern Nigeria, where a decoction of the whole plant is used as a remedy for diabetes, diarrhoea and pile. Previously, the proximate composition, mineral elements and anti-nutritional factors of *A. mannii* was reported by (Musa *et al.,* 2009, Osuntokun *et al.,* 2016b).

9. *Daniella oliveri* (*Caesalpiniacea sp.*)

Daniella oliveri is a plant found in the Amazon region and other parts of Southern Nigeria. It is an indigenous African tree found extensively in Nigeria. It produces liquid oleoresin which has been used as medicine by indigenous people of Nigeria for more than 400 years (Jones, 1996). The oleoresin is produced in the *Daniella oliveri* tree's trunk, stem, and leaves and it consists of large but varying amounts of volatile oils (primarily composed of sesquiterpene hydrocarbons usually including caryophyllene), non volatile resinous substances and small quantities of acids. The oleoresin is traditionally used as an anti-inflammatory agent and in the treatment of a variety of genito-urinary tract diseases and skin ailments. Moreover, it is used as an anti-rheumatic, antiseptic, antibacterial, diuretic, and hypotensive agent, and also as an expectorant, laxative, purgative, vermifuge and vulnerary.

The *Daniella oliveri* leaves are also used in folk medicine as an anti-diabetic agent. Modern scientific studies have authenticated some of these *Daniella oliveri* medicinal uses of oleoresin such as efficacy as an antibacterial, antiinflammatory, and anti-oxidant agent. The leaves are used traditionally to treat diabetes and yellow fever. The leaves were found to contain quercitrin, quercameritrin, rutin and the rare flavoured glycoside quercitin-3-methoxy 3-o-rhamnosylpranosyl,-β-d-Glycopyranoside (Narssine) isolated from N-butanol extract (Wijesekera, 1991).

The root is considered diuretic and a decoction is taken to treat veneral diseases, absence of menses, anxiety, insanity, food poisoning, and skin diseases. Leafy twigs are put in baths to treat fever and jaundice, and also as atonic. A decoction of the leafy twigs with salt is taken as a purgative, to treat constipation and stomach-ache (Zolfaghari and Ghannadi, 2000).

10. *Erythrina senegalensis*

Erythrina senegalensis is a thorny shrub small tree with common names that include coral tree English) and minjirya (Hausa, Nigeria). The stem bark and root bark are used by traditional healers to cure wide range of illnesses (Adomi, 2008). The leaves are used to treat malaria, gastrointestinal disorders, fever, dizziness, secondary sterility, diarrhoea, jaundice, nose bleeding and pain (Donfack *et al.,* 2010). The stem bark extract has been shown to have antimicrobial activity against *Staphylococcus aureus, Streptococcus pyogenes,*

Escherichia coli, Salmonella typhi, Pseudomonas aeruginosa, Aspergillus flavus, Aspergillus fumigatus, Candida albicans, Penicillium notatum and inhibitory activity against HIV-1 protease (Lee *et al.,* 2009; Ngadjui and Moundipa, 2005). The root ethanol extract has a strong activity against *Plasmodium falciparum*. The stem bark has shown to have hepatoprotective properties (Togola *et al.,* 2008).

Phytochemicals such as tannins, glycosides, alkaloids, cardiac glycosides, prenylated isoflavones and flavones have also been identified in the *E. senegalensis* stem bark (Contu, 2009). The leaf extract shown to contain many phytochemicals. Appreciable data has been reported on the stem and root bark, little is known about the leaf extract of *E. senegalensis* (Wandji *et al.,* 1994). In Nigeria, the leaves of *E. senegalensis* are pounded and added to soup to help treat sterility/ barrenness (Osuntokun *et al.,* 2016a). In Nigeria, preparations of *E. senegalensis* are made from different parts of plant and used to bath the body, used as fumigations or taken orally to treat Malaria, cough, pneumonia, fewer, gastrointestinal disorders, snake bites, jaundice, nose bleeding, venereal diseases, etc, and the root infusion is used to relieve toothache in Nigeria. *E. senegalensis* is used by traditional healers to treat amenorrhea, malaria, jaundice, infections, body pains and used for abortions (Doughari, 2010; Togola *et al.,* 2008, Kone *et al.,* 2011).

11. *Bridelia ferruginea*

Bridelia ferruginea is a shrub growing up to 8 meters tall, or a straggly tree with a crooked bole which can grow up to 15 meters tall. It sometimes has spiny branches. The tree is much utilized from the wild by local people that use it for medicine and many other commodities. Medicinal preparations are sold in local market. They are found in Western tropical Africa, Nigeria. It hangs in numerous clusters of more than dozen on the trees (Iwu, 2010). The Stem bark is sometimes added to palm-wine to strengthen it and enhance fermentation. The medicinal uses of *B. ferruginea are* as follows. The leaf-extract of *B. ferruginea* in saline solution has shown to produce a marked reduction of blood-sugar. Decoctions of the leaves, leafy twigs and stem bark are commonly used in the treatment of urethral discharges; dysentery and diarrhoea; fever and rheumatic pains (Iwu, 2010).

The grated stem bark may be taken mixed with tapioca flour to treat dysentery and the stem bark, and the bright red infusion from it, are commonly used as a mouth-wash and remedy for thrush in children. The Stem bark has a great reputation as an antidote against poisons and snake bite. Some parts of Nigeria uses *B. ferruginea* as an antidote for scorpion bite. The Stem bark is chewed and then applied to a wound caused by a poisoned arrow, after which the wound is sucked to remove any more poison.

The non-medicinal uses of *B. ferruginea* the roots are used by the Yoruba as chew-sticks, grind the wood to a fine powder for use as a dentifrice. It is said to be termite-proof, and is used to make granaries and primary structural timber *B. ferruginea* is recognized as a good firewood, indeed as a 'woman's firewood that is one good for the hearth and cooking place, long-lasting while the housewife is away on other chores, and picking up quickly from sleeping embers with a hot flame and minimal amount of smoke (Govaerts *et al.,* 2000). *B. ferruginea* are commonly used as a mouth-wash and remedy for thrush in children and to treat skin diseases, infections and eruptions. A tea of the leaves is taken to relieve stomach ache, various inflammatory condition and wound healing (Jansen, 2005).

12. *Alstonia boonei*

Alstonia boonei (Ahun in Yoruba) is a tall forest tree, which can reach 45 metres (148 ft) in height and 3m (9.8ft) in girth. *A. boonei* is a large evergreen tree is one of the widely used medicinal plants in Nigeria and beyond. Important plants of the genus *Alstonia* includes *A. scholaris*, viz. leaves, stem bark; root and inflorescences. It is not edible as food but possess roots, stem barks, leaves fruits, seeds, flowers, and latex which are has medicinal properties in some cultures. The stem bark of *A. boonei* is used in traditional medicine to treat fever, painful menses, insomnia malaria and chronic diarrhea, rheumatic pains, as anti-venom for snake bites and in the treatment of arrow poisoning (Tepongning *et al.,* 2011; Osuntokun *et al.,* 2017b).

A. congensis and *A. macrophylla* which have proved to be useful in various diseases (Amole and Ilori, 2010). All the parts of the plant are very useful but the thick bark cut from the matured tree is the part that is most commonly used for therapeutic purposes (Amole and Ilori, 2010). The bark of the tree is highly effective when it is used in its fresh form; however, the dried one could equally be used. Therapeutically, the stem bark has been found to possess antirheumatic, anti-inflammatory, analgesic /pain-killing, antimalaria/antipyretic, antidiabetic (mild hypoglycemic), anti helminthic, antimicrobial and antibiotic properties (Odugbemi *et al.,* 2007; Oliver, 2000).

A decoction of *A. boonei* could be sweetened with pure honey and be taken up to 4 times daily as an effective painkiller for the following conditions; Painful menstruation (dysmenorrhoea), when associated with uterine fibroid or ovarian cysts in women; lower abdominal and pelvic congestion associated with gynecological problems such as pelvic inflammatory diseases; to relieve the painful urethritis common with gonococcus or other microbial infections in men. *A. boonei* decoction also exerts a mild antibacterial effect in this case, relieving the aches and pains associated with malaria fever. *A. boonei* is taken in the form of preparations that exhibits antipyrexia and anti-malaria effects, to combat rheumatic and arthritic pains (Opoku and Akoto, 2015).

The decoction of *A. boonei* stem bark could be taken alone as an effective pain-killing agent. A cold infusion made from the fresh or dried Stem bark of *A. boonei* taken orally two to three times daily exerts a mild hypoglycemic effect on diabetic patients. The cold infusion is also administered orally for expelling around worms, threadworms, and other intestinal parasites in children (Fakae *et al.,* 2000).

The fresh Stem bark of *A. boonei* are used in preparing herbal tinctures; it is particularly useful as an effective antidote against snake, rat, or scorpion poison. It is also useful in expelling retained products of conception and afterbirth when given to women. Asthma can be treated with a drink prepared from parts of *Trema orientalis* and decoction of the bark of *A. boonei* mixed with the roots and Stem bark of cola and fruits of *Xylopia parviflora* with hard potash. The Stem bark decoction of *A. boonei* is used with other preparations in the treatment of fractures or dislocation, jaundice, and for inducing breast milk. Its latex is taken as a purgative. *A. boonei* is considered a sacred tree and worshiped in the forest and hence human beings in those countries do not eat its parts (Marjorie, 2009).

13. *Ceiba pentandra*

Ceiba pentandra (Araba in Yoruba) is a tropical tree of Order Malvales and the Family Malvaceae. Bark decoction of *C. pentandra* has been used as a diuretic, aphrodisiac, and to treat headache, as well as type II diabetes. It is used as an addictive in some versions of the hallucinogenic drink. The root forms part of preparations to treat leprosy. Pulverized roots and root decoctions are taken against diarrhoea and dysentery. Root decoctions are oxytocic. Macerations of the root bark are drunk against dysmenorrhoea and hypertension. The root and the stem barks are taken to treat stomach problems, diarrhoea, hernia, gonorrhoea, heart trouble, oedema, fever, asthma and rickets; they are also applied on swollen fingers, wounds, sores and leprous macules. Stem bark extracts of *C. pentandra* are considered emetic: they are drunk or applied as enema (Selvakumar *et al.*, 2011).

Young leaves of *C. pentandra* are warmed and mixed with palm oil to be eaten against heart problems. Pounded leaves of *C. pentandra* are applied as a dressing on sores, tumours, abscess and whitlows. Leaf sap is applied to skin infections and drunk to treat mental illness. Leaf macerations are drunk or used in bathes against general fatigue, stiffness of the limbs, headache and bleeding of pregnant women. Leaf preparations are used as an eye-bath to remove foreign bodies from the eye. In veterinary medicine a decoction of the leaf extract of *C. pentandra* is used to treat trypanosomiasis. The flowers are taken to treat constipation, and flowers and fruits are taken to with water against intestinal parasites and stomachache. Kapok fibre is used for cleaning wounds; the seed oil is rubbed in for treatment of rheumatism and applied to heal wounds (Selvakumar *et al.*, 2011).

14. *Chrysophyllum albidum* (African Star Apple)

Chrysophyllum albidum (African Star Apple) is one fruit of great economic value in tropical Africa, especially Nigeria, due to its diverse medicinal and therapeutic uses. *C. albidum* (African Star Apple) plant has become a crop of medicinal and commercial value in Nigeria. (Olasehinde *et al.*, 2016; Osuntokun *et al.*, 2017a). *C. albidum* is a plant which has been used in traditional/alternative medicine in Nigeria to treat health problem, various parts of this herb have been proved to have a wide range of therapeutic effects. Generally, the roots, barks and leaves of *C. albidum* are widely used as an application to sprains, bruises and wounds in southern Nigeria.

C. albidum stem bark is employed for the treatment of yellow fever and malaria. In Southern Nigeria, the roots and stem barks are employed in urinary infections. The leaf is used as an emollient and for the treatment of skin eruption, stomachache and diarrhoea (Mac-Donald *et al.*, 2014). The decocted leaves of *C. albidum* are administered as a cancer remedy and as pectoral. The bitter pulverized seed of *C. albidum* is taken as a tonic, diuretic, febrifuge and in the treatment of diarrhoea (Adisa, 2000). The seeds and roots extracts of *C. albidum* is used to arrest bleeding from fresh wounds, and to inhibit microbial growth of known wound contaminants, and also enhance wound healing process. In addition, its seeds are a source of oil, which is used for diverse purposes (Florence and Adiaha, 2015; Olorunnisola *et al.*, 2008, Oboh *et al.*, 2009).

15. *Polyalthia longifolia* (Indian mast tree)

Polyalthia longifolia is an evergreen plant commonly used as an ornamental street tree due to its effectiveness in combating noise pollution. *P. longifolia* exhibits symmetrical pyramidal growth with willowy weeping pendulous branches and long narrow lanceolate leaves with undulate margins. In traditional medicines, its various herbal preparations are being used for treating duodenal ulcers (Chen *et al.*, 2007).

Figure 2. *Polyalthia longifolia* leaves

The stem bark; roots and leaves have been studied for various biological activities, such as antibacterial, antidiabetic, anti-inflammatory, and antioxidant activity. In addition, the bioactive compounds of this plant and its pharmacological activities due to its geographic density are relatively new areas for investigation (Muller-Schwarze, 2006).

In traditional medicine, various herbal preparations are being used for treating duodenal ulcers. The plant has been used in traditional system of medicine for the treatment of fever, skin diseases, diabetes, hypertension and helminthiasis. The plant extract and isolated compounds have been studied for various biological activities like antibacterial, cytotoxicity, antifungal activity. The Stem bark, flower, leaf, root and fruit can be used as potential herbal samples in pharmacy as decoction (Chen *et al.,* 2007).

16. *Persea americana* (Avocado essential plant)

Avocado essential oil is widely used by all sections of the population either directly as folk remedy or indirectly in the preparation of pharmaceuticals. The essential oil is a medicinal plant used in the treatment of malaria and other ailments of human body.

Figure 3. Avocado leaves *(Persea americana)*

The essential oils from Avocado stem bark and seed are used as astringent against cold, cough and sore throat (Wang *et al.*, 2005, Olonisakin, 2014). The hexane extract of the fruit has been tested and found effective against isoniazid and ethambutol resistant *M. tuberculosis* strains *in vitro*. It also possesses antioxidant properties, the flavonoids present in this oil have hydroxyl radical scavenging activity, anticancer and antioxidant activity (Wang *et al.*, 2005).

The seed oil of *Persea americana* is highly effective as an antifungal agent against *lentinus sajor-caju*, which causes white rot in wood, hence it can be used as a preservative agent in the management of wood infection. The extracts from the seed has a characteristic antimicrobial, anti-inflammatory action against bacteria, especially *Mycobacterium tuberculosis*, *Streptococcus pyogenes*, *Staphylococcus aureus* and varieties of fungi (Yekeen *et al.,* 2014). The crushed and boiled seeds of *Persea americana* is used in treating toothache and mouth sores. *Persea americana* stem bark extract have been used as analgesic, anti-inflammatory, hypoglycaemic, anticonvulsant, anti-diabetic and vasorelaxant (Gomez-Flores, 2008).

17. *Parkia biglobosa* (Jacq.) Benth

Parkia biglobosa (Jacq.) Benth is an economic tree found in west Africa, Nigeria savannahs and dry forests. It belongs to the family Fabaceace – pea f.amily, of the order Fabaceae. It is popularly known as the Africa locust bean, Igba or Irugba (Yoruba), Dorowa

(Hausa) and Orgili (Ibo). It is an important multipurpose tree and is well known in many African countries. Apart from providing building materials, wood, food, fodder, weapons and other commodities, *P. biglobosa* plant is especially important as traditional medicine in Nigeria. In Nigeria west Africa, various identified uses of *P. biglobosa* were medicinal and therapeutic uses. The fermented seeds of *P. biglobosa* are used in all parts of Nigeria and indeed the West coast of Africa for seasoning traditional soups. The yellow pulp is a high energy giving food with up to 60% sugar content (Osuntokun *et al.,* 2018d).

The powdered pods of *P. biglobosa* are used to paint traditional Hausa buildings in northern Nigeria. Parkia biglobosa has been identified as source of tannin, saponins, gum, fuel and wood. Seeds of various species of *P. biglobosa* have also been investigated for their protein and mineral content. Indigenous healers in West Africa, Nigeria use different parts of the locust bean tree for health benefits. *P. biglobosa* is one of the highest cited plants used for treating hypertension (Wagenlehner and Naber, 2006), Ajaiyeoba, 2002). It is also on record that the plant is also listed as having real wound-healing properties, for the treatment of malaria fever (antimalarial) and typhoid fever (antibacterial properties). It is a potential source of compounds used in the management of bacterial infections (Koura *et al.*, 2011; Sourabie and Nikiema, 2013).

18. *Ageratum conyzoides* (Billy goat-weed)

Ageratum conyzoides (Billy goat-weed, Chick weed, Goat weed, White weed; Ageratum obtusifolium Lam., Cacaliamentrasto Vell) (Imi-esu in Yoruba) is native to tropical America, especially Brazil, but were found in Nigeria, West Africa.

Figure 4. *Ageratum conyzoides* (Billy goat-weed)

The essential oil of *A. conyzoides* contains 5% eugenol, which has a pleasant odour. The oil from Ageratum conyzoides plants growing in Africa has an agreeable odour, consisting almost entirely of eugenol (Okwori *et al.,* 2006). A decoction of the fresh *A. conyzoides* plant is used as a hair wash, leaving the soft, fragrant and dandruff-free.

19. *Cleistopholis patens* (Benth) Engl. and Diels

Cleistopholis patens (Benth) found distributed in various parts of tropical Africa, Nigeria. It is used in traditional medical practices in many parts of Nigeria where it has several applications (Osuntokun, 2018b). *C. patens* (Benth) are potent anti-fungal and antibacterial agents effective against *Klebsiella pneumonia* (Adonu *et al.*, 2013). Several of the applications in traditional medicine have been confirmed by research, e.g. antimicrobial, anthelmintic and antimalarial activities. This may offer opportunities for drug development (Talalay and Talalay, 2001). *C. patens* have exhibited significant activity against *Candida albicans, Aspergillus fumigatus*, and *Cryptococcus neoformans* isolated from root bark of *C. patens*. The antiplasmodial activity of non-volatile and volatile extract from the stembark of *C. patens* had been reported. According to ethnomedicinal report, the stem bark of the plant is used in the treatment of jaundice, infective hepatitis and stomach disorders. The roots are used as a vermifurge and the leaves are said to remedy fever (Stary and Storchova, 1991).

20. *Piptadeniastrum africanum*

Piptadeniastrum africanum is commonly used in traditional medicine, mostly the stem bark, sometimes roots and leaves. Stem bark decoctions are used internally to treat cough, bronchitis, headache, mental disorders, haemorrhoids, genitor-urinary infections, stomachache, dysmenorrhoea and male impotence, it is also been used as an antidote. Externally, they are applied to treat fever, toothache, pneumonia, oedema, skin complaints and rheumatism to expel worms, to dispel fleas, and as a purgative and abortifacient (Famobuwa, 2012, Osuntokun et al., 2019b).

The stem bark is used in arrow poison, and as ordeal poison and fish poison, mixed with rice it is used to poison mice. Pounded leaves and stem bark decoctions are applied as an enema to treat gonorrhoea and abdominal complaints. The stem bark fibre has been used to weave mats. Edible caterpillars feed on the leaves, and the flowers parts are sources of nectar for honeybees. *P. africanum* is considered a magic tree (Katende *et al.,* 2005).

21. *Adenopus breviflorus* (Benth) Fruit

 Adenopus breviflorus (Benth) (Tagiri in Yoruba) is a tree plant, commonly known as "*Lagenaria breviflora* Roberty*"*, belongs to the family of Cucurbitaceae (Gourd family) (Osuntokun *et al.,* 2019a). It is a perennial, seasonal creeping tendril climber. The fruit (bulb) appear green with cream-colored narrow blotches measuring 1-5 cm in length and its pulp is bitter. The stem bark when crushed has an unpleasant smell and a decoction from it is said to be used in Nigeria for headache and as a vermifuge (Ajayi *et al.*, 2002).

Figure 5. *Adenopus breviflorus* (Benth) fruit

Adenopus breviflorus (Benth) has a long medicinal history values for treating various conditions in Nigeria. The fruits are major important items of trade in Western region of Nigeria and they are of immense values in curative and preventive control measures against conditions such as measles, chickenpox, intestinal worms, enteritis (diarrhoea), *Diabetes mellitus,* Newcastle diseases, leather and wood preservatives (Adedeji *et al.*, 2017), as wound antiseptics (umbilical incision wound) and as depilatory agent. It has been reported that the methanol extract of its whole fruit has anti-implantation activity and abortifacient activity (Ajayi *et al.*, 2002). The ethanol extract of its whole fruit has been reported to have a broad spectrum antibacterial activity as well as anti-oxidant and anti-ulcerogenic effects. The ethanol extract of its whole fruit has been reported to cause increase in red blood cells, total white blood cell count and packed cell volume values, as well as caused electrolytes imbalances and spermatotoxic effect in rats (Banjo *et al.*, 2013).

Extracts from *A. breviflorus (Benth)* have been identified as potent anti-inflammatory agents (Akinyemi *et al.*, 2005), while fruits and seeds were reported to possess miracicidal and cercaricidal substances effective for controlling transmission of Schistosomiasis (Tomori *et al.,* 2007). In a related development, *A. breviflorus (Benth)* fruit applied as phytobiotics improved the growth performance of broiler and exhibited excellent control of *Eimeria oocyst* and *Ascaris galli* (Elujoba *et al.,* 1990, Oridupa and Saba, 2012). The extract demonstrated antioxidant activity by its ability to quench free radicals generated by nitric oxide and superoxide anion with a concomitant scavenging potential against DPPH-induced radical formation and enhance the recovery from oxidative stress (Saba *et al.,* 2012).

22. *Zingiber officinale* (Ginger)

Ginger is one of the most delicious and healthiest spices available on Earth. Not only it adds a tasty flavor to food, it is full of nutrients and bioactive compounds that are beneficial for brain and body. Ginger, mostly grows in warmer areas of Asia is popular to cure an upset stomach. Some other benefits includes treating motion sickness, cures chronic indigestion, helps in menstrual pain, lowers blood cholesterol, protect against Alzheimer and fights Infections (Mbaveng and Kuete, 2017; Ebrahimzadeh *et al.,* 2017).

23. *Allium Sativum* (Garlic)

Allium sativum commonly called garlic is a species of the Allium family, and is a common food ingredient all around the world. It has a unique flavor which makes it an indispensable element in many dishes. Ancient Greek physicians have referred it to be the father of western medicine. Not that it just eases people with heart and blood issues, it has several other health benefits too. It is used to combats sickness, Prevents Alzheimer, Increases life-span, improves athletic performance, Ends hair loss and clears skin impurities (Eja *et al.,* 2011, Katsa *et al.,* 2014; Johnson *et al.,* 2013).

24. *Tamarindus indica*

Tamarindus indica of the Fabaceae, sub-family Caesalpinioideae, is an important food in the tropics. It is a multipurpose tree of which almost every part finds at least some uses either nutritional or medicinal. *T. indica* is indigenous to tropical Africa (Nigeria). *T. indica* (Tamarind) seed has been found to be effective in the treatment of water. Therefore, it is recommended that local industries should consider using it for wastewater treatment as an alternative to chemical coagulant, because of its biological origin, cost affordability and availability (El-Siddig *et al.,* 2006, Ayangunna *et al.,* 2016).

Flour from the seed may be made into cake and bread. Roasted seeds are claimed to be superior to groundnuts in flavor. *T. indica* leaves and flowers can be eaten as vegetables and are prepared in a variety of dishes They are used to make curries, salads, stews and soups in many countries, especially in times of scarcity.

There is medical interest in the use of purified xyloglucan from tamarind in eye surgery for conjunctival cell adhesion and corneal wound healing Tamarind fruit is regarded as a digestive, carminative, laxative, expectorant and blood tonic, anti-hepatoxic, anti-inflammatory, anti-diabetic activities. *T. indica* combined with lime juice, honey, milk, dates fruit, spices or camphor, the pulp is considered to be effective as a remedy for biliousness and bile disorders (Martinello *et al.,* 2006).

25. *Tetrapleura tetraptera*

Tetrapleura tetraptera is a perennial tree which is commonly distributed along the Tropical regions of Africa. The common names of *T. tetraptera* are: Gum tree (English), Ìshíhí

(Oshoho), Aidan (Yoruba), Ighimiaka (Bini) and Edeminnangi (Efik) (Irondi, *et al.,* 2016). In Nigeria, it is used for numerous purposes. The powdered fruit is used as fish poison and in ointment for the treatment of skin diseases (Atawodi *et al.,* 2014). The intensive odour produced when the fruit is roasted is claimed to repel insects and snakes. The methanol extract of the fruit which was linked to their saponin content has been reported to have molluscidal property and its mechanism of action is by ultrastructural effects of the snail digestive system. The leaves, stem bark, roots, fruits and kernels are used for medicinal purposes (Atawodi *et al.,* 2014, Adesina *et al.,* 2016).

Tetrapluera tetraptera is used for the treatment of skin diseases, stem bark extract for the treatment of gonorrhoea. It is used for treatment of skin diseases, stem bark extract for the treatment of gonorrhoea. The stem bark of this plant has an inhibitory effect on the leutenizing hormone released by the pituitary gland. This suggests why this plant has a contraceptive property. Both the stem bark, leaves and fruits are used as concoction for managing convulsion, it has anticonvulsant properties. The *T. tetraptera* extract is known for its anti-inflammatory properties and this suggest its inhibitory impacts against certain human pathogens (Kren and Walterova, 2015; Jimmy and Ekpo, 2016).

The pods contains essential chemical compounds such as flavonoids, triterpenoid glycoside (aridanin) and phenols, which have been reported effective for healing wounds. The taub is an excellent source of antioxidants such as polyphenols, alkaloids, tannins and flavonoids. Antibacterial ability of the *T. tetraptera* plant has been revealed by researchers that water extract and alcoholic mixture of aridan fruit can inhibit the growth of *Staphylococcus aureus* (Akintola *et al.,* 2015). The presence of glycosides and tannins in ethanolic and water extract have been proven effective for inhibiting the growth of bacteria. It is also used for dermatological care as the fruit can be dried and blended into powdered form for producing dermatological products such as soap. The great attention drawn to the use of this plant for manufacturing soap is due to its high antimicrobial and antibacterial properties (Moukette *et al.,* 2015). It is worthy of note that the aridan plant helps to promote soap forming as well as its hardness. To make soap with aridan, the dried powdered herbs can be combined with shear butter, palm kernel oil or any other bases of choice.

26. *Vernonia amygdalina*

*Vernonia amygdalina (*Asteraceae) is a genus of about 1,000 species of forbs and shrubs of which *V. amygdalina* is the most prominent species, and one of the pan tropical tribes of the family Asteraceae. It is popularly called bitter leaves because of its bitter taste and is used as vegetables or as flavour decoction soups. The bitter taste of *V. amygdalina* is as a result of its anti-nutritional components such as alkaloids, saponins, glycosides and tannins (Egedigwe, 2010). The leaves are used for human consumption and washed before eating to get rid of the bitter taste. They are used as vegetable and stimulate the digestive system, as well as to reduce fever. *V. amygdalina* are used as local medicine against leech, which are transmitting bilharziose. Free living chimpanzees eat the leaves, if they have been attacked by parasites. *V. amygdalina* is also used instead of hops to make beer in Nigeria. *Vernonina amygdalina* found in homes, in villages as fence post and pot-herb. In general it has there been found that *V. amygdalina* have an astringent taste, which affects its intake. The bitter taste is due to anti-nutritional factors such as alkaloids, saponins, tannins and glycosides (Farombi and Owoeye, 2011).

The roots and leaves decoction of *V. amygdalina* are commonly used in ethno medicine to treat fevers, hiccups, kidney problems and stomach discomfort among other several uses It is also used in the treatment of diarrhoea, dysentery hepatitis and cough and as a laxative and fertility inducer. The leaves of *V. amygdalina* are also commonly used as a treatment against nematodes in humans and chimpanzees as well as other intestinal worms. *V. amygdalina* extracts of the plants have been reported to be used in Nigerian herbal homes as tonic, in the control of tick and treatment of hypertension. *V. amygdalina* is attributable to the complex active secondary plant compounds that are pharmacologically active (Abosi and Raseroka, 2003). The aqueous and alcoholic crude extracts of the *V. amygdalina* leaves, bark, stem and roots are reported to be widely used as antimalarial, for the treatment of eczema and as a purgative. The roots and the leaves of *V. amygdalina* are used in traditional medicine to treat fever, stomach discomfort, hiccups and kidney problems. *V. amygdalina is* known as quinine substitute because it is widely used for the treatment of fevers (Kigigha and Onyema, 2015).

The wood, particularly those from the root is a tooth cleaner (Chewing sticks), an appetizer, fertility inducer and also for gastrointestinal upset. The root infusion is taken in Nigeria for the treatment of intestinal worms as well as for enteritis and rheumatism. Other documented medicinal uses include the treatment of schistosomiasis, amoebic dysentery, treatment of malaria, wound healing, veneral diseases, hepatitis and diabetes (Olamide and Agu, 2013). Fresh leaves of *V. amygdalina* have been reported to have abortifacient and purgative activities. The chopped roots of *V. amygdalina* are used for the treatment of sexually transmitted diseases. The root of *V. amygdalina* is used for its antifertility effect and for the treatment of amenorrhoea (Muhammad *et al.,* 2014).

27. *Garcinia kola* Heckel

Figure 6. *Garcinia kola* Heckel

The principal application of *Garcinia kola* is for chew-sticks. In Nigeria, fruit and root are said to whiten the teeth, and thought to prevent dental caries. Pharmacological extracts of stem bark, roots and seeds have shown strong anti-hepatotoxic and hepatotropic activity (Theophine and Emeka, 2014).

The powdered stem bark is applied to malignant tumours, cancers, etc., and the gum is taken internally for gonorrhoea, and externally to seal new wounds. Stem bark decoction of *G. kola* is taken for female sterility and to ease child-birth, the intake being daily till conception is certain and then at half quantity throughout the term. The Stem bark is added to that of *Sarcocephalus latifolius* (Rubiaceae), a tisane of which has a strong reputation as a diuretic, urinary decongestant and for chronic urethral discharge. The Stem bark is also thought to be galactogenic, the bark is used with *P. guineense* (Piperaceae) and sap from a plantain stalk (Musaceae) to embrocate the breast for mastitis (Okonkwo *et al.,* 2014).

The leaves of *G. kola* have a bitter taste. They are used as a deterrant to fleas. A leaf-infusion is purgative. The fruits are edible, orange-sized, and contain a yellow pulp surrounding four seeds. The fruits are eaten in Nigeria as a cure for general aches in the head, back, etc., and as a vermifuge.

Since the emergence of sexually transmitted diseases, noble scientists have been battling night and day to bring lasting solution to the plague. Although, their battle may seem somewhat difficult as a result of the complicatedness of the diseases, results have shown that seed and leaf of bitter kola have antibacterial activity on clinical isolates of *Staphylococcus aureus, Escherichia coli, Salmonella typhi* and *Streptococcus pyrogens* (Cheek, 2004). The fibers and lung tissues are not only strengthened when bitter kola helps in the maintenance of good respiratory track as well as treating chest cold (Cheek, 2004). The stem, bark, and seeds of bitter kola are used to treat acute fever, inflammation of the respiratory tract and throat infections. Bitter kola has been found to be an amazing remedy for the eye, when bitter kola is eaten, at least twice a day, it could successfully reduce the eye pressure (Mac-Donald *et al.*, 2014).

28. *Uvaria afzelii* (Scott-Elliot)

Uvaria afzelii is a small tree or spreading shrub growing up to 5m tall. The tree is used locally, being harvested from wild for food and medicines (Iroabuchi, 2008). Locally, it is used in the treatment of cough, vaginal tumor, breast aches, swollen hands feet's, diabetes as well as leucorrhoea and gonorrhoea (Iroabuchi, 2008; Lawal and Adeniyi, 2015). A number of investigations carried out to ascertain the claimed uses of the plant includes its reported bactericidal activity against Gram-positive and acid-fast bacteria (Ogbole *et al.*, 2013) antihelminthic and antiparasitic activities. Other ethnomedicinal uses of the plant includes its benefit as a remedy for jaundice, infections of the liver, kidney, and bladder. Silymarin is a standardized extract of the milk thistle plant (*Silybum marianum*) which majorly contains flavonoids: silybin, silybinin, silydianin and silychristin (Akewele, 1990). Seeds of this plant have been used for years to treat liver and gall bladder disorders, including hepatitis, cirrhosis and jaundice and to protect the liver against poisoning from chemicals, environment toxins, snake bites, insect stings, mushroom poisoning and alcohol (Fasola and Egunyomi, 2002). The fruit is edible the leaves are used for treating fever locally (Groombridge, 2000) and boiled with pepper are taken in draught, or rubbed on the skin for yellow fever in Nigeria. The plant is held to be good for bronchial troubles and for stomach ache in Ivory Coast and in the Gagnoa area pulped leaves are eaten with oil-palm. In folk medicine, the stem and leaves extracts of *Uvaria afzelii* can be used in the preventing and treatment of hypertension (Ugbogu and Odewo, 2004).

Researchers agree that *Uvaria afzelii* is effective in preventing high blood pressure and for improving the oxidative position in salt model of hypertension patients. The stem and bark of *Uvaria afzelii* can also be used for preparing herbal medicines for treating diabetes. Being an excellent source of key vitamins such as potassium, iron, calcium, magnesium, and zinc, *Uvaria afzelii*, helps to strengthen our immune system. Iron helps to regenerate lost blood, zinc offers protection against viruses, especially those that can cause respiratory tract infections, while calcium and potassium helps to manage, prevent and control bones and muscles disorder Tapsell *et al.*, 2006).

Conclusion

Medicinal plants have becomes a mobile tools in the treatment of infectious diseases, and in the production of new and novel antibiotics in the pharmaceutical world. It is necessary to further explore its efficacy, medicinal and theraupeutic properties. The success can even serve as another foreign earning in a country that entirely depends solely on petroleum resource for its annual budget.

References

Abbiw, D.K. (1990) Useful plants of Ghana: West African uses of wild and cultivated plants. Intermediate Technology Publications, London and Royal Botanic Gardens, Kew, Richmond, United Kingdom, 1990. pp. 337.

Abila, B., Richens, A., Davies, J.A. (1993) Anticonvulsant effects of extracts of the West African black pepper, *Piper guineense. J. Ethnopharmacol.* 34: 261-1264.

Abosi, A.O. and Raseroka, B.H. (2003) *In vivo* antimalarial activity of *Vernonia amygdalina.* Br. J. Biomedical Sci., 60(2): 89-91.

Adedeji, G.A., Aiyeloja, A.A., Oyebade, B.A. and Uriel, T. (2017) Harnessing the instrumentalities of *Lagenaria breviflora* fruits for optimising wood protection. In: V.A.J. Adekunle, O.Y. Ogunsanwo and A.O. Akinwole (Eds.). Proceedings of the 39th Annual Conference of Forestry Association of Nigeria, 20-24 February, 2017, Ibadan, Oyo State, Nigeria 39: 947-953.

Adepoju, O.T., O.E. Oyewole (2008) "Nutrient Composition and Acceptability Study of Fortified Jams from Spondias Mombin (Hog Plum, Iyeye in Yoruba) Fruit Pulp" *Nigerian Journal of Nutritional Science* 29 (02): pp.180–189

Adeshina, G.O., Onaolapo, J.A., Ehinmidu, J.O., Odama, L.E. (2010) Phytochemical and antimicrobial studies of ethyl acetate extract of *Alcornea cordiforlia* leaf found in Abuja, Nigeria. J. Med. Plants Res. 4(8):649-658.

Adesina, S.K., Iwalewa, E.O. and Johnny, I.I (2016) *Tetrapleura tetraptera Taub* Ethnopharmacology, Chemistry, Medicinal And Nutritional Values- A Review. *British Journal of Pharmaceutical Research* 12(3): 1-22.

Adeyemi, T.O.A., Ogboru, R.O., Idowu, R.O., Owoeye, E.A., Isese, M.O. (2014) Phytochemical screening and health potentials of *Morinda lucida* Benth. International Journal of Innovation and Scientific Research.;11(2):515-519. Available:http://www.ijisr.issr-journals.org

Adisa, S.A. (2000) Vitamin C, protein and mineral content of African apple (*Chrysophyllum albidum*) in proceedings of the 18th Annual Conference of NIST. (Eds), 141-146.

Adomi, P.O. (2008) Screening of the leaves of three Nigerian medicinal plants for antibacterial activity. Afr. J. Biotechnol. 7 (15): 2540-2542

Adonu, C.C., Okechukwu, P.C., Esimone, C.O., Emmanuel, C., Abubakar, B. (2013) Phytochemical analyses of the menthanol, hot water and n-hexane extracts of the aerial parts of *Cassytha filiformis* (Linn) and leaves of *Cleistopholis patens* (Benth). Res J Pharma Biol Chem Sci 4: 1143-1149.

Aiyeloja, A.A. and O.I. Ajewole (2006) Non-timber forest products'marketing in Nigeria. A case study of Osun state" reproduction). Educational Research and Reviews, Academic Journals1 (2).;52–58. ISSN 1990-3839. OCLC 173185259. Retrieved 2009-06-25.

Ajaiyeoba, E.O. (2002) Phytochemical and antibacterial properties of *Parkia biglobosa* and *Parkia bicolor* leaf extracts. *Afr. J. Biomed. Res.*;5:125–129.

Ajayi GO, Awoja NC, Abulu LE (2002) The miracicidal and cercaricidal activity of the methanolic extract of *lageneria breviflora* Robert family Cucurbitaceae fruit in *Schistosoma mansoni. Nig Q J hosp Med.*;12(1):57-59.

Akewele, O (1990) Medicinal Plants and Primary Health care; an agenda for action. Essential Drugs Monitor. No. 10. World Health Organization, Geneva, p. 6-11.

Akinside, K.A., Olukoya, D.K. (1995) Vibrocidal activities of some local herbs. J. Diarhoeal Dis. Res. 13:127-129.

Akintola, O.O., Bodede, A.I. and Ogunbanjo, O.R. (2015) Nutritional and medicinal importance of *Tetrapleura tetraptera* fruits (Aridan). *African Journal of Science and Research* 4 (6): 33-38.

Akinyemi KO, Oladapo O, Okwara CE, Ibe CC, Fasure KA (2005) Screening of crude extracts of six medicinal plants used in South-West Nigerian unorthodox medicine for anti-methicillin resistant Staphylococcus aureus activity. BMC Complement Altern Med;5:6. Doi:10.1186/1472-6882-5-6

Amole, O.O., Ilori, O.O. (2010) Antimicrobial activity of the aqueous and ethanolic extracts of the stem bark of *Alstonia boonei*. Int. J. Phytopharmacol. 1(2):119–123

Anyanwu C. U. and Nwosu G. C. (2014) Assessment of the antimicrobial activity of aqueous and ethanolic extracts of *Piper guineense* leaves, *Journal of Medicinal Plant Research*, Vol. 8(10), pp. 337-439.

Anyanwu, C.U, and Nwosu, G.C (2014) Assessment of antimicrobial activity of aqueous and ethanolic extracts of *Piper guineense* leaves. J Med Res. 8 (10): 337 – 439.

Aregheore E. M. and Singh E., (2003) Seasonal variation of macro- and micro-mineral contents of someruminant browse species from five countries in the South Pacific Region. *Tropical Agriculture* 80 (2), 69-73.

Arvind, G., Bhowmik, D., Duraivel, S., Harish, G. (2013)Traditional and medicinal uses of *Carica papaya*, J Med Car Pap; 1(1):2320-3862.

Atawodi, S.E.,Yakubu, O.E., Liman, M.L. and Iliemene, D.U (2014) (Effect of methanolic extract of *Tetrapleura tetraptera* [Schum and Thonn] Taub leaves on hyperglycemia and indices of diabetic complications in alloxan-induced diabetic rats). Asian Pac. Journal. Tropical. Biomedical 4: 272-278

Ayangunna, R.R., S.O. Giwa and A. Giwa (2016) Coagulation-Flocculation Treatment of Wastewater Using Tamarind Seed Powder, International Journal, Journal of ChemTech Research. 9, (5) pp 771-780

Ayoka, A.O, Akomolafe, R.O, Iwalewa, E.O, Akanma, M.A & Ukponmwan. O.E (2006) Sedative,epileptic and antipsychotic effects of *Spondias mombin* L (Anacardiaceae) in mice and rats. *Journal of Ethnopharmacolology*, 103(2): 166-175. Doi: 10.1016/ j.jep.

Banjo, T.A., Kasim, L.S., Iwalokun, B.A., Mutiu, W.B., Olooto. W.E., Mba, N.G., James, E.S. and Shorunmu, T.O. (2013) Effects of different extraction methods on in-vitro antimicrobial properties of *Lagenaria breviflora* whole fruits. *New York Science Journal*. 2013;6(10):60-65.

Chaisson R.E. and N.E., Martinson (2008) Tuberculosis in Africa: Combating an HIV-driven crisis, N. Engl. *Journal of Medicine*, 3581089-1092.

Cheek, M. (2004) *Garcinia kola*. IUCN Red List of Threatened Species. Archived June 27, 2014, at the Wayback Machine Downloaded on 20 July 2007.

Chen, Chung-Yi; Fang-Rong Chang; Yao-Ching Shih; Tian-Jye Hsieh; Yi-Chen Chia; Huang-Yi Tseng; Hua-Chien Chen; Shu-Jen Chen; Ming-Chu Hsu; Yang-Chang Wu (2007) "Cytotoxic Constituents of *Polyalthia longifolia* var. *pendula*". *Journal of Natural Products*. **63** (11): 1475–1478. doi:10.1021/np000176e. PMID 11087586. S0163-3864(00)00176-2. Retrieved 2007-09-21.

Chukwujekwu J, Staden J and Smith P(2005) Antimicrobial, anti-inflamatory and antimalarial activities of some Nigerian medicinal plants;71:315-325

Coates N. J., Gilpin M. L.Gwynn M. N., Lewis D. E., Milner P. H., Spear S. R. and Tyler J. W. (1994) SB-202742 a novel beta-lactamase inhibitor isolated from *Spondias mom bin.Journal of Natural Products*. 57, 654 – 657.Corporate Ltd. Pp 310-367.

Contu, S. (2009) *Erythrina senegalensis*. Assessment using IUCN categories and criteria 3.1 (IUCN 2001). Royal Botanic Gardens, Kew.

Corthout J., Pieters, L. A., Claeys, M., Vanden Berghe, D. A. and Viletinck A. J. (1991) Anti   viral; *Ellagitannins* from *Spodias mombin.Phytochemistry* 30, 1190.

Cos, P., Hermans, N., De Bruyne, T., Apers, S., Sindambiwe, J. B., Vanden Berghe, D., PietersL., Vlietinck, A. J. (2002) Further evaluation of Rwandan medicinal plant extracts for their antimicrobial and antiviral activities. J. Ethnopharmacol. 79, 155– 163. doi: 10.1016/S0378-8741(01)00362-2

Dada A. A., Ifesan B. O. T. and Fashakin J. F. (2013) Antimicrobial And Antioxidant Properties Of Selected Local Spices Used In "Kunun" Beverage In Nigeria, *Acta Sci. Pol., Technol. Alignment,* issue 12, vol. 4, p.374

David, A.A, Emmanuel, O.A., Olayinka, A.A., Oluseun, F.A. and Anthony, I.O. (2014) Evaluation of Antibacterial and Antifungal Properties of *Alchornea laxiflora* (Benth.) Pax. & Hoffman Evidence-Based Complementary and Alternative Medicine 2015:1-6http://dx.doi.org/10.1155/2015/684839

Donfack, J.N., Njayou, F.N., Rodrigue, T.K., Chuisseu, D.D.P, Tchana, N.A., Finzi, V.P. (2010) Evaluation of antimicrobial potentials of stem bark extracts of *Erythrina senegelensis* DC. Afr. J microbiol Res.; 4(17):1836-1841.

Doughari, J.H. (2010) Evaluation of antimicrobial potential of stem bark extracts of *Erythrina senegalensis* DC, Afriacan Journal of Microbiology Research 4 (7): 1836-1841.

Ebrahimzadeh, A.V., Malek, M.A., Javadivala, Z., Mahluji, S., Zununi, V.S., Ostadrahimi, A. (2017) A systematic review of the anti-obesity and weight lowering effect of ginger (*Zingiber officinale* Roscoe) and its mechanisms of action. Phytotherapy research : PTR. 2017;Epub ahead of print. https://doi.org/10.1002/ptr.5986.

Egedigwe, C.A. (2010) Effect of dietary incorporation of *Vernonia amygdalina* and *Vernonia colorata* on blood lipid profile and relative organ weights in albino rats (Thesis). Department of Biochemistry, MOUAU, Nigeria

Eja, M.E., Arikpo, G.E., Enyi-Idoh, K.H., Ikpeme, E.M. (2011) An evaluation of the anti-microbial synergy of garlic (*Allium sativum*) and utazi (*Gongronema latifolium*) on *Escherichia coli* and *Staphylococcus aureus*. Malays J Microbiol.;7:49–53.

Ekanem, A.P., Udoh, F.V and Oku, E.E. (2010) Effects of ethanol extract of *P guineense* seeds on the conception of mice (*Mus musculus*). *Afr. J. Pharm Pharmacol*, 4 (6): 362-367.

El-siddig, K., Gunasena, H.PM., Prasa, B.A., Pushpakumara, D.K., N.G. Ramana, K.V.R. Vijay and. P., Williams, J.T (2006) *Tamarindus indica* L. fruits for the future. Southampton center for underutilized crops, Southampton, UK 188p.

Elujoba, A.A., Fell, A.F., Linley, P.A. and Maitland, D. (1990) Triterpenoid saponins from the fruits of *Laganaria breviflora* Robert. *Phytochemistry* 29:3281-3285.

Fakae, B.B., Campbell, A.M., Barrett, J. (2000) Inhibition of glutathione S- transferases (GSTs) from parasitic nematodes by extracts from traditional Nigerian medicinal plants. Phytotherapy Res. 14(8):630–634.

Famobuwa, O. E. (2012) Phytochemical screening and invitro antimicrobial effect of methanol bark extract of *Piptadeniastrum africanum*. *Journal of science and science education.* 3: 93-96

Farombi E. O. and Owoeye O. (2011) Antioxidative and Chemopreventive Properties of *Vernonia amygdalina* and *Garcinia biflavonoid*, International Journal of Environ Res Public Health, 8(6): 2533–2537.

Farombi, E.O., Ogundipe, O.O., Uhunwangho, E.S., Adeyanju, M.A. & Moody, J.O. (2003) Antioxidant properties of extracts from *Alchornea laxiflora* (Benth) Pax and Hoffman. Phytotherapy Research 17(7): 713–716.

Fasola, T. and Egunyomi, R. (2002) Bark extractism and uses of some medicinal plants. Journal of botany.; 15:26-36.

Fennell, C.W., Lindsey, K.L., McGaw, L.J., Sparg, S.G., Stafford, G.I., Elgorashi, E.E., Grace, O.M. and van Staden, J. (2004) Assessing African Medicinal plants for efficacy and Safety: Pharmacological screening and toxicity. J. Ethnopharmacol., 94:205-217

Florence, A.B. and Adiaha, A.H. (2015) Storage effects and the postharvest quality of African star apple fruits (*Chrysophyllum africanum*) under ambient conditions. African Journal of Food Science and Technology, Vol. 6(1): 35-43.

Fred-Jaiyesimi, A.K. Abo and R. Wilkins (2009*) α-Amylase inhibitory effect of 3 β-olean-12-en-3-yl (9 Z)-hexadec-9-enoate isolated from Spondias mombin leaf. Food Chemistry* ,16: 285-288.

Gomez-Flores R (2008) Antimicrobial activity of *Persea americana* (Lauraceae) (avocado) seed and stem bark extract. Am-Euras J Sci Res 3: 188-194.

Govaerts, R., Frodin, D.G. & Radcliffe-Smith, A. (2000) World Checklist and Bibliography of Euphorbiaceae (and Pandaceae) 1-4: 1-1622. The Board of Trustees of the Royal Botanic Gardens, Kew

Graw-Hill, M.C. (1980): Encyclopedia of Food, Agriculture and Nutrition, (repairs) vol 4. Published by Mc.Graw- Hill/Inc USA. pp. 629,

Groombridge, B. and Jenkins, M.D. (2002) World Atlas of Biodiversity: Earth's Living Resources in the 21st Century. Berkely, CA: University of California Press.

Hess, S. C., Brum, R. L., Honda, N. K., Cruz, A. B., Moretto, E., Cruz, R. B., Messana, I., Ferrari, F., Filho, V. C., Yunes, R. A. (1995) Antibacterial activity and phytochemical analysis of *Vochysia divergens. Journal of Ethnopharmacology* 47, 97-100.

Igwe, G.O.C., Onyeze, V.A., Onwuliri, C.G., Osuagwu and A.O. Ojiako (2010) Evaluation of the chemical compositions of the leaf of *Spondias Mombin* linn from Nigeria, *Australian Journal of Basic and Applied Sciences* , 4(5) , 706-710

Iroabuchi, F. (2008) Phytochemical constituents of *Clerodendron splendens* (A. cheval) and *Uvaria chamae* (P. Beauv) M.Sc., Thesis, Michael Okpara University of Agriculture, Umudike, Nigeria. 2008;3-16

Irondi, E.A., Oboh, G., Agboola, S.O., Boligon, A.A. and Athayde, M.L (2016) Phenolics extract of *Tetrapleura tetraptera* fruit inhibits xanthine oxidase and Fe2+-induced lipid peroxidation in the kidney, liver, and lungs tissues of rats *in vitro*). Food Science and Human Wellness. 5: 17-23.

Iwu, M.W., Duncan, A.R. and Okunji, C.O. (1999) New antimicrobials of plant origin. In: Perspectives on New Crops And New Uses. J. Janick (Ed). ASHS Press, Alexandria VA.;457– 462

Iwu, M.M. (2010) Handbook of African medicinal plants (2nd ed). Taylor and Francis group LLC. Pp 216-217.

Jackson, D.D (1989) Searching for medicinal wealth in Amazonia. Smith sonian, pp.95-103.

Jansen, P. C. M. (2005) *Dyes and Tannins*. PROTA (Plant Resources of Tropical Africa). p. 102. ISBN 9057821591. Retrieved 2013-06-19

Jimmy, E.O., and Ekpo, A.J. (2016) Upgrading of lethal dose of *Tetrapleura tetraptera* extract enhances blood cell values. *Journal of Hematological & Thromboembolic Dis*, 3(4): 256-269.

Johnson, O.O., Ayoola, G.A., Adenipekun, T. (2013) Antimicrobial activity and the chemical composition of the volatile oil blend from *Allium sativum* (Garlic clove) and *Citrus reticulata* (Tangerine fruit) Int J Pharm Sci Drug Res. 2013;5:187–93.

Jones, F. (1996) Herbs – useful plants. Their role in history and Today. Euro J Gastroenterol Hepatol.;8:1227-123.

Juliani, H.R., Koroch, A.R. Koroch, L. Giordano, L. Amekuse, S. Koffa, J. Asante-Dartey and J.E. Simon (2013) *Piper guineense* (Piparaceae) Chemistry, Traditional uses and Functional Properties of West African Black Pepper. Vol II. Discoveries and challenges in chemistry, Health and Nutrition. ACS Symposium Series.1127: 3 – 48.

Katende, A. B., Birnie, A. and Tengnas, B. (2005) Useful trees and shrubs for Uganda: identification, propagation and management for agricultural and pastoral communities. Technical Handbook 10. Regional Soil Conservation Unit, Nairobi, Kenya. p: 710-734

Katsa, M., Gyar, S.D., Reuben, C. (2014) *In vitro*: Antibacterial activity of *Allium roseum* (Wild Allium or Rosy garlic) against some clinical isolates. J Microbiol Res Rev. 2014;2:30–3.

Kigigha, LT, Onyema E. (2015) Antibacterial activity of bitter leaf (*Vernonia amygdalina*) soup on *Staphylococcus aureus* and *Escherichia coli*. Sky J Microbiol Res. 2015;3:41–5.

Kone, W.M., Solange, K.E., and Dosso, M. (2011) Assessing Sub- saharian Erythrina for efficacy: traditional Uses, biological activities and phytochemistry. *Pak. J. Biological Sci.* 14(10):560- 571.

Koura, K., Ganglo, J.C., Assogbadjo, A.E. (2011) Ethnic differences in use values and use patterns of *Parkia biglobosa* in Northern Benin. *J Ethnobiol Ethnomed*.;7:42.

Kren, V. and Walterova, D.S. (2015) Silybin and silymarin- new effects and applications *Biomedical Paper. Medical Faculty* University Palachy Olomouc Czech Republic 149 (1): 29-41

Lawal, T.O., and Adeniyi, B.A. (2015) *In vitro* susceptibility of *Mycobacterium tuberculosis* to extracts of *Uvaria afzelii* Scott Eliot and *Tetracera alnifolia* wild. *Africa Journal BiomedRes* 14: 17-21.

Lee, J., Wk O.H, Ahn, J.S., Kim, Y.H., Mbafor, J.T., Wandji, J. *(*2009) Prenyliso flavonoids from *Erythrina senegalensis* as novel HIV-1 protease inhibitors. *Planta Med.* 2009; 75:268-27.

MacDonald, I., Nosa, O.O., Emmanuel, O.O. and Joseph, O.E. (2014) Phytochemical and antimicrobial properties of *Chrysophyllum albidum*, *Dacryodes edulis*, *Garcinia kola* chloroform and ethanolic root extracts Intercult Ethnopharmacol Journal, 3(1):15-20.

Marjorie, C. (2009) Plant Products as Antimicrobial Agents. Clin. Microbiol. Rev.12:564-582.

Martinello, F., Soares, S.M., Franco, J.J., Santos. A.C, Sugohara., A., Garcia, S.B., Curti, C., Uyemura,S.A. (2006) Hypolipidemic and antioxidant activities from *Tamarindus indica* L. *pulp fruit extract in hypercholesterolemic hamsters*. Food and Chemical Toxicology, 44, 810-818.

Martinez, M., Lazaro, R., Olmo L. and Benito, P (2008) Anti-infectious activity in the anthemideae tribe. In: Atta-ur- (Ed.) *Studies in Natural Products Chemistry,* Vol. 35. Elsevier. pp 445-516.

Mbaveng, A.T., Kuete, V. (2017) In: Medicinal Spices and Vegetables from Africa, 2017

Moronkola D. O., Adeleke A. K. and Ekundayo O. (2003) Constituents of the *Spondias mombin* Linn and the comparison between its fruit and leaf essential oils. *Journal of Essential Oil Bearing Plants*, 6 (3): 148 – 152.

Moukette, B.M., Pieme, A.C., Biapa, P.C., Njimou, J.R., Stoller, M., Bravi, M. and Yonkeu, N.J. (2015) *In vitro* ion chelating, antioxidative mechanism of extracts from fruits and barks of *Tetrapleura tetraptera* and their protective effects against fenton mediated toxicity of metal ions on liver homogenates. *Evid.-Based Complement. Altern. Med,* 2:14-16

Muhammad, A.B., Doko, H.I., Ibrahim, A., Abdullahi, B., Yahaya, H., Sharfadi, R.S. (2014) Studies on inhibitory effects of extract of *Vernonia amygdalina* used in traditional poultry farming against some bacteria isolated from poultry dropping. Sch J Appl Med Sci. ;2:3185–92.

Muller-Schwarze D (2006) Chemical ecology of vertebrates. Cambridge University Press, UK. p. 287.

Musa, A. M., Aliyu, A. B, Yaro A. H., Magaji, M. G., Hassan, H. S. and Abdullahi, M. I. (2009) Preliminary phytochemical, analgesic and anti inflammatory studies of the methanol extracts of *Anisopus mannii*(N.E. Br) (Asclepiadaceae) in rodents. African Journal of Pharmacy and Pharmacology Vol. 3(8). pp. 374-378

Ngadjui, T.B. and Moundipa, F.P. (2005) Study of hepatoprotective and antioxidant fraction from *Erythrina senegalensis* stem bark extract in vivo pharmacology online 1:120-130.

Nwankwo, C.S., Ebenezer, I. A., Ikpeama, A.I. and Asuzu, F.O. (2014) The Nutri tional and anti-nutritional values of two culinary herbs – Uziza Leaf (*Piper guineense*) and Scent Leaf (*Ocimum gratissium*) *Popularly used in Nigeria Intern J. Sci Engineering Res,* 5 (12): 1160-1163.

Nworu, C.S., Akah, P.A., Okoli, C.O., Okoye, T.C. (2007) Oxytocic activity of leaf extract of *Spondias mombin*. *Pharmaceutical Biology* 45: 366-371.

Nwozo S. O., Ajagbe, A.A., Onyinloye, B.E. (2012) Hepatoprotective effect of *Piper guineense* aqueous extract against ethanol induced toxicity in male rats. J. Exp. Integrative Med. 2 (1): 71 – 76.

Oboh, I.O, Aluyor, E. O and Audu, T.O.K. (2009) Use of *Chrysophyllum albidum* for the removal of metal ions from aqueous solution. Scientific Research and Essay Vol. 4(6):632-635.

Odugbemi, T.O., Akinsulire, O.R., Aibinu, I.E., Fabeku, P.O. (2007) Medicinal plants useful for malaria therapy in Okeigbo, Ondo State, Southwest Nigeria, Afr. J. Traditional, Complementary and Alternative Med. 4:191-198.

Offiah, V.N., Anyanwu I.I. (1989) Abortifacient activity of an aqueous extract of *Spondias mombin*leaves. Journal of ethnopharmacology 26(3): 317-20.

Ogbole, O.O., Adeniji, J.Λ., Λjaiyeoba, O.E., Adu, D.F. (2013) Anti-poliovirus activity of medicinal plants selected from the Nigerian ethnomedicine. Afri. J. Biotech.; 12(24):3878-83.

Ogundipe, O.O., Moody, J.O., Houghton, P.J. and Odelola, H.A. (2001) Bioactive chemical constituents from *Alchornea laxiflora* (Benth.) Pax and Hoffman. Journal of Ethnopharmacology 74(3): 275–280.

Okigbo, K.N. and Igwe, D. I(2007) Antimicrobial effects of *P.guineense*"Uziza" and phylantrusamarus on *Candida albican*s and *Streptococcus faecalis*. Acta microbiologica et immunologica Hungarica.54(4):353-366.

Okonkwo, H. O., Koyejo, O. A., Osewa, S. O. and Babalola, O. T. (2014) Techniques for Improvement of *Garcinia kola* (Heckel) Seeds Germination. International Journal of Applied Research and Technology. 3(8): 80 – 86

Okwori, A., Dina, C., Junaid, S., Okeke, I., Adetunji, J., Olabode, A. (2006) Antibacterial activities of *Ageratum conyzoides* extracts on selected bacterial pathogen. The Internet Journal of Microbiology. 4:1-5.

Okwu, D.E. (2004) Chemical composition of *Spondias mombin* Linn. Plant parts. J. Sustain. Agric Environ. 2004;6 (2):140-147.

Oladele, O.I. (2008) Contribution of Neglected and Underutilized Crops to Household food security and Health among Rural Dwellers in Oyo State, Nigeria. Symposium Proceedings, online publication of presented papers. International Symposium, Underutilized plants for food, nutrition, income and sustainable development, Arusha Tanzania 3–7 March 2008. Colombo, Sri Lanka: International Centre for Underutilised Crops (ICUC);.

Olamide, S.O., Agu, G.C. (2013) The assessment of the antimicrobial activities of *Ocimum gratissimum* (wild basil) and *Vernonia amygdalina* (bitter leaf) on some enteric pathogen causing dysentery or diarrhea. Int J Eng Sci. 2013;2:83–96.

Olasehinde, G.I., Okolie, Z.V., Oniha, M.I., Adekeye, B.T. and Ajayi, A.A. (2016) In vitro antibacterial and antifungal activities of *Chrysophyllum albidum* and *Diospyros monbuttensis* leaves. Journal of Pharmacognosy and Phytotherapy Vol. 8(1): 1-7.

Olawale, K.O., Fadira, S.O., Taiwo, S.S. (2011) Prevalence of hospital acquired enterococci infections in two primary care hospitals in Osogbo, Southwestern Nigeria. Afr. Jour. Infect. Dis. 5(2): 45-46.

Oliver, B. (2000) Medicinal plants in Nigeria 1st edition. Nigerian College of Arts, Science and Technology, Ibadan. Pp43-53.

Olonisakin, A. (2014) Comparative study of essential oil composition of seed and stem bark of avocado essential life. J of Scie 16:2.

Olorunnisola, D.S., Amao, I.S., Ehigie, D.O. and Ajayi, Z.A.F. (2008) Anti-Hyperglycemic and Hypolipidemic Effect of Ethanolic Extract of *Chrysophyllum albidum* Seed Cotyledon in Alloxan Induced-Diabetic Rats. Res. J. Appl. Sci, Vol. 3: 123-127.

Olugbuyiro, J.O., Moody, M.T., Hamann (2013) Phytosterols from *Spondias mombin* Linn.with antimycobacterial activities, *African Journal of Biomedical Research* , 16 (1)182- 186

Omodamiro, O.D. and Ekeleme, C.M. (2013) Comparative study of *in vitro* antioxidant and anti-microbial activities of *Piperguineense, Cormuma longa, Gongronems latifolium, Allium sativum, Ocimum gratissi mum.World J. Med. Medical Sci*. 1 (4): 51 – 69,

Opoku, F. and Akoto, O. (2015) Antimicrobial and phytochemical properties of *Alstonia boonei* extracts. Organic Chemistry Current Research

Oridupa, O.A. and Saba, A.B. (2012) Relative anti-inflammatory and analgesic activities of the whole fruit, fruit bark, pulp and seed of *Lagenaria breviflora* Roberty. Journalof Pharmacology and Toxicology. 7(6):288-297. Doi: 10.3923/jpt.2012

Orwa C., A Mutua., Kindt R., Jamnadass R., S Anthony. (2009) Agroforestree Database:a tree reference and selection guide version 4.0 (http://www.world agro forestry.org/ sites /treedbs/treedatabases.asp).

Osuntokun, O.T. (2018a) Evaluation of Inhibitory Zone Diameter (IZD) of crude *Spondias mombin* (Linn.) extracts (root, leaf, and stem bark) against thirty infectious clinical and environ mental isolates. J Bacteriol Infec Dis. 2(1):8-16

Osuntokun, O.T. (2018b) Evaluation of Inhibitory Zone Diameter (IZD), Phytochemical Screening, Elemental Composition and Proximate Analysis of Crude *Cleistopholis patens* (Benth) on Infectious Clinical Isolates. J Mol Biomark Diagn Volume 9: 385. doi: 10.4172/2155-9929. 1000385,ISSN:2155-9929,Pp 1-9.

Osuntokun O.T. and Ajayi, A.O. (2014) Antimicrobial phytochemical and proximate analysis of four Nigerian Medicinal plants on some Clinical Microorganisms.;5:457-461. ISSN:2320-2246.

Osuntokun O.T., Olajubu F.A (2014) Comparative study of phytochemical and proximate analysis of seven Nigerian medicinal plants. *Applied Science Research Journal.App. Sci. Res.J Pon* Publishers. Ekpoma Edo state Nigeria ; 2(1):10-26.

Osuntokun, O.T., Ojo, R.O., Ogundeyi, S.B. (2014) Chewing sticks and oral health care in owo Local Government Ondo State. Journal Archives of Biomedical Science and the Health Pon Publishers Ekpoma Edo state Nigeria. 2(1):68-74

Osuntokun, O.T., and O.A. Thonda (2016) Comparative Study of the Antibacterial and antifungal Spectrum, Phytochemical screening and Antioxidant potentials of *Alchornea laxiofolia and Piliostigma reticulatum* Leaf on Pathogenic Isolates *The Pharmaceutical and Chemical Journal* 3(2): 1-11.

Osuntokun, O.T., Ajayi, A.O, Olorunnipa, T.A., Thonda, O.A. and Taiwo, O.V. (2016a) Phytochemical screening and antimicrobial properties of partially purified ethyl acetate extracts of *Erythrina senegalensis* leaf and bark against selected clinical isolates, *Journal of Medicinal Plants Studies* 4(3): 259-269.

Osuntokun, O.T., Akanji O.C. and A.A. Joshua (2016b) Antimicrobial activity and phytochemical composition of crude ethyl acetate extracts of *Anisopus mannii* on some selected clinically important microorganisms. *International Journal of Pharma Sciences and Research* 7(5): 231-239.

Osuntokun, O.T., Ayegbusi, B.L., Yusuf-Babatunde, A.M., Ige, O.O. (2017a) Assessment of Antimicrobial, Phytochemical Screening and Gas Chromatography-Mass Spectrophotometric Profile of Crude *Chrysophyllum albidum* Essential Oil, *Chemistry Research Journal,* 2017, Volume 2(5):68-85.

Osuntokun, O.T., Bamidele O, Aladejana O.M and Ogunlade A.O (2017b) Evaluation of Antimicrobial Activity, Synergistic Efficacy, Qualitative and Quantitative Phytochemical Determination of *Alstonia boonei* Leaf and Stem Bark on Selected Clinical Isolates, *Triple A Research Journal of Multidisciplinary* (JMD) 1(1): Pp; 001- 014.

Osuntokun, O.T., Baraldi, C., Gamberini, M.C. (2018a) Evaluation of Quantitative Elemental Compositions and Antioxidant Potentials of *Spondias mombin* Extracts (Linn) ,A Precursor Against Infectious Diseases. World Journal of Pharmacy and Pharmaceutical Sciences, 7(3): 3: 964-985.

Osuntokun, O.T., Idowu, T.O., and Gamberini, M.C. (2018b) Bio-guided Isolation, Purification and Chemical Characterization of Epigallocatechin; Epicatechin, Stigmasterol Phytosterol from of Ethyl Acetate Stem Bark Fraction of *Spondias mombin* (Linn.) *Bio chem. Pharmacol* (Los Angel,USA) 7:240.doi: 10.4172 /2167 0501 .10002 40,Volume 7 • Issue 1 • 1000240, Pp.1-9.

Osuntokun, O.T., Ige, O.O., Idowu T.O, Gamberini, M.C. (2018c) Bio-activity and Spectral Analysis of Gas Chromatography/Mass Spectroscopy (GCMS) Profile of Crude *Spondias mombin* Extracts. SF J Anal Biochem 2:1.SF J Anal Biochem; 2(1) 1000001,ISSN:xxxx-xxxx SFJAB, an open access journal,page 1- 12.

Osuntokun, O.T., Jemilaiye, T.A., Akinrodoye, A.R. (2018d) Comparative study between the effect of *Parkia biglobosa* (JACQ) benth and conventional antibiotics against multiple antibiotic resistant uropathogenic bacteria (MARUB). *MOJ Bioequiv Availab.* 2018;5(4):200–212. DOI: 10.15406/mojbb.2018.05.00103

Osuntokun, O.T., Ibukun, A.F., Yusuf-Babatunde, A.M., Abiodun, S. (2019a) Pre/Post-Plasmid ProfileAnalysis, Killing- Kinetics and Secondary Metabolites Screening of *Adenopus breviflorus (Benth)* Fruit Extract Against Multiple Drug Resistant Isolates Using *Staphylococcus aureus (MDRSA)* as a Case Study. J Adv Res Biotech 4(1):1-17. www.symbiosis online publis hing.com ISSN Online: 2374-8362.

Osuntokun, O.T., Jemilaiye, T.A, Adefemiwa, A.T., Thonda, A.O. (2019b) Zones of Inhibition and Molecular Docking (In-silico Approach) of *Piptadeniastrum africanum* Extracts on Clinical and Multiple Antibiotic Resistant Isolates. *International Journal of Microbiology and Application.* 5, No. 4, 2018, pp. 70-80.

Pal D. and Verma P. (2013) Flavonoids: A Powerful And Abundant Source Of Antioxidants, *International Journal of Pharmacy and Pharmaceutical Sciences*, Vol. 5, Issue 3, p.97.

Parle M., and Gurditta, G. (2011) Basketful benefits of Papaya, Int Res J Pharm 2011; 2(7):6-12.

Prior, M. P. (2007) Helping you lose weight deliciously Alternative-medicine online. Retrieved ; 2007

Rodrigues, K. F. and Hasse M. (2000) Antimicrobial activities of secondary metabolites produced by endophytic fungi from *Spondias mombin. Journal of Basic Micro biology* 40, 261 – 267.

Saba, A.B., Onakoya, O.M. and Oyagbemi, A.A. (2012) *J.Basic Clin. Physiol. Pharmacol.* 2012;23(1):27-32.

Sara, H.F., Jens, S., Anna, K.J. (2015) Medicinal plants used as excipients in the history in Ghanaian herbal medicine. Journal of Ethno pharm. 174:568

Selvakumar, K., Madhan, R., Srinivasan, G., Baskar, V. (2011) Antioxidant assays in pharmacological research. Asian J Pharm Tech 1(4): 99- 103.

Sofowora A(1993) Medicinal plants and traditional medicines in Africa. Chic HesterJohn Willey &Sons. 1993;256.

Sofowora A., (1982) Medicinal Plants and Traditional medicine in Africa. 1st ed., New York: John Wiley and Sons Ltd. pp. 168-171.

Soladoye, M.O., Amusa, N., Raji-Esan, S., Chukwuma, E., Taiwo, A. (2010) Ethnobotanical survey of anti cancer plants in Ogun State, Nigeria. Ann. Biol. Res. 1:261-73.

Soobitha, S., Tan, C.C., Kee, CC, Thayan, R., Mok, B.T. (2013) *Carica papaya* leaves juice significantly accelerates the rate of increase in platelet count among patients with dengue fever and dengue haemorrhagic fever, Evi Bas Compl Alter Med, 2013, 148-151.

Sourabie, K.Y. and Nikiema, J.B. (2013) Ethnobotanical Survey of Medicinal Plants Used By the Traditional Medical Healers in the Villages of Bérégadougou and Fabédougou (Cascades Region, Burkina Faso). *IOSR Journal of Pharmacy*. 2013;3(7):38–45.

Stary F, Storchova H (1991) A mutual guide to medicinal herbs and plants. Tiger Books International, UK. pp. 44.

Talalay, P., Talalay, P (2001) "The importance of using scientific principles in the development of medicinal agents from plants".*Academic Medicine* 76 (3): 238–47. doi: 10.1097/000 018 88-200103000-00010. PMID 11242573

Tapsell, L.C., Hemphill, I., Coblac, L. (2006) Health benefits of herbs and spices. "The past, the present and the future" *Med. J. Aust.* 185 (4 Suppl): 54 – 24.

Tepongning, R.N., Lucantoni, L., Nasuti, C.C., Dori, G.U., Yerbanga, S.R., Lupidi, G. (2011) Potential of a *khaya ivorensis- Alstonia boonei* extract combination as antimalarial prophylactic remedy. J Ethnopharmacol. 2011; 137: 743-751

Theophine, C.O., Emeka, K.O. (2014) In: Toxicological Survey of African Medicinal Plants, 2014

Togola, A., Austarheim, I., Theis, A., Diallo, D., Paulsen, B.S. (2008) Ethno-pharmacological uses of *Erythrina senegalensis*: a comparison of three areas in Mali, and link between traditional knowledge and modern biological science. J Ethnobiol. Ethnomed. 2008; 4:6. DOI: 10.1186/1746- 4269-4-6

Tomori, O.A., Saba, A.B., Dada-Adegbola, H.O. (2007) Antibacterial activity of ethanolic extract of whole fruit of *Lagenaria breviflora* Robertys. *J Anim Vet Adv.*;6:752-757.

Ugbogu, O.A., Odewo, P. (2004) Some medicinal plants in the traditional Medicare of Nigeria. Journal of Forestry Research and Management 1(1&2):29-34.

Uhegbu, F.O., Chinedu, I., and Amadike, E.U. (2015) Effect of Aqueous extract of *Piper guineense* seeds on some Liver enzymes, antioxidant enzymes and some Hematological Parameters in Albino rats. *Intern J. plant Sci. Ecology.* 1 (4): 167 – 171.

Wagenlehner, F.M., Naber, K.G. (2006) Treatment of bacterial urinary tract infections: presence and future. *Eur Urol.*;49(2):235–244.

Wandji, J., Fomum, Z.T., Tillequin, F., Baudouin, G., Koch, M. (1994) Expoxyiso flavones from *Eryhtrina senegalensis*. Phyto chem. 1994; 35:1573-1577.

Wang, Y., Wong, M., Roquejo, C., McGhie, T., Woolf, A. (2005) Recent research on the health components in oil pressed avocado oil.

Watt, J.T., Berger-Brandwijk *B.S.* (2002) Cosmetic containing plant extracts, Official Gazette of US patents and trademark Office Patents 1259(3).

Wendakoon C., Calderon P., Gagnon D. (2012) Evaluation of selected medicinal plants extracted in different ethanol concentrations for antibacterial activity against human pathogens. *J Med Active* Pl 2012 ; 1(2): 60-68.

Wijesekera, R.O.B. (ed.) The medicinal plant industry. CRC Press, Boca Raton; 1991

Yekeen, M, Ajala O, Alarape A (2014) Antifungal activities of *Citrus sinensis* seed oil against *Lentinus sajor-caju*. Adv Appl Sci 2: 397-409.

Zolfaghari B, and Ghannadi, A. (2000) Research in Medical Sciences. Journal of Isfahan University Medical Science. 6, 2000, 1-6.

CHAPTER 10

Contribution of Microbial Inoculants in Sustainable Maintenance of Human Health, including Test Methods and Evaluation of Safety of Microbial Pesticide Microorganisms

by

Omena Bernard Ojuederie, Chinenyenwa Fortune Chukwuneme, Oluwaseyi Samuel Olanrewaju, Modupe Ayilara, Tope Taofeek Adegboyega, and Olubukola Oluranti Babalola*
Food Security and Safety Niche Area,
Faculty of Natural and Agricultural Sciences,
North-West University,
Private Mail Bag X2046, Mmabatho 2745, South Africa
*Correspondence email: olubukola.babalola@nwu.ac.za

Abstract

The continuous application of chemical fertilizers and pesticides for crop improvement is a major threat to human health and the environment. Depending on the level of exposure, it could cause neurological, carcinogenic, gastrointestinal, respiratory or reproductive effects and, in some cases, death. Since pesticides and herbicides have detrimental impact on the well-being of man, it, therefore, becomes essential that a cost-effective and eco-friendly approach towards food production should be considered by various governments and policymakers, to safeguard human health. Thus, there is an urgent need for nations to migrate towards using a bio-based approach in crop management and production through the use of microbial inoculants. This review discussed the potential benefits of microbial inoculants in crop production while sustaining human health, as well as the mechanisms utilized in biocontrol of phytopathogens and the bottlenecks involved in its application and possible commercialization. Increasing interest in microbial inoculant bioentrepreneurship necessitates more investment in microbiome research. We also elaborated on the prospects in the use of microbial inoculants to safeguard human health.
Keywords: Bioentrepreneurship; Bioremediation; Food safety; Nano-formulations; Microbial pesticides; Human health

Introduction

Food and health are two important factors important to human existence and sustainability. Availability and access to highly nutritious diet can help to maintain balanced health. With the postulated increase in human population by 2050 and a decrease in land availability for agriculture, there is increased importance and urgency for improving food availability and accessibility. Agriculture has always played an essential role in the food and nutritional well-being of man. There will always be the need to improve food production since the global population keeps on increasing at a startling rate, thus the need to explore more efficient ways of enhancing food production to meet the daily needs of the populace. In a bid to increase crop yield, farmers often apply chemical fertilizers as well as pesticides often at high rates, to get rid of pests and diseases and improve the nutrients in the soil. This, however, is an unhealthy practice since these agrochemicals degrade the environment and eliminate a number of beneficial insects in the food web, in addition to been hazardous to human health.

The soil as a resource needs to be managed appropriately, not only because we depend on it to promote food security, but also because it is an essential part of our ecosystems (Me'lanie and Filion, 2013). Coupled with increased disease occurrence in humans, runoff from the fields into water bodies leads to eutrophication with the resultant decrease in dis-solved oxygen and death of aquatic life, due to disturbances in the hydro-biome from toxic chemicals. In the early ages, farmers practiced mixed cropping on the same piece of land, which aided the control of insect pests as some pests from a particular crop fed on others. Nonetheless, they still had problems with weeds and crop yield. The advent of the Green Revolution in the 1930s doubled food production, and emphasis was placed on mono-cropping. However, mono-cropping reduces the ability of the soil to naturally eliminate pests and replenish nutrients (Alori *et al.*, 2017). The Green Revolution witnessed an in-creased use of chemical fertilizers and pesticides to ensure that the crops were pest and disease-free, and the high yields maintained. During the Second World war, there was a remarkable increase in the development of synthetic pesticides such as DDT (chlorinated organic pesticide) aldrin, dieldrin, endrin, parathion, and 2,4-D, to boost agriculture not minding the potential risk to human health and the environment (Bernardes *et al.*, 2015). Moreover, DDT was also used as a treatment against vectors such as the female anopheles' mosquito carrier of the *Plasmodium falciparum* parasite that causes malaria in Africa, as well as other pathogen carrying insects. More farmers applied high levels of DDT on their farms on a broad scale to get rid of these agricultural pests affecting their sole-cultivated variety of crops. Although pest reduction was achieved with increased crop yield, it had a negative impact on the food chain, consequently leading to the death of fishes, birds and mammals whose prey were contaminated with the pesticide.

In Africa, the subsistence farmers are often uneducated and not well kitted to spray pesticides and herbicides on their farms. This exposes them to a lot of health hazards such as respiratory diseases, and other ailments after consumption of foods containing residues of such pesticides. Human health is of utmost importance to the survival of humanity; hence stringent measures should be in place to ensure that the activities of man do not interfere with his health. To this end, the use of microbial inoculants as agents of biocontrol and biofertilizers for plant growth enhancement is gradually becoming the method of choice for enhanced food production.

In recent times, much emphasis has been given to the utilization of plant growth-promoting rhizobacteria as microbial inoculants towards improving plant health and nutrition, and as biopesticides or biocontrol agents to control disease-causing phytopathogens (Me'lanie and Filion, 2013). The microbial inocula are eco-friendly compared to the conventional chemicals, offering an added advantage of being renewable sources of nutrients that restores the fertility and quality of the soil (Timmusk *et al.,* 2017). Besides wading off phytopathogens, these tiny organisms also relieve plants from abiotic stresses such as heavy metal contamination, drought, excessive temperature or salinity which negatively affects crop production (Timmusk *et al.,* 2015; Timmusk *et al.,* 2014; Sharma *et al.,* 2016; Bharti *et al.,* 2016; Timmusk *et al.,* 2017). Some biological control methods can actually prevent economic damage to agricultural crops.

Biopesticides that contain microorganisms as their active ingredient are the essential substitute for chemical pesticides in shielding crops from pests, pathogens, and weeds (Moazami, 2011). Utilization of microbial inoculants in biopesticides production will greatly reduce the incessant application of conventional pesticides at high doses, which in the long run, has an adverse effect on man and the environment. Biopesticides could either be microbial pesticides which contain microbes as the active ingredient and controls an array of pests; for example, mycopesticides produced from entomopathogenic *Beauveria bassiana* and *Metarhizium anisopliae* have been used in the control of nymphs of the blacklegged tick, *Ixodes scapularis*, or plant-incorporated protectants which are pesticidal substances that plants produce from genetic material that has been added to the plant using genetic engineering techniques often from microbes (introduction of the Bt. pesticidal protein from *Bacillus thuringiensis* to maize, cowpea, and cotton).

Despite the benefits derived from microbial inoculants in enhancing food production and indirectly human well-being, it is limited by certain factors which will be discussed in this review. There are safety issues involved in the use of microbial inoculants as biopesticides, bioherbicides, and biofertilizers, which should be adhered to during formulation, trials, and commercialization. Thus, risk assessment studies are indispensable to confirm that the microbial-based inoculants are eco-friendly with no possible harm to human health. In this review, we elaborated on the beneficial use of microbial inoculants as biopesticides and biofertilizers, towards enhancing food production and safeguarding human health. The potentials for bio-entrepreneurship and commercialization, as well as methods used for formulation and testing, were also highlighted.

Microbial inoculants in sustainable maintenance of human health

Microbial inoculants, also known as bioinoculants are mainly living microorganisms that are usually applied to either agricultural soils, seeds or plant surfaces to promote plant growth (Mohammadi and Sohrabi, 2012; Bashan *et al.,* 2014). These organisms form beneficial relationships with plants and also assist in the development of plants. Microbial inoculant effects on plants vary according to the microbes that are involved. Inoculants make use of various mechanisms to improve plant health; some are directly elicited effects while some are elicited indirectly (Olanrewaju *et al.,* 2017). Some microbes elicit direct effects on plant growth promotion by synthesizing 1-aminocyclopropane-1-carboxylate (ACC) deaminase which reduces ethylene stress in plants, fixing atmospheric nitrogen, enhancing nutrient uptake in plants via phytohormone synthesis and phosphate solubilization, thereby reducing the rate of demand for chemical fertilizers (Alori and Babalola, 2018; Olanrewaju *et al.,* 2017). Other microbes promote plant growth indirectly by suppression of plant pathogens through resource competition, and the production of compounds that inhibit the growth of or possibly kill the pathogens such as siderophore production, induction of systemic resistance, antibiotics production, bacteriophages, production of cellulose wall degrading enzymes, and cyanogenesis (Alori and Babalola, 2018; Olanrewaju *et al.,* 2017). Certain fungi can actively transfer immune activation signals from stressed plants to neighboring plants. Microbes also assist in the protection of plants against environmental stresses like flooding, drought, extreme temperatures, and salinity, as well as assist in the reduction of the negative effects of pollution (Meena *et al.,* 2017). Bioinoculants can be classified as bacteria and fungi and subcategorized as intracellular and extracellular for bacteria; and root-associated fungi (RAF), ectomycorrhizas (EcM), and arbuscular mycorrhizas (AM) in the case of fungi (Adesemoye *et al.,* 2009; Patil and Solanki, 2016).

Utilization of microbial inoculants in healthy crop production

The incessant use of chemical-based fertilizers in agriculture has resulted in numerous environmental challenges, including decreased soil quality, pollution of water bodies, global warming and imbalance of soil microbial flora (Usta, 2013). Beating these challenges requires the application of microbial inoculants, which have been discovered to be the most suitable and effective alternative (Alori and Babalola, 2018). Biofertilizers are majorly gotten from bacteria and fungi, which are majorly isolated from the soil. Some are gotten from the rhizosphere, and some are from the endophytes of plants (Babalola, 2010). These organisms greatly assist in the improvement of soil fertility and quality and, therefore, provide natural means of nutrient mobilization in the soil (Ahemad and Kibret, 2014). Plant growth-promoting microbes generally consist of different naturally occurring microorganisms, which, when inoculated to the soil, improves soil health, soil microbial diversity, soil physicochemical properties, plant growth, development, and productivity. Many articles have published great progress and landslide achievements in the use of these microorganisms as biofertilizers and disease control agents (Alori *et al.,* 2019; Cheng *et al.,* 2012). Crops such as maize (Adjanohoun *et al.,* 2011), rice (Amprayn *et al.,* 2012), chickpea (Gopalakrishnan *et al.,* 2015), millet (Umesha *et al.,* 1998), wheat (Majeed *et al.,* 2015), soybean (Kumawat *et al.,* 2019), and other essential food crops have been improved using microbial inoculants. In recent studies, microorganisms have been applied successfully to improve soil fertility and crop yield (Bhattacharyya and Jha, 2012; Pieterse *et al.,* 2014). Phosphorus and nitrogen are essential elements of plant growth; therefore, to improve these nutrients in plants, the use of microbial inoculants formulated with phosphate solubilizing and nitrogen-fixing microbes is encouraged. Several soil bacteria, particularly those of *Bacillus*, *Streptomyces*, *Agrobacterium*, *Trichoderma*, *Pseudomonas*, *Penicillium*, *Serratia* sp. etc., are able to solubilize phosphate and release organic acids from adjacent soils (Alori and Babalola, 2018; Patil and Solanki, 2016; Ahemad and Khan, 2011; Valverde *et al.,* 2007).

Wani and Khan (2010) reported an improved protein of up to 86%, dry weight, grain yield, and the number of nodules in chickpea (*Cicer arietinum*) that were inoculated with *Mesorhizobium* sp. RC3 as compared to control plants, as well as more nitrogen contents in treated plants. Gholami *et al.* (2009) and Braud *et al.* (2009) also recorded significant improvement in maize growth by the nitrogen-fixing bacteria *Azospirillum brasilense*, compared to the controls. Nitrogen-fixing bacteria have also improved the growth of several plants including lentil (Ahemad and Khan, 2011), Soybean (Gupta *et al.,* 2005), common bean (Remans *et al.,* 2008), green gram (Wani *et al.,* 2008), pea (Ahemad and Khan, 2011), *Lupius leteus* (Dary *et al.,* 2010), etc. A study by Habibi *et al.* (2011) showed that mixed strains of biofertilizers and half dose of organic fertilizer resulted in the highest oil and grain yield in medicinal Pumpkin. It was also suggested that biofertilizers could take the place of the 50% phosphorus and nitrogen fertilizers needed as they improve the utilization of available phosphorus and nitrogen provided, as well as reduce the number of chemical fertilizers applied. In this way, environmental pollution as a result of incessant chemical fertilizer applications on plants is prevented, leading to the sustainable improvement of human and animal health. The highest growth, height, and grain yield was observed in Canola plants inoculated with phosphate-solubilizing bacteria, *Trichoderma* sp., and farm-yard manure, compared to the control (Mohammadi, 2010). Microbial inoculants increase plant nutrition, help to control pathogens, and ultimately improve plant health, thereby sustaining human health through the availability of safe and healthy food production (Adesemoye and Egamberdieva, 2013).

Microbial inoculants as an environmentally friendly alternative to conventional pesticides and fertilizers

Microorganisms in the soil aid in the decomposition of organic matter and soil nutrient recycling (Adesemoye *et al.,* 2009). They are also able to establish relationships with the roots of plants and inhabit both the interior and exterior plant parts, providing certain benefits to the plants. These benefits may include resistance to plant diseases and insects, heat tolerance, and drought tolerance (Pieterse *et al.,* 2014). The desire to increase plant yield for use by the ever-increasing population has led to the widespread use of chemically formulated pesticides and fertilizers and in very high amounts. Too much use of these pesticides and chemicals has adversely affected the atmosphere and directly or indirectly causing serious health issues for humans and animals (Bhattacharyya and Jha, 2012). The detrimental effects of the high application of chemical fertilizers on living organisms and degradation of the environment, though with improvement in food production, were reiterated by Suyal *et al.* (2016). The residues of these chemicals were transferred to man through the food chain. It took a period of 35 years before it was banned in the United States due to research findings by Müller's that it was detrimental to human health and the environment. After years of banning DDT, some children were deformed at birth as the toxic effect of the chemicals was transferred from mother to child before delivery. Despite the fact that some of these toxic chemicals have been prohibited in the United States and other European countries, several of them are still being applied to crops indiscriminately in different regions of the world, especially in the developingcountries (Baez-Rogelio *et al.,* 2017).

Microbial inoculants have, therefore, proven to be the effective and efficient solution to these problems and are seen by life scientists as the possible alternatives to the conventional agriculture that is chemically based (Patil and Solanki, 2016). In recent times, a lot of countries have adopted the use of microbial inoculants for the maintenance of food safety (Patil and Solanki, 2016). The use of microbial inoculants ensures the raising of unpolluted crops by utilizing biofertilizers, biopesticides, and bio-manures that control pests and pathogens attacks as well as aid in the provision of ideal nutrients to plants. Therefore, for better sustenance of human and animal health, the use of microbe-dependent techniques in agriculture to produce healthy crops as well as a safe and clean environment should be greatly encouraged.

The role of microbial inoculants in bioremediation

Pollution of the environment has drastically increased in many regions of the world as a result of industrialization and the misuse of natural resources (Brusseau *et al.,* 2019). These pollutants, including pesticides, solvents, surfactants, hydrocarbons, plastics, silicones, polycyclic aromatic hydrocarbons, etc. have led to the contamination of both water bodies and soil globally. Toxic metals are released into our environments in mine sites and on agricultural soils. These heavy metals can also find their way into water bodies, which are sources of drinking water to communities. By drinking this, toxic metals are taken into the body system of humans and animals, resulting in the alteration of DNA and protein degradation, which could later result in cancer and other genetic diseases. To remove these pollutants from the environment, the use of specific chemical, physical and biological methods that are specific to the pollutants and the soil properties are essential (Tripathi *et al.,* 2013). The traditional methods commonly used to remove chemicals from the environment include electrochemical treatment, precipitation, evaporation recovery, chemical reduction and ion exchange (Leitão, 2009). On the other hand, these methods present numerous disadvantages, including incomplete removal, high cost of recovery, generation of other toxic by-products, and high chemical and energy consumption. Bioremediation is capable of providing a cheaper alternative compared to other methods. Bioremediation is the process of using microbes to degrade organic contaminants both in the soil and water bodies. This is an alternative method of protecting the environment from the harmful effects of chemical fertilizers and has become an interesting topic in sustainable agriculture and biosafety programs. Thus, microbial inoculants are being used as mop-ups for these toxic metals and also have the potential to remove pollutants from water bodies and the atmosphere. They are also used to biologically transform organic substances like transforming propylene to epoxypropane and the production of chiral alcohols (Gai *et al.,* 2009;

Stępniewska and Kuźniar, 2013). Various studies have shown the importance and use of microbial inoculants for heavy metal remediation (Ayangbenro *et al.*, 2018; Fashola *et al.*, 2016). Human health is therefore enhanced and sustained by these beneficial microbial inoculants, which aid in the remediation of polluted environments.

Current trends in the use of microbial inoculants

One of the major roles of microbial inoculants is the ability to make nutrients available to plants. Phosphorus, nitrogen, iron, zinc, magnesium, and other minerals are made readily accessible for plant use by inoculants. They aid in the solubilization of phosphorus to be accessible by plants. In like manner, they also fix atmospheric nitrogen in legumes (Babalola *et al.*, 2017). Rhizobium is the major group of rhizobacteria that carry out this function (Dwivedi *et al.*, 2015). They assist in trapping atmospheric nitrogen and converting it to the form needed by the plants. Through the activities of siderophore, iron and other essential metals are made available to plants (Saha *et al.*, 2015). Siderophores are used to absorb these essential minerals, which are then used by the plants.

Currently, the major and most evolved use of microbial inoculants has come from the area of plant growth promotion either as biofertilizer or as a biocontrol agent. As discussed above, various mechanisms are applied by microbial inoculants to serve this purpose. Researches have a focus on the physiological improvement of plant growth. This has confirmed the ability of these inoculants to improve plant growth and many inoculants have been made into final products for use in this regard. This aspect has been substantially discussed above. Microbial inoculants have also been a source of antibiotics production. They make use of these antibiotics in pathogen control. Plant pathogens and human pathogens have been controlled using antibiotics from microbial inoculants (Lu *et al.*, 2018). Like antibiotics, microbial inoculants are currently one of the go-to areas when it comes to enzyme production. Industrially and medically important enzymes are presently gotten from microbial inoculants, especially inoculants formulated from the *Streptomyces* genus (Čihák *et al.*, 2017; Olanrewaju and Babalola, 2018).

Evaluation of safety of microbial pesticide

Pests are a major challenge hindering agricultural productivity. Pests could destroy crops on the field or cause post-harvest damage. Thus, the use of pesticides in preventing losses from agriculture is a common practice (Damalas and Koutroubas, 2016). Pesticides could be classified into synthetic and organic pesticides according to the source of their content. Synthetic pesticides are more commonly or generally used compared to organic pesticides.

Though they come with a lot of adverse effects. Humans, animals, soil, air, water bodies, as well as soil organisms are exposed to toxicity risks (Nawaz *et al.,* 2016). Due to these detrimental effects, organic pesticides came into limelight. Organic pesticides could be from plants and microbial origin. Microbial pesticides are highly degradable, and they pose less threat to the environment compared to the synthetic counterparts. This goes a long way in enabling sustainable development in agriculture. In the last few years, there has been a debate on whether microbial pesticides are safe or not, though contradictory reports are available on this. Where some researchers have proved that microbial pesticides pose less threat to the environment (Sarwar and Technology, 2015), some researchers have identified some potential threats (Nawaz *et al.,* 2016). The proposed threats include the accidental invention of a dangerous species and the introduction of resistant species. The search for more alternative microbes that could be used to counter the proposed problems will go a long way in enhancing the use of microbes as pesticides. Hence, it could be safe to say that microbial pesticides are effective, less toxic to the environment and are a better alternative to the use of synthetic pesticides.

Pests can also be classified according to the type of organism responsible for plant destruction. For instance, we can classify pests as insect pests, animal pests, weed, and microbial pathogen pests. Insect pests can further be classified into tubular sucking and chewing mouthparts based on their mouthparts (Flint, 2018). (Figure 1).

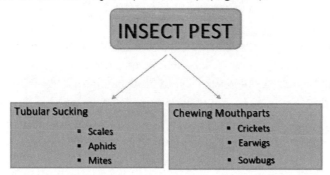

Figure 1: Classification of Insects according to mouthparts

Classification of Pesticides

Pesticides can also be classified according to the pest they control and according to their origin. Based on origin, pesticides could be synthetic or organic; this classification is based on the raw materials of the pesticides (Table 1). According to the pests they control, pesticides could be classified into rodenticides, bactericides, nematicides, miticides (acaricides), virucides, algicides, insecticides, avicides, fungicides, molluscicides, and herbicides (Uqab *et al.,* 2016).

Table 1: Differences between organic and synthetic pesticides

Organic Pesticides	Synthetic Pesticides
They are made from plant, animal, naturally occurring rock sources (copper, sulfur and lime sulfur) or microbial sources (Nawaz *et al.,* 2016; Flint, 2018; Ashishie and Ashishie, 2018).	Produces from organophosphorus, halogenated and carbate raw materials (Nawaz *et al.*, 2016).
They are highly specific in their mode of actions and have a slower rate of activity (Alori and Babalola, 2018).	They are always very active and have fast actions.
They are less toxic to other inhabitants of the environment (Ashishie and Ashishie, 2018).	They are toxic to organisms, soil, plants, animals, water bodies, as well as humans (Damalas and Koutroubas, 2016).
Organic pesticides are degradable and do not stay long in the environment; it is also difficult for them to be produced in large quantities (Sarwar and Technology, 2015). Therefore the chances of them being overused are slim.	Overuse and inappropriate use of these pesticides are capable of altering the genetic formation of existing pests, leading to the evolution of new species, which could be resistant to synthetic pesticides (Nawaz *et al.*, 2016).
They are highly degradable, and majorly do not have a persistent effect on the environment (Flint, 2018; Martínez *et al.*, 2018).	Their presence in food in quantities that exceed the acceptable limits leads to the rejection of food products during exportation (Damalas and Koutroubas, 2016).

Microbial Pesticides

The present-day agriculture has reflected the use of various advanced agricultural practices to ensure practical and environmentally friendly approaches towards sustainable agriculture. In the past, organic farming was used to produce high yields of crops with superior quality (Patil and Solanki, 2016). It was considered to be an ecofriendly way of promoting soil biological activity and biodiversity. However, in recent times, the use of biopesticides have replaced organic farming practices as the most accepted and environmentally friendly farming practice in evolving agriculture (Thakore, 2006). Biopesticides that are obtained from beneficial microbes that possess the ability to control pests are known as microbial pesticides and are used as a better and advanced alternative to manage plant diseases (Pérez-García *et al.,* 2011). Microbial pesticides are produced from nematodes, bacteria, protozoa, fungi, viruses, algae, and their derivatives (Nawaz *et al.,* 2016). The use of biopesticides in the management of pests in plants began after it was discovered that pesticides with chemical origin resulted in the development of resistance in insects which in the long run, affects the health of humans and animals, and raising environmental concerns (Anwer, 2017; Usta, 2013). Biopesticides have been successfully applied in agriculture worldwide, using members of the genera: *Pseudomonas, Agrobacterium, Streptomyces,*

Bacillus, *Trichoderma*, *Beauveria*, etc. Biopesticides can be used in Integrated Pest Management Programs to reduce the application of chemical-based pesticides, thereby, increasing crop yields (Alori and Babalola, 2018; Patil and Solanki, 2016; Trabelsi and Mhamdi, 2013).

Most of the research carried out on organic pesticides focused on insects and microbial pathogens. Organisms are capable of producing metabolites that drive away pests and insects feeding on the plant seeds, leaves, and fruits, thereby leading to death or reduced yield in plants (Olanrewaju *et al.*, 2017). Microbial pesticides have no toxic residual effect, they break down easily, and they promote sustainable development of agriculture. Furthermore, microbial pesticides are less toxic to non-target organisms and the environment (Nawaz *et al.*, 2016; Flint, 2018; Martínez *et al.*, 2018; Alori and Babalola, 2018). (Alori and Babalola, 2018) and are highly specific in eliminating target pests and insects, while indirectly enabling the survival of beneficial microbes in treated plants (Patil and Solanki, 2016). Furthermore, microbial pesticides, unlike conventional pesticides, maintain their efficacy when applied in small quantities, decomposing easily, thereby avoiding environmental pollution in addition to having lower exposure periods. These pesticides are safe to be applied to plants at any time because they do not leave any hazardous residue (Alori and Babalola, 2018).

These microbial pesticides are also generally innocuous to man, animal or other organisms except the target pests (Alori and Babalola, 2018; Anwer, 2017). Biopesticides are effective even in low safe to be applied to plants at any time because they do not leave any hazardous residue (Alori and Babalola, 2018). As good as this sounds, the disadvantage is that eggs that hatch after their application as well as new insects that come into the environments after this period, will not be affected (Flint, 2018). Presently, more than 3000 microbes that are capable of causing diseases in insects have been discovered (Nawaz *et al.*, 2016). It is, therefore, necessary to carry out more research to unravel novel organisms that can have the same functions, as this will help to make an alternative ready in cases where pests become resistant to present microbial pesticides. Microbial pesticides are specific in their mode of actions; each pesticide is specific for a particular pest (Nawaz *et al.*, 2016). The pest ranges from insects, plants, animals to other microbes. More than 100 species of bacteria have been identified to cause diseases or death in insects, out of which *Bacillus thuringiensis* (Bt.) is well known (Nawaz *et al.*, 2016). Over 1000 species of Viruses, 800 species of fungi, and 1000 species of protozoa have entomopathogenic effects (Nawaz *et al.*, 2016). There are majorly two groups of nematodes that attack insects; they are 15 species of *Steinernema* and 12 species of *Heterorhabditis* (Koul, 2011).

Bt. is a popular bacterial pesticide and it has been established to be environment-friendly, it is not toxic to aquatic insects, Arachnida, soil organisms, parasitoid, *Crustecea*, honeybees, Collembola, salamanders, predators, Mollusca, birds, humans and other mammals (Vimala Devi *et al.*, 2012). Humans lack receptor sites that bind to the toxin released by

Bt, also it degrades rapidly in the digestive systems of humans. Different strains of Bt are specific for various pests, and this is because each pest has a different endotoxin receptor sites on their gut wall. For instance, *Bacillus thuringiensis* var. *israelensis affects* larvae of flies (fungus gnats), *Bacillus thuringiensis* var. *san diego* affects beetles larvae (Colorado potato beetles and elm leaf beetles) while *Bacillus thuringiensis* var. *kurstaki* affects moths and butterflies caterpillars (Crickmore, 2006; Ruiu *et al.,* 2013; Sarwar and Technology, 2015).

Other bacteria (*Trichoderma viride* Pers., *Pseudomonas fluorescens*), fungi (*T. harzianum Rifai*), and viruses (*Migula Baculoviruses)* has also been found to be capable of managing plant diseases considering environmental safety (Vimala Devi *et al.,* 2012). Fungi such as *Paecilomyces, Isaria, Beauveria, Tolypocladium, Lecanicillium*, and *Metarhizium* have pesticidal effects, the species *Beauveria bassiana* has the ability to attack a wide range of arthropods (Sarwar and Technology, 2015). Protozoa have also proven to be active pesticides against insects. The protozoa *onidospora, sporozoa*, and subphrla are able to reduce the life span, activities and reproduction of insects (eg. Mosquitoes, corn borer, grasshoppers and Lepidoptera) (Sarwar and Technology, 2015). Asides the entomopathogenic microbes, microbes are also used against the microbial pathogen. These groups of microbes are referred to as plant growth-promoting microbes (PGPM). They produce antimicrobials, volatile compounds or outcompete with pathogens, thereby protecting the plants against diseases (Enebe and Babalola, 2019).

Mechanism of action of microbial pesticides

The organisms which are used as biocontrol agents can establish themselves in a population of the pests, thereby preventing generation to generation or in subsequent season (Nawaz *et al.,* 2016). The mode of action of each pesticide differs. Some microbes are inhibiting the growth, feeding and reproductive process of insect pests (Ashishie and Ashishie, 2018). Entomopathogens could enter the host insects through the guts or integument; after the invasion, they multiply, and this leads to the death of the host insect (Sarwar and Technology, 2015). Herbicides can be produced from microbial exudates. These exudates alter the defensive system of weeds and eliminate death (Mosttafiz *et al.,* 2012). For instance, the bacterium, *Bacillus thuringiensis* (Bt.), reproduce by producing spores along with endotoxins, which, if ingested by an insect, get attached to a receptor site on its gut wall and breaks down the gut lining (Sarwar and Technology, 2015). This allows the insect to get access to the bloodstream of the host, become virulent, and subsequently leads to the death of the host insect (Sarwar and Technology, 2015). Some viruses can also be transferred from one insect to another through mating or production of spores, which attaches, germinates and penetrates the cuticle of the host-pathogen (arthropods) (Sarwar and Technology, 2015).

Assessment methods for microbial inoculants

Microbial inoculants are major stakeholders in the biofertilizer and biopesticides indus-tries. There is a high demand for the production of effective soil microbial inoculants con-taining agricultural relevant qualities needed to boost agricultural production as safe and nutritious food demands will likely increase in the coming decades. However, care must be taken in the formulation of potential inoculants before they are commercialized for farmers' use. Utilization of these beneficial soil microbes for plant growth promotion while maintaining the stability of the environment is on the increase. Even though several bacte-ria and fungi naturally perform plant growth promotion and biocontrol of phytopathogens in the environment, the prospects for their successful application are still limited, since introduced cells do not always survive and perform well in the soil ecosystem (Van Elsas *et al.,* 1998). The application of biofertilizers and biocontrol agents in agricultural practice is still hindered by several factors such as standardization of the products, the quality of which can be affected the inoculant viability and persistence in soil, as well as the possi-bility of locating the inoculant in the environment (Malusá and Vassilev, 2014; Bashan *et al.,* 2014; Canfora *et al.,* 2016). Assessment of the persistence of inoculants and their trace-ability in the soil is essential to the improvement of their efficacy, as it gives us a mecha-nistic understanding of their behavior in the crop environment, and to modify the method of application (Torsvik and Øvreås, 2002; Fierer *et al.,* 2003; Canfora *et al.,* 2016).

Changes in the structure of the soil microbiome may also be altered when the soil is inoc-ulated, or seed bacterization is carried out. This is important, especially as regards the safety of the environment when these microbes are introduced. Before these soil microbes are utilized either as growth-promoting, stress resistance or biocontrol agents, several steps are taken to identify the most suitable isolates to be used. In this regard, the culture-de-pendent and culture-independent methods have been used over the years, each with its limitation and successes (Figure 2). For the culture-dependent method, the microbial iso-lates are obtained from the rhizosphere soil samples or if endophytes from the tissues of the host plants. The isolates are cultured on defined medium or selective medium after serial dilution to obtain pure cultures before been subjected to biochemical tests and mi-croscopy to identify the isolates. To develop PGPR as microbial inoculants, the PGPR should have the ability to grow well in culture, be easily propagated, have a positive influ-ence on plant growth and should be safe for humans and the environment (Martínez-Hi-dalgo *et al.,* 2018). After pure single colonies are obtained, isolates are selected based on their morphology and Gram staining results and then evaluated *in vitro* for plant growth-promoting, abiotic stress-tolerant or biocontrol traits; (phosphate solubilization, ACC de-aminase activity, siderophore production, indole-3-acetic acid production, ammonia pro-duction, nitrogen fixation, exopolysaccharide production, and hydrogen cyanide production). The best performing isolates are subjected to DNA extraction and PCR amplification using 16SrRNA universal bacteria primers or ITS primers in the case of fungi, before being sent for Sanger sequencing.

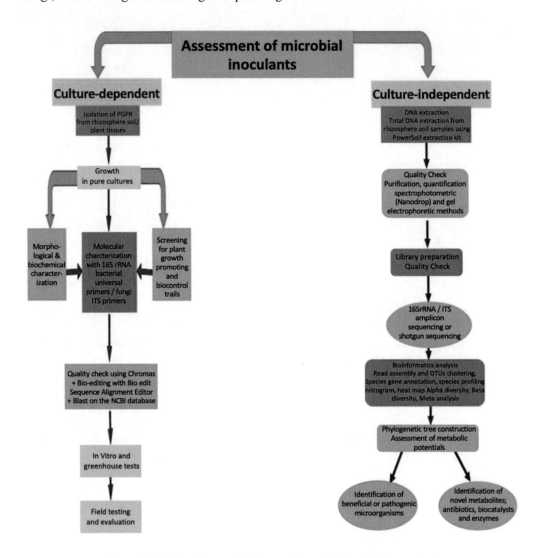

Figure 2. Methods used in the testing of microbial inoculants

In some cases, selected strains are also screened using specific primers for these traits to determine if the genes are amplified when run in a thermocycler using a defined cocktail. However, the culture-dependent method has the disadvantage of not being able to capture the un-culturable viable microbial cells in the soil and only attracts the culturable microbes (Van Elsas *et al.*, 1998). It is also noteworthy that the isolates to be applied as biofertilizers or biocontrol agents should be evaluated for environmental and human safety. To this end,

Vílchez *et al.* (2016) developed a biosafety test method for plant growth-promoting bacteria (PGPB) environmental and human safety index (EHSI) which proposed a scoring system to assess the safety of potential isolates for PGPB/PGPR within the limitations of the different assays they used. They made use of two risk groups. Risk Group 1 had *Pseudomonas putida* KT2440 and *Burkholderia cepacia* CC-Al74 as Risk Group 2 PGPR. *Pseudomonas aeruginosa* PA14 was used as Risk Group 2 representative for some animal tests (Tan *et al.,* 1999). Furthermore, strains *Rhizobium legominosarum* IABRL05, *Pseudomonas fluorescens* IABPF05, *Bacillus subtilis* IABBS05, and *Azotobacter vinelandii* were used as PGPB and *Serratia marcescens* 615 (Almaghrabi *et al.,* 2013), *Serratia entomophila* A1 (Johnson *et al.,* 2001), *Serratia proteamaculans* 28151 (Bai *et al.,* 2002), and *P. aeruginosa* PA14 as potential PGPB and pathogenic strains were included in order to validate the Index. A scale of values from 0-100 was made for the EHSI (Figure 3).

Values at a cut-off score of 50 ± 0.5 indicated the need for further safety tests before the potential strain could be considered safe for use bioinoculants whereas, at higher values, it indicated that the bacterial strain has a greater chance of been safe for use as a bioinoculant (Vílchez *et al.,* 2016). The EHSI conducted was based on tests of mortality (M) which is the major factor utilized in the determination of the pathogenicity of bacterial strains (Navas *et al.,* 2007), reproduction (R), and development (D) of target organisms (Vílchez *et al.,* 2016).

The persistence of inoculated microbes in the soil can be detected using PCR-based methods which could be used to monitor and identify the inoculated strains from the indigenous microbes in the soil (Castrillo *et al.,* 2007; Savazzini *et al.,* 2008; López-Mondéjar *et al.,* 2010; Canfora *et al.,* 2016). However, DNA fingerprinting methods are mainly qualitative in nature. Advancement in molecular technologies has offered different methods of detection of microbial inoculants used in plant growth promotion, which is a culture-independent method. The DNA of the soil is extracted, and specific DNA markers employed. Recently, Canfora *et al.* (2016) developed a culture-independent method based on simple sequence repeat (SSR) markers and qPCR for the direct, discriminant and simultaneous detection and quantification of 2 bioinoculants in soil using multilocus simple SSR genotyping.

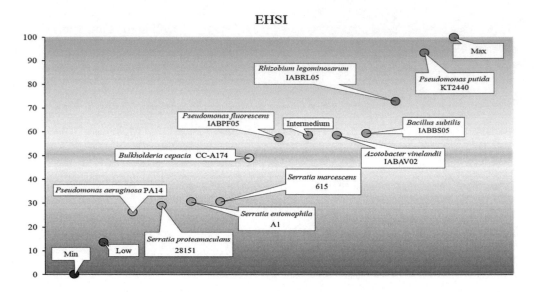

Figure 3 Environmental and human safety index (EHSI). Scores in the green zone show that the strains can be considered safe for use as a PGPB if they appear in the green zone. Additional tests are required for scores in the red zone before the strains can be considered safe as bioinoculants while strains in the yellow zones are in a transition region of uncertainty. All values presented are the mean and standard deviation of three measurements. (Source: Vílchez *et al.* (2016).

With this protocol, the authors were able to trace and quantify two entomopathogenic fungi, *Beauveria bassiana* (Bals.-Criv.) Vuill. and *B. brongniartii* (Sacc.) Petch used as biocontrol agents of *Melolontha melolontha* L. (European cockchafer) in field experiments. The metabolic potentials of the soil microbiome can also be harnessed by using high-throughput sequencing technologies now available such as Illumina sequencing or shotgun amplicon sequencing. Scientists are better able to determine the functionality of the different organisms in the soil and their metabolic potentials as it relates to plant growth promotion and biocontrol of phytopathogens using metagenomics approach.

Challenges and environmental effects associated with microbial pesticides
A number of challenges are currently faced with the use of microbial pesticides, these range from the inability to produce in large scale, slow action, to high specificity in their mode of action (Sarwar and Technology, 2015). To solve these challenges, the use of recombinant technology has been adopted (Karabörklü *et al.,* 2017). Though a number of challenges have been proposed against this, for instance, Ashishie and Ashishie (2018) suggested that recombinant technology might lead to the accidental invention of a

dangerous strain of microbe, which can be very problematic. Although this has not been scientifically proven, it will be necessary to carry out more researches to ascertain this. If this is ascertained, the exemption of the problematic strains as microbial pesticides should be done, and the bioremediation of any effects they might have caused should also be carried out. The proposed problematic strains should, therefore, not be an excuse to neglect the potential of the very useful microbial pesticide strains. Therefore, it could be safe to state that microbial pesticides are safe to use until research proves otherwise.

It has also been reported that the use of recombinant technology could also lead to the transmission of virulence traits (Keen, 2012) and antibiotic resistance in bacteria species due to horizontal gene transfer (Koonin *et al.*, 2001; Kay *et al.*, 2002; Gyles and Boerlin, 2014; Ashishie and Ashishie, 2018). Microbes are very diverse in nature as regards their functions and abilities. Hence, the discovery of more organism which can naturally perform the traits that are sought using recombinant technology will be important. The effects of microbial inoculants are often not easily observed as chemical herbicides or fertilizers (Babalola and Glick, 2012a). Furthermore, microbial experiments that involve the development of plant-incorporated protectants must be carried out under containment to prevent issues regarding possible ecological effects of transgenic microbe releases (Babalola and Glick, 2012a). Also, the indiscriminate use of microbial pesticides should be prevented to avoid resistance of pests to microbial pesticides. Furthermore, more researches should be carried out on the simultaneous use of more than one organism having different desired traits and activities.

Role of microbial inoculants in food processing

In food processing industries, microbial inoculants play major roles in enhancing the nutritional values and properties of foods, including the texture, taste shelf-life, and aroma (Nissar *et al.*, 2017; Sundarraj *et al.*, 2018). They also play roles in food preservation and fermentation (Borneman *et al.*, 2013), as well as in the production of products like dyes, enzymes, flavoring agents, vitamins (Vitorino and Bessa, 2017; Dikshit and Tallapragada, 2018; Kang *et al.*, 2012), and key compounds used in the pharmaceutical industries (Vitorino and Bessa, 2017). Some microbes can produce specific flavoring agents and polyunsaturated fatty acids, amino acids like glutamic acid and lysine, and complex carbohydrates that are commonly used in food formulations. Using microbial inoculants in the processing of food promotes the safety, yield, product quality and consistency of food products. Studies have reported the effective use of microbes in food and drug processing industries. For instance, Mojsov (2016) reported the production of citric acid by *Aspergillus niger* for the preservation of food. Also, *Aspergillus* sp. is used to produce alcoholic beverages (Alori and Babalola, 2018). Microbial inoculants can be an eco-friendly alternative to the chemicals used in food processing because they offer more advantages over their chemical counterparts. Some advantages of microbes in food processing include being able to be genetically modified, able to be produced on a large scale in industrial fermenters, lower costs of production, microbial enzymes yield fast culture development than those from animals and vegetables (Lipkie *et al.,* 2016).

Advantages and disadvantages of microbial inoculants

Microbial inoculants offer a lot of advantages to human health. Some of these advantages include:

- Increased food productivity.
- Improved nutrient cycling and environmental detoxification. When inoculants increase the survival rate of crop plants, they indirectly enhance environmental cleansing and nutrient cycling. Toxic air pollutants are being recycled by plants, and oxygen is released through photosynthesis and respiration processes. This, in turn, improves human survival.
- Contaminants in soil reduce soil availability for agriculture. The use of microbial inoculants has been successfully applied to address this issue, although work is still ongoing for full implementation. This is a positive step in the right direction towards reclaiming back the land that have been lost to heavy metal pollution as a result of human activities.
- Microbial inoculants are good sources of antibiotics for various disease eradication. Most antibiotics and vaccines are synthesized by these microbes (Olanrewaju and Babalola, 2018).

Despite the advantages highlighted above, inoculants could be disadvantageous. Some of these disadvantages are:

- Development of antimicrobial resistance. Resistance to antibiotics by pathogens is rapidly increasing. Pathogens are adapting and modifying their physiology and metabolism to adjust to the antibiotics being used to treat them (Olanrewaju and Babalola, 2018). As a result, antibiotics that usually work have lost their efficacy. In some cases, combinations of antibiotics have to be used before they can be effective. These have led to great concern among some researchers that the continuous use of microbial inoculants might increase antibiotic resistance in pathogens.
- Concerns are also being raised about the safety of microbial inoculants. Unethical activities can accidentally lead to the release of inoculants that are not environmentally safe.
- Most studies do not confirm the establishment of these inoculants on the field. Further work still needs to be done in this regard.

Limitations/bottlenecks

Acceptance is one of the limitations of microbial inoculants. The same issue with genetically modified crops (GMO) crops is also affecting inoculant acceptability. Many see microbes as pathogens and do not agree that they can be of any good purpose. Making people change their belief is a little bit difficult, although a lot of progress has been made concerning this. Another limitation is the lack of funding for the necessary technology needed for thorough research. Reduced price of next-generation sequencing has led to tremendous improvement in microbial inoculant studies, but the majority of researchers still cannot afford the cost. Funding by various organizations and governments is presently not enough as most funding is directed towards other aspects of agriculture. Getting enough funding for microbial inoculants for researchers especially those in the developing world is a big challenge.

Microbial Inoculants bioentrepreneurship

Amidst global projected population increase, regions such as sub-Saharan Africa and South-east Asia characterized by extreme environmental conditions and burden of the high population may be the worst hit. Hence, the need to explore microbial inoculant use for food increased production and environmental sustainability, as suggested by Sessitsch *et al.* (2018).

Microbial inoculants have proven to possess several plant growth-promoting traits such as control of microbial infections, replenishment of soil nutrients and other plant stimulating attributes (Babalola and Glick, 2012b). Besides, the appropriate maintenance of the soil is crucial for meeting the growing need for agricultural produce in the face of the global population and climate change. Active microbes that are present in biofertilizers may serve as a viable alternative approach to the strategies of increasing food production sustainably without necessarily interfering with the natural ecosystems and human health as currently being witnessed with the use chemical fertilizers (Alori *et al.*, 2017). Agricultural yield loss due to abiotic stress, climate change scenarios, development of resistance to pathogens and pests against chemical treatments are some of the drivers of microbial inoculants bioentrepreneurship. Globally, several companies have emerged using microbial inoculants for plant beneficial activities (Pavela and Benelli, 2016; Marrone, 2014; Yakhin *et al.*, 2017).

In 2014 alone, more than 800 million USD was realized from the global biopesticides market (Yakhin *et al.*, 2017). About 3 billion USD was made from the global biocontrol market, with projections of over 10 billion USD in six years from now. More than 50% of these were microbial products with a projected annual rate of over 15 % expected before 2020 (Caradonia *et al.*, 2019). Outside India and Asia, there are more than 300 microbial inoculant companies as reported by Sessitsch *et al.* (2018), with surplus returns. Due to the increased interest in microbial inoculant bioentrepreneurship, most major global seed companies have made substantial investments in microbiome research, and they are currently exploring new commercial opportunities as an alternative to agrochemicals. Small and medium enterprises can benefit from the global biocontrol market as individuals/groups can be trained on local production and utilized to increase income for smallholder farmers and also improve poor soils predominant in SSA.

In South America, for example, *Azospirillum* has been shown to possess flourishing potentials. Different levels of scientists have done field testing and recommended known regimens. Consequently, several products containing *Azospirillum* strains with over 50 firms with a special focus on maize, wheat, and soybean cultivation have been commercialized (Cassán and Diaz-Zorita, 2016). Furthermore, the total microbial inoculant production is projected to be over 200 Billion USD in 2023. This shows the extent of the financial inflow derivable from inoculant bioentrepreneurship (Prakash *et al.*, 2013; Singh *et al.*, 2016). Consequently, the overall contribution of microbial inoculant bioentrepreneurship cannot be overemphasized (Aleti *et al.*, 2015; Hassani *et al.*, 2018).

Commercial outlets/industries/products

Globally, several companies, including but not limited to BayerCrop Science, Agrium US Inc., Plant Health Care, Inc, Biosynthetica Pty. Ltd and IITA BIP are producing, distributing and marketing various brands. For a rapid increase in the market distribution of microbial inoculants, aspects of cost-effectiveness, efficiency on the application, as well as a high return on yield and storage flexibilities must be well considered (Babalola and Glick, 2012b). Tables 2-4 describes selected commercially available microbial biofertilizers, mycorrhizal inoculants, and ectomycorrhizal fungi inoculants currently being marketed across the globe while selected pictures are shown below.

Nodumax™

Aflasafe™

MycoTree®

MycoApply®

Table 2: Commercially available microbial products

Product	Microbial constituent (s)	Attributes	Reference
Actinovate Plus	*Streptomyces lydicus* strain WYEC 108	Control of soil-borne diseases	Su and Marrone (2019)
Aflasafe	*Aspergillus flavus*	Reduces aflatoxin in maize and groundnuts	Gbadamosi and Kolawole (2017)
AgBio	*Streptomyces griseoviridis* strain K61	Reduces the effects of common pathogens such as *Fusarium* spp.	Goertz and Es-Sayed (2017)
AQ-10	*Ampelomyces quisqualis* M-10	Against powdery mildews	Khasa (2017)
AtEze	*Pseudomonas chloroaphis* strain 6328	Suppresses the growth of *Rhizoctonia solani* and *Pythium* spp	Özer and Coşkuntuna (2016)
Biofox	*Fusarium oxysporum*	Helps against *Fusarium moniliforme*	Sachdev and Singh (2016)
Bioplus-2B (pellet)	*Bacillus licheniformis* (DSM 5719) *Bacillus subtilis* (DMS 5750)	Probiotic feed additive	EFSA (2016)
Bioshiel	*Serratia entomophila*	Reduces the actions of soil-habiting grass grub larvae	Vandana *et al.* (2017)
BiotaMax	*Bacillus subtilis; megaterium; licheniformis; pumilu and laterosporus, Paenibacillus polymyxa, Trichoderma harzianum, T. viride, T. polysporum, T. koningii*	Produces plant growth hormones and processed nutrients more efficiently.	Vandana *et al.* (2017)
Companion	*B. subtilis*	Reduction of infections caused by *Rhizoctonia, Fusarium* and *Phytophtora.*	Jabs *et al.* (2016)
Custom B5	*Bacillus subtilis; Laterosporus; licheniformis; megaterium* and *pumilus*	Increases the performance of soil	Vandana *et al.* (2017)
Custom N2	*Paenibacillus polymyxa*	Enhances nitrogen nutrition in plants	Vandana *et al.* (2017)
EcoGuard	*Bacillus licheniformis* SB3086	Good in the control of fungal diseases	Rouabhi (2010)
JumpStart	*Penicillium bilaiae*	Support growth in the root, assist nitrogen fixed, it is low cost and greatly enhances seed growth.	Steckler *et al.* (2012)
Mycostop	*Streptomyces griseoviridis* Strain K61	Reduces wilt diseases	Keswani *et al.* (2016)
Nodumax	USDA 110	Boosts yield of soybean by 30-40%	Gbadamosi and Kolawole (2017)
Ovalis Rhizofertil	*Pseudomonas putida I-4163*	Stimulates plant growth and enhances soil performance	Celador-Lera *et al.* (2018)
PONCHO/ VOTiVO	*Bacillus firmus* mixed with clothianidin	Provide plant protection against insects and nematodes	Wilson and Jackson (2013)

Recharge	*Azospirillum brasilense* CD	Support the growth of roots and resuscitate degraded soils.	Babalola and Glick (2012b)
Rhapsody	*Bacillus subtilis*	Bacterial and fungal diseases	Taranta *et al.* (2018)
RhizUp	Rhizobium bacteria	Improves soil porosity	Bougoure *et al.* (2016)
YieldPlus	*Cryptococcus albidus*	Useful against postharvest fruit diseases	Droby *et al.* (2016)

Table 3: Selected commercially available mycorrhizal products

Product	Nature of mycorrhiza	Reference
AgBio-Endos	Endomycorrhizal inoculant	Hall *et al.* (2000)
AM 120	Microbial inoculant	Siddiqui and Kataoka (2011)
Bio/Organics	Endomycorrhizal inoculant	Akram *et al.* (2016)
Biogrow	Hydo-sol Endomycorrhizae	Madsen *et al.* (2014)
BioVAm	Mycorrhizal powder	Sukiman *et al.* (2015)
BuRize	VAM inoculant	Mishra and Arora (2016)
Cerakinkong	VA mycorrhizal fungi	Parihar *et al.* (2019)
Diehard	Endodrench	Reid (2013)
Endorize	Mycorrhizal product	Bati *et al.* (2015)
MYCOgold	AM fungi	Mohammed and Yusop (2016)
Mycor	Endo/ectomycorrhizae	Mirabelli *et al.* (2007)
Mycosym	Mycorrhiza vitalizer	Bhromsiri and Bhromsiri (2010)
Pro-Mix 'BX'	Endomycorrhizal fungi	Iheshiulo *et al.* (2017)
Rhizanova	Endomycorrhizae	Pal *et al.* (2016)

Source: Siddiqui and Kataoka (2011)

(Disclaimer: The authors declare no association with the selected products and manufacturers and those not listed here)

Table 4: Selected commercially available ectomycorrhizal fungi products

Product label	Manufacturer	Reference
BioGrow Blend	Terra Tech, LLC, Eugene, Oregon, America	Siddiqui and Kataoka (2011)
Ectomycorrhiza Spawn	Sylvan Spawn Laboratory, Inc. (China, Europe and USA)	Siddiqui and Kataoka (2011)
MycoApply	Mycorrhizal Applications, Inc. Oregon, America	Siddiqui and Kataoka (2011)
Mycobead	Biosynthetica Pty. Ltd, Western Australia	Siddiqui and Kataoka (2011)
Mycor Tree	Plant Health Care, Inc. Raleigh, USA	Siddiqui and Kataoka (2011)
MycoRhiz	Abbott Laboratories, Chicago, USA	Siddiqui and Kataoka (2011)
Somycel PV	INRA-Somycel S.A. France	Siddiqui and Kataoka (2011)

Source: Siddiqui and Kataoka (2011)

Most of the items listed in the above tables have been applied in the forms of liquids or dry formulations with huge successes recorded, but have challenges on some large scale applications. This is because a large amount is required for optimum performance of the inoculants (O'Callaghan, 2016). However, with renewed interest and partnerships among relevant stakeholders in Research and Development, these success-limiting challenges are surmountable.

Future prospects

The continuous application of chemical fertilizers and pesticides to enhance crop production is not a sustainable approach due to the devastating effect it has on human health and the environment. We have seen the great potentials of microbial inoculants as a veritable alternative to the conventional system. However, before microbial inoculants are formulated for use as biofertilizers, bioherbicides or biopesticides, the potential microbe has to be characterized phenotypically as well as evaluated for plant growth-promoting traits and most importantly to ascertain if they are non-pathogenic. It is possible that some strains of the same bacteria species may be mutualistic to plants, while others may have some pathogenic tendencies. Keep in mind that some of these beneficial bacteria may be phylogenetically related to species that are virulent or opportunistic human pathogens detrimental to human health (Berg *et al.*, 2005; Martínez-Hidalgo *et al.*, 2018) A good example is the genus *Pseudomonas* which has been widely used as microbial inoculant for crop production; *P. fluorescens, P. putida, P. pseudoalcaligenes, P. stutzeri, and P. putrefaciens* although these strains are also found in clinical samples (Martínez-Hidalgo *et al.*, 2018). For example, *Pseudomonas aeruginosa,* which has been used as biofertilizers, also has strains that cause respiratory tract infection in individuals with cystic fibrosis. It is often not easily distinguishable from the mutualistic species using culture-based methods. Hence the use of the culture-independent method is becoming useful in the characterization of total organisms found in a given soil sample. This is achievable by using the metagenomic analysis, either 16S amplicon or shotgun approach. This will enable comparative and functional analyses of the organisms found in the soil, of which bacteria are the majority, and give us a better understanding of their gene structure. Moreover, other omic technologies such as proteomics, transcriptomics, and metabolomics will give reliable information on the type of proteins produced under different environmental conditions, which could enhance the use of the microbe in biotic or abiotic stress tolerance in plants. Additionally, the genes up-regulated under diseased conditions could enable the identification of strains with potential as biocontrol agents. Metabolomics approach is essential in having a mechanistic understanding of the metabolic pathways elicited by microorganisms. Hence, it will be much easier to differentiate pathogenic from non-pathogenic microbes, permitting only the beneficial ones to be used for sustainable agricultural productivity and sustain human health. With the omic technologies now in place, bacteria strains that are able to degrade pesticides that contaminate the soil can be identified and utilized in bioremediation or phytoremediation of contaminated agricultural land. By so doing, the transfer of chemical residues to plants consumed by man will be greatly reduced, thereby promoting the healthy well-being of man. Furthermore, with the advancement in nanotechnology, there is likely to be a great reduction in food contamination as the technology will aid early detection of disease-causing phytopathogens before they spread and also improve the efficacy and stability of microbial inoculants as biofertilizers, biopesticides or bioherbicides through the process of nano-encapsulation. However, these nanoparticles may also affect the functionality of the soil microbiome hence its activities in a given environment must be evaluated before they are utilized as components for biofertilizers or biocontrol agents.

Conclusion

Environmental pollution arising from the use of synthetic pesticides and chemical fertilizers has called for the incorporation of organic and sustainable practices in Agriculture. Microbial pesticides have been used as an alternative to synthetic fertilizers for years. They are active and less toxic, which has promoted their use compared to their synthetic counterparts. Though some environmental threats are being proposed against the use of microbial pesticides, their use as an alternative to synthetic fertilizers and pesticides is on the increase. The generation of resistant, as well as virulent strains, are seen as major challenges. To avoid and ameliorate these challenges, microbial pesticides should not be used indiscriminately, and more research studies should be carried out to discover microbes that can be used as alternatives in case a species or strain becomes problematic, or a pest become resistant to it.

Proper handling of microbial pesticides should also be ensured as mishandling can lead to contamination by pathogenic organisms or render the product inactive. In addition, more research should be carried out to discover other microbial species that have the potentials of performing better than the presently available ones. Since presently, there are no reports of microbial pesticides that pose environmental threats, it's necessary to register products to ensure proper regulations and quality control of microbial pesticides before being released into the commercial market. This will help to prevent future occurrences of environmental pollution arising from the use of microbial pesticides. Finally, more researches should be carried out to evaluate the potency and compatibility of mixed microbial exudates or microbes as pesticides, as it will help to ensure a wide spectrum of action

References

Adesemoye A, Torbert H, Kloepper J (2009) Plant growth-promoting rhizobacteria allow reduced application rates of chemical fertilizers. Microb Ecol 58 (4):921-929

Adesemoye AO, Egamberdieva D (2013) Beneficial Effects of Plant Growth-Promoting Rhizobacteria on Improved Crop Production: Prospects for Developing Economies. In: Maheshwari DK, Saraf M, Aeron A (eds) Bacteria in Agrobiology: Crop Productivity. Springer Berlin Heidelberg, pp 45-63. doi:10.1007/978-3-642-37241-4_2

Adjanohoun A, Allagbe M, Noumavo P, Gotoechan-Hodonou H, Sikirou R, Dossa K, GleleKakaï R, Kotchoni S, Baba-Moussa L (2011) Effects of plant growth promoting rhizobacteria on field grown maize. J Anim Plant Sci 11:1457-1465

Ahemad M, Khan M (2011) Toxicological assessment of selective pesticides towards plant growth promoting activities of phosphate solubilizing Pseudomonas aeruginosa. Acta Microbiol Immunol Hungar 58 (3):169-187

Ahemad M, Kibret M (2014) Mechanisms and applications of plant growth promoting rhizobacteria: current perspective. J King Saud Uni-Sci 26 (1):1-20

Akram M, Rizvi R, Sumbul A, Ansari RA, Mahmood I (2016) Potential role of bio-inoculants and organic matter for the management of root-knot nematode infesting chickpea. Cogent Food Agric 2 (1):1183457

Aleti G, Sessitsch A, Brader G (2015) Genome mining: prediction of lipopeptides and polyketides from Bacillus and related Firmicutes. Computat Struct Biotechnol J13:192-203

Almaghrabi OA, Massoud SI, Abdelmoneim TS (2013) Influence of inoculation with plant growth promoting rhizobacteria (PGPR) on tomato plant growth and nematode reproduction under greenhouse conditions. Saudi J Biol Sci 20 (1):57-61

Alori ET, Babalola OO (2018) Microbial inoculants for improve crop quality and human health. Front Microbiol 9:2213

Alori ET, Babalola OO, Prigent-Combaret C (2019) Impacts of Microbial Inoculants on the Growth and Yield of Maize Plant. Open Agric J 13 (1):1-8

Alori ET, Dare MO, Babalola OO (2017) Microbial inoculants for soil quality and plant health. In: Lichtfouse E. (ed) Sustainable Agriculture Reviews 181–308. Springer Berlin, pp 281-307 doi: 10.1007/978-3-319-48006-0

Amprayn K-o, Rose MT, Kecskés M, Pereg L, Nguyen HT, Kennedy IR (2012) Plant growth promoting characteristics of soil yeast (Candida tropicalis HY) and its effectiveness for promoting rice growth. Appl Soil Ecol 61 (0):295-299

Anwer MA (2017) Biopesticides and Bioagents: Novel Tools for Pest Management. CRC Press pp 402

Ashishie P, Ashishie CJ (2018) Biopesticide, their Ecological and Toxicological Effects. Int J Sci 7 (08):21-25

Ayangbenro AS, Olanrewaju OS, Babalola OO (2018) Sulfate-reducing bacteria as an effective tool for sustainable acid mine bioremediation. Front Microbiol 9:1986

Babalola OO (2010) Beneficial bacteria of agricultural importance. Biotechnol Lett 32 (11):1559-1570

Babalola OO, Glick BR (2012a) Indigenous African agriculture and plant associated microbes: current practice and future transgenic prospects. Sci Res Essays 7 (28):2431-2439

Babalola OO, Glick BR (2012b) The use of microbial inoculants in African agriculture: current practice and future prospects. J Food Agric Environ 10 (3&4):540-549

Babalola OO, Olanrewaju OS, Dias T, Ajilogba CF, Kutu FR, Cruz C (2017) Biological Nitrogen Fixation: In Panpatte, Y Jhala, R Vyas & H Shelat (eds), The Role of Underutilized Leguminous Plants. In: Microorganisms for Green Revolution. Springer Singapore, pp 431-443 DOI:10.1007/978-981-10-6241-4_20.

Baez-Rogelio A, Morales-García YE, Quintero-Hernández V, Muñoz-Rojas J (2017) Next generation of microbial inoculants for agriculture and bioremediation. Microb Biotechnol 10 (1):19-21

Bai Y, Souleimanov A, Smith DL, Joeb J (2002) An inducible activator produced by a Serratia proteamaculans strain and its soybean growth-promoting activity under greenhouse conditions. Can J Microbiol 53 (373):1495-1502

Bashan Y, de-Bashan LE, Prabhu S, Hernandez J-P (2014) Advances in plant growth-promoting bacterial inoculant technology: formulations and practical perspectives (1998–2013). Plant Soil 378 (1-2):1-33

Bati CB, Santilli E, Lombardo L (2015) Effect of arbuscular mycorrhizal fungi on growth and on micronutrient and macronutrient uptake and allocation in olive plantlets growing under high total Mn levels. Mycorrhiza 25 (2):97-108

Berg G, Eberl L, Hartmann A (2005) The rhizosphere as a reservoir for opportunistic human pathogenic bacteria. Environ Microbiol 7 (11):1673-1685

Bernardes MFF, Pazin M, Pereira LC, Dorta DJ (2015) Impact of pesticides on environmental and human health. Toxicology Studies-Cells, Drugs and Environment (Andreazza C y Scola G Eds) InTech, Croacia:195-233

Bharti N, Pandey SS, Barnawal D, Patel VK, Kalra A (2016) Plant growth promoting rhizobacteria Dietzia natronolimnaea modulates the expression of stress responsive genes providing protection of wheat from salinity stress. Sci Rep 6:34768

Bhattacharyya PN, Jha DK (2012) Plant growth-promoting rhizobacteria (PGPR): emergence in agriculture. World J Microbiol Biotechnol 28 (4):1327-1350

Bhromsiri C, Bhromsiri A (2010) The effect of plant growth promoting Rhizobacteria and Arbuscular mycorhizal fungi on the growth, development and nutrient uptake on different vetiver ecotypes. Thai J Agric Sci 43 (4):239-249

Borneman AR, Schmidt SA, Pretorius IS (2013) At the cutting-edge of grape and wine biotechnology. Trends Genet 29 (4):263-271

Braud A, Jézéquel K, Bazot S, Lebeau T (2009) Enhanced phytoextraction of an agricultural Cr-and Pb-contaminated soil by bioaugmentation with siderophore-producing bacteria. Chemosphere 74 (2):280-286

Brusseau ML, Pepper IL, Gerba C (2019) Environmental and Pollution Science. Academic Press pp 662

Bougoure J, Flavel R, Glaser A, Koehle H, Mainwaring D, Mark T, Mathes F, Murphy D, Murugaraj P, Pearce JD (2016) Mixtures comprising a superabsorbent polymer (sap) and a biopesticide. Google Patents,

Canfora L, Malusà E, Tkaczuk C, Tartanus M, Łabanowska B, Pinzari FJ (2016) Development of a method for detection and quantification of B. brongniartii and B. bassiana in soil. Sci Rep 6:22933 doi:10.1038/srep22933

Caradonia F, Battaglia V, Righi L, Pascali G, La Torre A (2019) Plant biostimulant regulatory framework: prospects in europe and current situation at international level. J Plant Growth Regul 38 (2):438-448

Cassán F, Diaz-Zorita M (2016) Azospirillum sp. in current agriculture: From the laboratory to the field. Soil Biol Biochem 103:117-130

Castrillo LA, Thomsen L, Juneja P, Hajek AE (2007) Detection and quantification of Entomophaga maimaiga resting spores in forest soil using real-time PCR. Mycol Res 111 (3):324-331

Celador-Lera L, Jiménez-Gómez A, Menéndez E, Rivas R (2018) Biofertilizers based on bacterial endophytes isolated from cereals: potential solution to enhance these crops. In: Meena V. (eds) Role of Rhizospheric Microbes in Soil. Springer, pp 175-203

Cheng Z, Woody OZ, McConkey BJ, Glick BR (2012) Combined effects of the plant growth-promoting bacterium Pseudomonas putida UW4 and salinity stress on the Brassica napus proteome. Appl Soil Ecol 61 (0):255-263.

Čihák M, Kameník Z, Šmídová K, Bergman N, Benada O, Kofroňová O, Petříčková K, Bobek J (2017) Secondary Metabolites Produced during the Germination of Streptomyces coelicolor. Front Microbiol 8 (2495).

Crickmore NJJoAM (2006) Beyond the spore–past and future developments of Bacillus thuringiensis as a biopesticide. J Appl Microbiol 101 (3):616-619

Damalas C, Koutroubas S (2016) Farmers' exposure to pesticides: toxicity types and ways of prevention. Toxics 4(1): 1.

Dary M, Chamber-Pérez M, Palomares A, Pajuelo E (2010) "In situ" phytostabilisation of heavy metal polluted soils using Lupinus luteus inoculated with metal resistant plant-growth promoting rhizobacteria. J Hazard Mater 177 (1-3):323-330

Dikshit R, Tallapragada P (2018) Comparative study of natural and artificial flavoring agents and dyes. In: Natural and Artificial Flavoring Agents and Food Dyes. Elsevier, pp 83-111

Droby S, Wisniewski M, Teixidó N, Spadaro D, Jijakli MH (2016) The science, development, and commercialization of postharvest biocontrol products. Postharvest Biol Tech 122: 22-29.

Dwivedi SL, Sahrawat KL, Upadhyaya HD, Mengoni A, Galardini M, Bazzicalupo M, Biondi EG, Hungria M, Kaschuk G, Blair MW, Ortiz R (2015) Chapter One - Advances in Host Plant and Rhizobium Genomics to Enhance Symbiotic Nitrogen Fixation in Grain Legumes. In: Donald L S (ed) Advances in Agronomy, vol Volume 129. Academic Press, pp 1-116. doi:10.1016/bs.agron.2014.09.001

EFSA (European Food Safety Authority) (2016) Scientific opinion on safety and efficacy of BioPlus 2B (R) (Bacillus subtilis DSM 5750 and Bacillus licheniformis DSM 5749) as a feed additive for sows, piglets, pigs for fattening, turkeys for fattening and calves. doi: 10.2903/j.efsa.2016.4558

Enebe MC, Babalola OO (2019) The impact of microbes in the orchestration of plants' resistance to biotic stress: a disease management approach. J Applied Microbiol Biotechnol 103 (1):9-25

Fashola M, Ngole-Jeme V, Babalola O (2016) Heavy metal pollution from gold mines: environmental effects and bacterial strategies for resistance. Int J Environ Res Public Health 13 (11):1047

Fierer N, Schimel JP, Holden PA (2003) Variations in microbial community composition through two soil depth profiles. Soil Biol Biochem 35 (1):167-176

Flint ML (2018) Pests of the garden and small farm: a grower's guide to using less pesticide, vol 3332. UCANR Publications http://ipm.ucanr.edu/IPMPROJECT/ADS/ manual_gardenfarms.html

Gai CS, Lacava PT, Quecine MC, Auriac M-C, Lopes JRS, Araújo WL, Miller TA, Azevedo JL (2009) Transmission of Methylobacterium mesophilicum by

Bucephalogonia xanthophis for paratransgenic control strategy of citrus variegated chlorosis. J Microbiol 47 (4):448-454

Gbadamosi J, Kolawole P (2017) Agricultural science and technology research to support food security. Journal of Emerging Trends in Educational Research and Policy Studies 8 (2):103-106

Gholami A, Shahsavani S, Nezarat S (2009) The effect of plant growth promoting rhizobacteria (PGPR) on germination, seedling growth and yield of maize. Int J Biol Life Sci 1 (1):35-40

Goertz A, Es-Sayed M (2017) Binary fungicidal composition. Google Patents

Gopalakrishnan S, Srinivas V, Alekhya G, Prakash B, Kudapa H, Rathore A, Varshney RK (2015) The extent of grain yield and plant growth enhancement by plant growth-promoting broad-spectrum Streptomyces sp. in chickpea. SpringerPlus 4 (1):1-10

Gupta A, Rai V, Bagdwal N, Goel R (2005) In situ characterization of mercury-resistant growth-promoting fluorescent pseudomonads. Microbiol Res 160 (4):385-388

Gyles C, Boerlin P (2014) Horizontally transferred genetic elements and their role in pathogenesis of bacterial disease. Vet Pathol 51 (2):328-340

Habibi A, Heidari G, Sohrabi Y, Badakhshan H, Mohammadi K (2011) Influence of bio, organic and chemical fertilizers on medicinal pumpkin traits. J Med Plants Res 5 (23):5590-5597

Hall K, Lamboy J, Rusinek T, MacAvery S, Lobdell E, Daughtrey M (2000) Implementation and Demonstration Report in Integrated Pest Management Microbial Products for Poinsettia Disease Suppression. New York State IPM Program Report https://ecommons.cornell.edu/handle/1813/46647

Hassani MA, Durán P, Hacquard S (2018) Microbial interactions within the plant holobiont. Microbiome 6 (1):58

Iheshiulo EM-A, Abbey L, Asiedu SK (2017) Response of Kale to single-dose application of k humate, dry vermicasts, and volcanic minerals. Int J Veg Sci23 (2):135-144

Jabs T, Seevers K, Reinot E (2016) Synergistic compositions comprising a Bacillus subtilis strain and a biopesticide. Google Patents,

Johnson V, Pearson J, Jackson T (2001) Formulation of Serratia entomophila for biological control of grass grub. J New Zealand Plant Protect 54:125-127

Kang Z, Zhang J, Zhou J, Qi Q, Du G, Chen J (2012) Recent advances in microbial production of δ-aminolevulinic acid and vitamin B12. Biotechnol Advance 30 (6):1533-1542

Karabörklü S, Azizoglu U, Azizoglu Z (2017) Recombinant entomopathogenic agents: a review of biotechnological approaches to pest insect control. World J Microbiol Biotechnol 34 (1):14

Kay E, Vogel TM, Bertolla F, Nalin R, Simonet PJ (2002) In situ transfer of antibiotic resistance genes from transgenic (transplastomic) tobacco plants to bacteria. Appl Environ Microbiol 68 (7):3345-3351

Keen EC (2012) Paradigms of pathogenesis: targeting the mobile genetic elements of disease. Front Cell Infect Microbiol 2:161

Keswani C, Bisen K, Singh V, Sarma BK, Singh HB (2016) Formulation technology of biocontrol agents: present status and future prospects. In: Arora N., Mehnaz S., Balestrini R. (eds) Bioformulations: for Sustainable Agriculture. Springer, New Delhi. pp 35-52

Khasa YP (2017) Microbes as biocontrol agents. In: Kumar V., Kumar M., Sharma S., Prasad R. (eds) Probiotics and Plant Health. Springer, Singapore, pp 507-552

Koonin EV, Makarova KS, Aravind L (2001) Horizontal gene transfer in prokaryotes: quantification and classification. Annu Rev Microbiol 55 (1):709-742

Koul O (2011) Microbial biopesticides: opportunities and challenges. CAB Reviews: Perspectives in Agriculture, Veterinary Science, Nutrition and Natural Resources 2011 6, No. 056 6:1-26

Kumawat K, Sharma P, Sirari A, Singh I, Gill B, Singh U, Saharan K (2019) Synergism of Pseudomonas aeruginosa (LSE-2) nodule endophyte with Bradyrhizobium sp. (LSBR-3) for improving plant growth, nutrient acquisition and soil health in soybean. World J Microbiol Biotechnol 35 (47)

Leitão AL (2009) Potential of Penicillium species in the bioremediation field. Int J Environ Res Public Health 6 (4):1393-1417

Lipkie TE, Ferruzzi M, Weaver CM (2016) Bioaccessibility of vitamin D from bread fortified with UV-treated yeast is lower than bread fortified with crystalline vitamin D2 and bovine milk. The FASEB Journal 30 (1_supplement):918.916-918.916

López-Mondéjar R, Antón A, Raidl S, Ros M, Pascual JA (2010) Quantification of the biocontrol agent Trichoderma harzianum with real-time TaqMan PCR and its potential extrapolation to the hyphal biomass. Bioresour Technol 101 (8):2888-2891

Lu C, Li J-M, Qi H, Zhang H, Zhang J, Xiang W-S, Wang J-D, Wang X-J (2018) Two new lankacidin-related metabolites from Streptomyces sp. HS-NF-1178. Journal Antibiotics 71 (3):397

Madsen K, Bertelsen I, Askegaard M (2014) Fertilizer placement and competitive ability of spring barley varieties–Results from two years of organic field trials. In: 10 th EWRS Workshop on Physical and Cultural Weed Control Alnarp, Sweden.

Majeed A, Abbasi MK, Hameed S, Imran A, Rahim N (2015) Isolation and characterization of plant growth-promoting rhizobacteria from wheat rhizosphere and their effect on plant growth promotion. Front Microbiol 6:198.

Malusá E, Vassilev N (2014) A contribution to set a legal framework for biofertilisers. Appl Microbiol Biotechnol 98 (15):6599-6607.

Marrone PG (2014) The market and potential for biopesticides. Europe 12 (9.5):13.608

Martínez-Hidalgo P, Maymon M, Pule-Meulenberg F, Hirsch AM (2018) Engineering root microbiomes for healthier crops and soils using beneficial, environmentally safe bacteria. Can J Microbiol 65 (2):91-104

Martínez A, Camarero S, Ruiz-Dueñas F, Martínez M (2018) Biological lignin degradation. Lignin Valorization: Emerging Approaches 19:199

Me'lanie MP, Filion M (2013) Engineering the rhizosphere for agricultural and environmental sustainability. In book: Applications of microbial engineering, Publisher: CRC Press, Editors: VK Gupta et al, pp.251-271

Meena KK, Sorty AM, Bitla UM, Choudhary K, Gupta P, Pareek A, Singh DP, Prabha R, Sahu PK, Gupta VK (2017) Abiotic stress responses and microbe-mediated mitigation in plants: the omics strategies. Front Plant Sci 8:172

Mirabelli C, Tullio M, Pierandrei F, Rea E (2007) Effect of arbuscular mycorrhizal fungi on micropropagated hazelnut (Corylus avellana L.) plants. In: III International Symposium on Acclimatization and Establishment of Micropropagated Plants pp 467-472

Mishra J, Arora NK (2016) Bioformulations for plant growth promotion and combating phytopathogens: A sustainable approach. In: Bioformulations: for Sustainable Agriculture. Springer, pp 3-33

Moazami, N (2011) Biological Control. Comprehensive Biotechnology (Second Edition) 3.61: 731-739

Mohammadi K (2010) Ecophysiological response of canola (Brassica napus L.) to different fertility systems in crop rotation. Ph.D thesis. Agronomy Department. Tarbiat Modares University, Tehran, Iran,

Mohammadi K, Sohrabi Y (2012) Bacterial biofertilizers for sustainable crop production: a review. J Agric Biol Sci 7:307-316

Mohammed A, Yusop MAS (2016) Effects of Light Intensity and Mycorrhiza Association on the Growth Performance of Capsicum annum. In: Regional Conference on Science, Technology and Social Sciences (RCSTSS 2014). Springer, pp 455-462

Mojsov K (2016) Aspergillus enzymes for food industries. In: New and Future Developments in Microbial Biotechnology and Bioengineering. Elsevier, pp 215-222

Mosttafiz SB, Rahman M, Rahman M (2012) Biotechnology: role of microbes in sustainable agriculture and environmental health. Int J Microbiol 10 (1)

Navas A, Cobas G, Talavera M, Ayala JA, López JA, Martínez JL (2007) Experimental validation of Haldane's hypothesis on the role of infection as an evolutionary force for Metazoans. PNAS 104 (34):13728-13731

Nawaz M, Mabubu JI, Hua HJ (2016) Current status and advancement of biopesticides: microbial and botanical pesticides. J Entomol Zool Studies 4 (2):241-246

Nissar J, Ahad T, Nazir F, Salim R (2017) Applications of Biotechnology in Food Technology. Int J Eng Technol Sci Res Volume 4, Issue12

CHAPTER 10 – Contribution of Microbial Inoculants in Sustainable Maintenance of Human Health,
including Test Methods and Evaluation of Safety of Microbial Pesticide Microorganisms

Ojuederie *et al.*

O'Callaghan M (2016) Microbial inoculation of seed for improved crop performance: issues and opportunities. Appl Microbiol Botechnol 100 (13):5729-5746

Olanrewaju OS, Babalola OO (2018) Streptomyces: implications and interactions in plant growth promotion. Appl Microbiol Biotechnol:1-10

Olanrewaju OS, Glick BR, Babalola OO (2017) Mechanisms of action of plant growth promoting bacteria. World J Microbiol Biotechnol 33 (11):197

Özer N, Coşkuntuna A (2016) The biological control possibilities of seed-borne fungi. In: Kumar P., Gupta V., Tiwari A., Kamle M. (eds) Current Trends in Plant Disease Diagnostics and Management Practices. Fungal Biology. Springer, Cham, pp 383-403.

Pal S, Singh HB, Farooqui A, Rakshit A (2016) Commercialization of Arbuscular Mycorrhizal Technology in Agriculture and Forestry. In: Agriculturally Important Microorganisms. Springer, pp 97-105

Parihar M, Meena VS, Mishra PK, Rakshit A, Choudhary M, Yadav RP, Rana K, Bisht JK (2019) Arbuscular mycorrhiza: a viable strategy for soil nutrient loss reduction. Arch Microbiol:1-13

Patil HJ, Solanki MK (2016) Microbial inoculant: Modern era of fertilizers and pesticides. In: D.P. Singh et al. (eds.), Microbial inoculants in sustainable agricultural productivity. Springer, India pp 319-343.

Pavela R, Benelli G (2016) Essential oils as ecofriendly biopesticides? Challenges and constraints. Trends Plant Sci 21 (12):1000-1007

Pérez-García A, Romero D, De Vicente A (2011) Plant protection and growth stimulation by microorganisms: biotechnological applications of Bacilli in agriculture. Curr Opinion Biotechnol 22 (2):187-193

Pieterse CM, Zamioudis C, Berendsen RL, Weller DM, Van Wees SC, Bakker PA (2014) Induced systemic resistance by beneficial microbes. Ann Rev Phytopathol 52:347-375

Prakash D, Nawani N, Prakash M, Bodas M, Mandal A, Khetmalas M, Kapadnis B (2013) Actinomycetes: a repertory of green catalysts with a potential revenue resource. BioMed Res Int Volume 2013, Article ID 264020, 8 pages

Reid JP (2013) Rainfall variability and mycorrhizal associations affect nitrogen retention in tree mesocosms. Non-equilbrium dynamics of ecosystem processes in a changing world:31

Remans R, Beebe S, Blair M, Manrique G, Tovar E, Rao I, Croonenborghs A, Torres-Gutierrez R, El-Howeity M, Michiels J (2008) Physiological and genetic analysis of root responsiveness to auxin-producing plant growth-promoting bacteria in common bean (Phaseolus vulgaris L.). Plant Soil 302 (1-2):149-161

Rouabhi, R., 2010. Introduction and toxicology of fungicides. In Fungicides. IntechOpen. DOI: 10.13140/RG.2.1.2099.9125

Ruiu L, Satta A, Floris I (2013) Emerging entomopathogenic bacteria for insect pest management. Bullet Insectol 66 (2):181-186

Saha M, Sarkar S, Sarkar B, Sharma BK, Bhattacharjee S, Tribedi P (2015) Microbial siderophores and their potential applications: a review. Environ Scie Pollut Res Int 23 (5):3984-3999

Sachdev, S., Singh, R.P., 2016. Current challenges, constraints, and future strategies for development of successful market for biopesticides. Clim Change Environ Sustain 4, 129-136.

Sarwar M, Technology AR (2015) Microbial insecticides- an ecofriendly effective line of attack for insect pests management. Int J Eng Adv Res Technol 1 (2):4-9

Savazzini F, Longa CMO, Pertot I, Gessler C (2008) Real-time PCR for detection and quantification of the biocontrol agent Trichoderma atroviride strain SC1 in soil. J Microbiol Methods 73 (2):185-194

Sessitsch A, Brader G, Pfaffenbichler N, Gusenbauer D, Mitter B (2018) The contribution of plant microbiota to economy growth. Microb Biotechnol 11 (5):801

Sharma S, Kulkarni J, Jha B (2016) Halotolerant rhizobacteria promote growth and enhance salinity tolerance in peanut. Front Microbiol 7:1600

Siddiqui ZA, Kataoka R (2011) Mycorrhizal inoculants: progress in inoculant production technology. In: Ahmad I, Ahmad F, Pichtel, J (eds.) Microbes and Microbial Technology. Springer New York, pp 489-506

Singh R, Kumar M, Mittal A, Mehta PK (2016) Microbial enzymes: industrial progress in 21st century. 3 Biotech 6 (2):174

Steckler, S., Caldwell, C., Priest, K., Greenshields, D., 2012. A fluorescent mutant of the phosphorus-solubilizing fungus Penicillium bilaiae to image rhizosphere growth. Soils and Crops Workshop. http://hdl.handle.net/10388/9080

Su, H., Marrone, P., 2019. Compositions containing anthraquinone derivatives as growth promoters and antifungal agents. Google Patents

Taranta C, Bork T, Schreieck J, Müller H, Riediger N, Klein CD, Willis R, Sikuljak T, Mertoglu M (2018) Solid agroformulations prepared from a melt of pesticide and polyalkoxylate, optionally containing liquid adjuvant complexed with polycarboxylate. Google Patents,

Vandana UK, Chopra A, Bhattacharjee S, Mazumder P (2017) Microbial biofertilizer: A potential tool for sustainable agriculture. In: Panpatte D., Jhala Y., Vyas R., Shelat H. (eds) Microorganisms for Green Revolution. Microorganisms for Sustainability, vol 6. Springer Singapore, pp 25-52

Wilson MJ, Jackson TA (2013) Progress in the commercialisation of bionematicides. BioControl 58:715–722.

CHAPTER 11

Present Views, Status and Updates in Biopesticide Usage

by
Emmanuel Okrikata
Department of Biological Sciences,
Federal University Wukari,
Taraba State, Nigeria.
Email:*okrikata@fuwukari.edu.ng*

Abstract

For centuries, suppressing the myriads of agricultural and public health pests has remained a huge challenge for the exponentially growing human population. Over reliance and faulty use of synthetic chemical pesticides for 'controlling the pests' is impacting so negatively on the environment and human health, leading to stricter regulations on pesticide usage. The utilization of biopesticides under the framework of integrated pest management is thus advocated by experts. This chapter examines the current standing of biopesticides from a global perspective vis-à-vis it's hitherto status, while charting the way forward for its optimum development and utilization. The chapter delineates biopesticides and presented a brief on its history while emphasizing the magnitude of its current development, acceptance and utilization with theoretical projections for the future. Factors, responsible for the current status of biopesticides are highlighted. The role of technologies (such as molecular-based technologies, high throughput screenings and nanoformulations) and government policies in the development and use of biopesticides are also discussed.

Keywords: Biopesticides; Integrated pest management; Nanoformulations; Pests; Synthetic chemical pesticides

Introduction

Aside public health pests which impacts negatively on human health, suppressing agricultural pests to enhance food production and sustainability is an important and evolving challenge confronting humanity. The magnitude of the challenges in these regard is to say the least, very obvious as it is projected that the world population will hit 9.2 billion by 2050, and 11 billion by 2100 as against the current 7 billion (FAO, 2009; Lade *et al.*, 2019). No doubt, food productions need to match food needs of our fast growing population. However, in achieving this, it is necessary for crop producers and those within the value chain who pay their bills from agriculture, to do so without endangering the socio-economic and environmental livelihood of upcoming generations or constraining the ecosystem and its services.

It has been shown that, aside 20% postharvest losses, about 40% of global possible crop yields are ravaged before harvest by a combination of invertebrates, pathogens and weeds numbering about 67000 different species (Chandler, et al., 2011; Kumar and Kalita, 2017; Arjjumend and Koutouki, 2018). These yield losses are no doubt high. Hence, an improvement in pest management strategies is important to increasing access to food. To achieve this, from 1950s up to late 1960s; the green revolution era, crop yields were increased dramatically by the use of synthetic chemical pesticides, among other interventions. In recent times, consumption of chemical pesticides have significantly increased with an estimated annual global usage of 2 million tones and a colossal consumption in Europe (45%), followed by the United States (25%), and 25% in the rest of the world (Aktar et al., 2009; Lade et al., 2019). This outrageous and faulty use of synthetic chemical pesticides has attracted significant negative effects to both man and the environment. These effects include; high biodiversity loss, soil degradation, water and air pollution, distorted ecological balance, pesticide resistance, pest resurgence, rising invasive species aside, serious acute and chronic health hazards and increased crop losses and the awareness has increased in recent years (Koul, 2012; Kumar, 2013; Okrikata and Ogunwolu, 2017). Hence, it can rightly be said that, the advent of pesticide technology which was aimed at securing food availability, has hardly been successfully and sustainably achieved.

Policy makers worldwide are now placing tighter safety regulations on imported commodities and stricter regulations on the amount of pesticide residues on products, with the intent of suppressing the over reliance on conventional chemical pesticides and promoting environmental and health safety. However, with the rising human population; the need to increase food production as was the case in the green revolution era is very obvious. The use of Biopesticides as an alternative to synthetic chemical pesticides in the context of Integrated Pest Management (IPM) is thus being advocated. Some successes have been recorded in this direction. We will explore some of these in this chapter. An important question to deal with first is; what is a biopesticide?

What is Biopesticide?
There exists varying definitions of biopesticide, this is because the term is viewed from different perspectives. However, for a pesticide to be termed 'a biopesticide', it must exhibit low, if any risk to the environment, wildlife, non target organisms including mammals and should be of biological origin. Biopesticide embraces products used to suppress noxious vertebrates or invertebrates, plants or pathogens. It ranges from whole living organisms to near synthetic molecules (Sundh and Goettel, 2013). A widely acceptable definition assumes that after the initial application, the bioactive agent will not persist for

long nor replicate (in the case of living biopesticidal agents) to such an extent that it controls the target pest without re-application. The term can be restrictively used to refer to live organisms (Kiewnick, 2007) or, broadened to encompass metabolites and extracts from those organisms. It may also be used for any naturally occurring compound that has a pesticidal action. Therefore, that biopesticide is a mass-produced, biologically based agent manufactured from a living microorganism or a natural product, which is sold for the control of pests is a widely acceptable definition of biopesticide. It is a formulation rooted on the action of a natural agent with pesticidal activity. The term recognizes the natural agent to be of biological than synthetic origin (Gupta and Dikshit, 2010; Chandler *et al.*, 2011; Sporleder and Lacey, 2013).

The U.S. Environmental Protection Agency (EPA) defined biopesticides as including naturally occurring substances or microorganisms that control pests. Included also are pesticidal substances produced by plants on addition of genetic materials - plant-incorporated protectants/PIPs. (https://www.epa.gov/pesticides/biopesticides#what, accessed November 11, 2019). The Food and Agriculture Organization of the United Nations (FAO) on the other hand accepts biopesticides to be biocontrol agents which are passive (botanicals, secondary metabolites/semiochemicals, genes/gene products and micro-organisms) and may also include those that actively search for the pest as in the case of predators, parasitoids and many species of entomopathogenic nematodes (www.fao.org/docs/eims/upload/agrotech/2003/global_perpective.pdf, accessed October 29, 2019). That some endophytic microorganisms can be regarded as biopesticides was stated by Glare *et al.* (2012). Such microorganisms inhabit a plant without harming it but confer pesticidal and/or pest-deterrent properties on the host plant.

From the foregoing, it is difficult or a bit controversial to delineate which organisms or even substances qualifies as biopesticidal agents. For example, plant extracts and pheromones were not included in the list of some authors and regulators and, controversies also exists as to whether natural products with toxic mode of action can be classified as biopesticides. However, from whatever perspective it is looked at, one can safely conclude that, a biopesticide is a pesticidal formulation which is based on the activity of a natural agent. It has the capacity to either inactivate or kill the pest or to obstruct the pest's reproductive and/or physiological functions such as impulse transmission, respiration or mating. For agricultural purposes, biopesticides are used for insect, pathogen, weed and nematode control. They are also used to regulate plant physiology and productivity. It is therefore obvious that, the value of biopesticide is to effectively manage pest with less environmental impact.

Categories/classes of biopesticides
The 3 widely acceptable classes of biopesticides are:

1. Microbial biopesticides: Microorganisms (dead, alive or spores) have been used as biopesticides to target specific invertebrate and vertebrate pests and, pathogens. Many of them have been commercialized. Pathogenic microbes and nematodes have the capacity to infect other organisms and scientists are recruiting this ability for developing microbial pesticides for pest management. The active ingredients are micro-organisms like bacteria species such as *Pseudomonas, Rahnella, Serratia, Xanthomonas*, but notably *Bacillus thuringiensis* (*Bt*) which produces a protein crystal *Bt* δ-endotoxin during spore formation which kills susceptible insects by rupturing their gut cells within 2 days of consumption. Other bioactive agents include some fungus eg., *Beauveria, Phytophthora, Trichoderma* and *Verticillium* species, virus (mainly cytoplasmic polyhedrosis virus – CPV, granulosis virus – GV and nuclear polyhedrosis virus – NPV) that apparently do not replicate in homeothermic animals, or protozoans (Sundh and Goettel, 2013; Seiber *et al.*, 2014; Lengai and Muthomi, 2018; Ruiu, 2018). Some microbial pesticidal agents, eg., *Pseudomonas* and *Trichoderma* sp. are fungicidal while, others eg., *Alternaria alternate, Puccinia chondrillina* and *Colletotrichum gloeosporioides* are herbicidal. Formulations of Bacteriophages have also been used to suppress bacterial pathogens (Roach *et al.*, 2008; Glare *et al.*, 2012). Nematodes such as *Heterarhabditis* and *Steinernama* species are entomopathogenic.

Microbial products may be made up of the microbes and/or their metabolites. The microorganism used may be natural, living or dead, or engineered genetically. Microbial pesticides constitute about 90% of all biopesticides and are largely sourced from agricultural fields where they live with other beneficial and pathogenic microbes, and are applied as either sprays, granules or dusts. More than 400 fungal species and over 90 bacterial species have been shown to infect insects (Sundh and Goettel, 2013; Kachhawa, 2017; Kumar *et al.*, 2019).

2. Biochemical biopesticides: These are pesticidal natural or synthetic chemicals/compounds. For the synthetic, their structural and functional identities are similar to their natural equivalents. They include plant products such as essential oils, and various compounds synthesized by other organisms such as chitin and chitosan. Unlike synthetic chemical pesticides that contain synthetic molecules that directly kill the pest, biochemical pesticides control pests by non poisoning mechanisms. They can be classified into the following biologically distinct functional classes:

a. *Semiochemicals:* Semiochemicals (eg., pheromones, allomones, kairomones and other groups of behavior modifying molecules) are plant-insect or insect–insect communication chemicals which can be used in pest management. They are considered an environmentally safe option, as they are usually very target-specific. Very low quantities are required and, as volatiles, they have low persistence (Rosell *et al.*, 2008). The most widely used semiochemical for pest management are insect pheromones. They are used in traps for either mass trapping, mating disruption or lure-and-kill techniques.

b. *Plant extracts and oils*: The use of botanical pesticides is as old as 17th century when nicotine and tobacco leaves were used against plum beetles (O'Neal and Dara, 2018a). Plants' auto-synthesized chemicals are used for defense against pathogens, weeds, insects and other herbivores. Well over 800 species of plants have been found to exhibit biocidal activity against pests and pathogens both pre- and post-harvest with active compounds having varying modes of action such as insect growth regulators, attractants, antifeedants, fumigants, repellents, deterrents, confusants and insecticidal activities (El-Wakeil, 2013; Hikal *et al.*, 2017). The active compounds/chemicals in plants include phenols, quinones, alkaloids, steroids, terpenes, alcohols and saponins such as azadirachtin or urushiol. It also include oils (particularly, essential oils) like rosemary, citrus and eucalyptus oils which can be extracted from plant leaves, seeds and flowers containing them; depending on the plant's family, varying antimicrobial components such as α- and β-phillandrene, β-caryophyellene, camphor, linalool, linalyl acetate and limonene. Pesticides formulated with these natural products (botanical pesticides) have generally little or no toxic impact on the environment.

c. *Plant growth regulators (PGRs):* These are natural and/or synthetic analogs of natural chemicals that affect key physiological functions of plants. They can inhibit, promote or alter the physiological characters of a plant. They are used to enhance plant agronomic performance and quality, and to improve consistency in productivity and conquer abiotic and genetic constraints to plant yields. Major PGRs are abscisic acid, auxins, cytokinins, ethylene and gibberellins (BPIA, 2014). They exhibit no pesticidal activity. However, since they are regulated by the EPA as biopesticides, they are so classified.

d. *Insect growth regulators (IGRs):* Are chemical substances that interfere with the life cycle of an insect by regulating maturation and reproduction. They are basically used as bioinsecticides to suppress populations of noxious insects. They mimic juvenile hormones and disrupt growth and reproduction in insects thereby inhibiting their proliferation eg., hydroprene and pyriproxifen. Included also, are the chitin synthesis inhibitors eg., diflubenzuron. The high specificity of IGRs on target pests vis-à-vis preservation of beneficial arthropods makes them important candidate for IPM. Generally, IGRs are hardly fatal to adults and less toxic to humans hence, largely labeled as 'reduced risk' by EPA. They are relatively expensive and slow in activity (Mondal and Parween, 2000).

3. Plant incorporated protectants (PIPs): Plant incorporated protectants are pesticidal substances that plants produce from genetic materials that have been incorporated into them via recombinant DNA technology. The inserted gene for a specific pesticidal protein modifies the crop (Genetically modified crop – GM crop) to exhibit resistance to pest attacks (as in *Bt* Maize) or, renders them tolerant to herbicide application. Additionally, crops can be engineered or "stacked" to express multiple characters. For example, a crop may be genetically manipulated to express resistance to multiple herbicides or to be resistant to herbicide application as well as insecticidal. All of such depends on the success of the genetic engineering process. The EPA stipulates that the protein and its genetic material are regulated and not, the plant (EPA, 2019). Aside Maize; Cotton, Plums, Potatoes and Soybeans have been genetically modified. More recently, *Bt* pod borer (*Maruca*) resistant Cowpea was approved for commercial release in Nigeria (Peace *et al.*, 2019).

The use of PIPs (as observed in the rising number of transgenic plants) is increasing, even though the technology remains controversial in many countries particularly because genes are widely sourced from organisms other than plants. An estimated 1.7 million hectares (mha) of land was cultivated with genetically modified crops globally in 1996, but this has recently risen to about 190 mha by 17 million farmers in 24 countries indicating an appreciable level of increase in global acceptability (ISAAA, 2017). To some extent, GM plants functions in a pattern similar to that of pesticidal endophytes in which case the pesticidal property is produced within the plant under the influence of a foreign bioagent (endophyte). It is expected that the use of biopesticides in agriculture and health programs will enhance environmental and health safety. However, suffice it to mention here that, the value of biopesticides may be eroded by continuous synthesizing and reliance on synthetic biocides as against natural biocides.

Current Views, Status and Updates in Biopesticide Usage

i. Brief history of biopesticides
Biopesticides have been used for pest control for centuries. Plant based pesticides were apparently the earliest biopesticides as it is on record that nicotine was used to control plum beetles in the 17th century (O'Neal and Dara, 2018a). However, an increase in the number of biopesticidal researches, beginning with today's widely used *B. thuringiensis* (*Bt*), sprung during the era of agricultural revolution of the 20th century (O'Neal and Dara, 2018a). The Japanese insect pathologist, Shigetane Ishiwata was the first to isolate and name *Bt* in his study of the causal agent of a disease afflicting silkworms in 1901. About a decade later, a German scientist; Ernst Berliner rediscovered it in a diseased flour moth caterpillar leading to its first commercial production in 1938 under the name Sporeine in France (Roh *et al.*, 2007; BPIA, 2019). However, as far back as 1835, an Italian

entomologist named Agostine Bassi experimentally showed that the fungus, *Beauveria bassiana* could be used to infect silkworm. A number of researches using different mineral oils in pest management were also recorded in the 19th century. Within the 1980s and 1990s, the commercial use of *Agrobacterium radiobacter* to suppress crown gall on tree crops and *Pseudomonas fluorescens* to prevent fire blight in orchards came to light (Escobar and Dandekar, 2003; Stockwell *et al.*, 2010; O'Neal and Dara, 2018a; BPIA, 2019).
The advances made in the 1990s and early 2000s in biopesticide formulation technology, as well as biochemical and microbiological researches greatly improved biopesticides. These triggered the development of new solutions with better efficiency (target and site-specific), improved application, conservation and shelf life.

ii. Current trends in biopesticide development and usage
Huge success have been achieved as an estimated over 3,000 tons of biopesticides are globally produced annually, and the increase is rapid. Recent report reveals that over 430 biopesticide active ingredients and 1320 products are registered and available in the USA, and it is on record that some new biopesticidal active ingredients were approved or launched within 2017 and 2018 (Kumar and Singh, 2015; EPA, 2018; Agrow, 2018). In Brazil, there are 166 approved commercial products out of which 37 were approved in 2017 and 9 as at May, 2018 (Agrow, 2018). Overall, over 225 microbial biopesticides are now manufactured within 30 Organization for Economic Cooperation and Development (OECD) countries, and an estimated 90% are derived from *Bt*. Of the total biopesticides sold globally, about 45% are used in the USA, Mexico and Canada, while Asia uses only 5% (Hubbard *et al.*, 2014; Lade *et al.*, 2019).

Reliable current statistics for Africa are hard to find due to low development and usage of biopesticides, the amendment and implementation of policies to minimize the usage of traditional pesticides and, promotion of the use of biopesticides by some countries has advanced the registration and launching of newer biopesticidal products in recent times across the world (Table 1).

Table 1. Some new active ingredients approved or launched within 2017 and 2018 across the world

Year	Company's name	Active ingredient	Use	Status
	AgBiome Innovations/SePRO Corporation	*Pseudomonas chlororaphis* strain AFS009 [biofungicide]	Food crops, turf and ornamentals	Approved in US as Zio and Howler
	Arysta LifeScience (Platform Specialty Products)	*Beauveria bassiana* strain 147 [bioinsecticide]	Ornamental palm trees	Approved in EU
	BASF	*Beauveria bassiana* strain NPP111B005 [bioinsecticide]	Bananas and ornamental, palm trees	Approved in EU
2017	BASF/Agrauxine	Cerevisane [disease resistance activator]	Grapevines	Approved in France as Romeo
	Bayer Crop Science	*Bacillus firmus* [bionematicide]	Cotton, maize and soybeans	Approved in Brazil as Oleage
	Eden Research	eugenol/geraniol/thymol [biofungicide]	Grapevines	Approved in France and Portugal as 3AEY and Mevalone
	Ihara	*Bacillus amyloliquefaciens* strain D747 [biofungicide]	Various	Launched in Brazil as Eco-Shot
	Marrone Bio Innovations	*Bacillus amyloliquefaciens* strain F727 [biofungicide]	Various including grapevines, leafy, greens, potatoes, carrots and onions	Approved in US as Stargus
	MosquitoMate	*Wolbachia pipientis* ZAP strain [bioinsecticide]	Mosquitoes	Approved in US as ZAP Males
	Novozymes BioAg	*Streptomyces lydicus* strain WYEC108 [biofungicide]	Vegetables, turf and ornamentals	Approved in Australia as Actinovate
	Nufarm	*Aureobasidium pullulans* strain DSM 14940 + A *pullulans* strain DSM 14941 [biofungicide]	Grapevines	Approved in Australia as Botector
	Stockton	*Melaleuca alternifolia* extract [biofungicide]	Cucumbers and Courgettes	Approved in Spain as Timorex Gold
	Syngenta (owned by ChemChina)	*Pasteuria nishizawae* [bionematicide] Soybeans	Soybeans	Approved in Brazil as Clariva PN
	Valent USA (subsidiary of Sumitomo Chemical)	*Bacillus amyloliquefaciens* strain PTA-4838 [bionematicide]	Soybeans	Launched in US as Aveo EZ
	Vitae Rural Biotecnologia	*Spodoptera frugiperda* multiple nucleopolyhedrovirus [bioinsecticide]	Maize	Launched in Brazil as CartuchoVIT
	BASF	*Beauveria bassiana* strain PPRI 5339 [bioinsecticide/acaricide]	Greenhouse ornament and vegetables	Approved in Canada as Velifer
2018	Bayer	*Bacillus amyloliquefaciens* strain QST 713 [biofungicide]	Grapevines	Approved in Australia as Serenade Opti
	Monsanto	Lipochitooligosaccharide SP104 [plant growth regulator]	Maize and Canola	Approved in US as Acceleron B-360 ST

Source: Agrow, 2018

256

Other biopesticide registrations/approvals or launchings across the world within 2017 and 2018 as reported by Agrow (2018) are;

1. Biofungicide, Integral Pro (*Bacillus amyloliquefaciens* MBI 600). Approved for use in France and the UK as seed application for oilseed rape by BASF.

2. Biofungicide, Velondis (*B subtilis* strain BU1814). Approved for use in US as seed treatments for soybeans, small-grain cereals and maize by BASF.

3. Biofungicide, Serifel (*B amyloliquefaciens* strain MBI600). Expanded the marketing with additional uses in US for mushrooms and Thailand for specialty crops by BASF.

4. Disease resistance activators, Bastide/Blason (COS-OGA). Launched against downy mildew and powdery mildew in high-value crops such as grapevines and greenhouse vegetables by Syngenta.

5. Bionematicide, BioAct Prime DC (*Paecilomyces lilacinus* 251). Launched in Greece in 2017 and granted regulatory approval in Spain in 2018.

Overall, microbial pesticide is the largest and fastest growing biopesticide segment and is expected to constitute about 60% of the market by 2025 with bioinsecticides and biofungicides dominating the market and bioherbicides remaining the least developed (Bullion, 2018). Majority of entomopathogenic fungi products used as biopesticides are based on the ascomycetes *Beauveria bassiana* or *Metarhizium anisopliae*. In Brazil for example, commercial biopesticide based on *M. anisopliae* was used annually against spittlebugs on about 250000 hectare of grassland and 750000 hectare of sugarcane (Li *et al.*, 2010). Tables 2-5 show some registered commercial microbial, botanical and mineral pesticides for controlling insects, mites, plant pathogenic fungi, and plant parasitic nematodes across the world.

Table 2. Microbial insecticides and acaricides

Active Ingredients	Type	Target Pest	Trade name	Manufacturer
Bacillus thuringiensis subsp. Aizawai	Microbial, Bacteria	Diamondback moth, armyworm	Xentari®	Valent BioSciences, Certis USA
Bacillus thuringiensis subsp. Kurstaki	Microbial, Bacteria	Broad range of caterpillars	Dipel®, Deliver®, Javelin®	Valent Bioscinces
Burkholderia rinojensis	Microbial, Bacteria	Broad range sucking and chewing insects, mites and flies	Venerate®	Marrone Bio Innovations
Chromobacterium subtsugae	Microbial, Bacteria	Broad range sucking and chewing insects, mites and flies	Grandevo®	Marrone Bio Innovations
Isaria fumosorosea Strain Apopka 97	Microbial, Fungus	Broad range sucking insects, mites and black vine beetles	PFR-97®	Certis USA
Metarrhizium anisopliae	Microbial, Fungus	Thrips, Mites, Whiteflies	Met52®, GreenGuard®, Green Muscle®	Novozymes, BASF

Source: O'Neal and Dara, 2018b

Table 3. Botanical insecticides and acaricides

Active Ingredient	Type	Target Pest	Trade name	Manufacturer
Azadirachtin	Plant extract	Broad range of sucking and chewing insects	Aza-direct®	Gowan
Chenopodium ambrosioides	Terpenes (synthetically made) from plant extracts	Sucking insects and mites	Requiem®	Bayer Crop Science
Citrus oil solution	Plant extract	Broad range of sucking insects	Oroboost®	OroAgri
Neem oil	Biochemical, soaps/ fatty acids	Broad range of sucking insects	Trilogy®	Certis USA

Source: O'Neal and Dara, 2018b

Table 4. Microbial and non-microbial fungicides

Active Ingredients	Type	Trade name	Manufacturer
Bacillus amyloliquefaciens D747	Microbial, Bacteria	DoubleNickle® 55	Certis USA
Bacillus pumilus 2808	Microbial, Bacteria	Sonata®	Bayer (Wilbus Ellis)
Bacillus mycoides Isolate J	Microbial, Bacteria	LifeGardTM	Certis USA
Gliocladium virens	Microbial, Fungi	SoilGard®	Cerits USA
Trichoderma harzianum T-22	Microbial, Fungi	RootShield® WP, PlantShield® HC	Bioworks
Trichoderma asperellum and *Trichoderma gamsii*	Microbial, Fungi	BIO-TAM 2.0®	Isagro (Marrone Bio)
Extract of *Reynoutria Sachalinensis*	Biochemical, Plant extract	Regalia®	Marrone Bio Innovations
Paraffin oil	Biochemical	Styletoil®	JMS
Potassium bicarbonate	Biochemical	Kaligreen®, Milstop®	Otsuka (Brandt), Bioworks

Source: O'Neal and Dara, 2018b

Table 5. Bionematicides

Active Ingredients	Type	Trade name	Manufacturer
Bacillus firmus	Microbial, Bacteria	Votiro® (Seed treatment)	Bayer Crop Science
Burhoderia rinojensis	Microbial, Bacteria	Majestene®	Marrone Bio Innovations
Myrothecium verrucaria	Microbial, Fungi	Ditera®	Valent BioSciences
Pasteuria nishizawae	Microbial, Bacteria	Clariva® (Seed treatment)	Syngenta
Purpureocillium lalacinus	Microbial, Fungi	MeloCon®	Bayer Crop Science
Saponins of *Quillaja saponaria*	Biochemical, Plant extract	Nema-Q®	Brandt

Source: O'Neal and Dara, 2018b

To ascertain the safe use of bacterial strains and any other microbes for that matter, it is important to conduct a comprehensive risk assessment (Oluwatobi, 2018). As is the case with synthetic chemical pesticides, the first and vital step to start with is the identification and characterization of the microorganism (in the case of microbial biopesticides). This is necessary to distinguish them from closely-related pathogenic variants. There is limited guidance in the EU on the characterization of microbes used as active ingredients in plant protection products (EFSA, 2018; Oluwatobi, 2018). Literature citations which may be inaccurate, misleading or insufficient have largely been relied on in characterizing biological agents (Agrow, 2018). Though still challenging, particularly for novel strains from taxonomic groups for which much is not known about, advanced gene technologies have substantially helped in designating novel strains of microbes with pesticidal potential to the correct taxonomic groups, and clearly distinguishing them from very closely-related pathogenic strains (Aiuchi *et al.*, 2008; Agrow, 2018; Oluwatobi, 2018). Better methods of characterization are however, still required.

In the past, aside from *Pyrethrum* which is globally commercialized due to its broad spectrum insecticidal efficacy, very few botanical insecticides were developed. However, the last three decades have been characterized by the discovery of phytochemicals capable as serving as alternatives to synthetic pesticides due to their environmental and health friendliness (Koul, 2012). Investigations by various researchers in different parts of the world have confirmed their efficacies against field and stored pests (Okrikata and Ogunwolu, 2019). However, variations in extraction methods, differences environmental conditions and plant ecotypes have been shown to influence the quality and quantity of constituent active compounds, thereby creating differences in responses by pathogens and pests to botanical pesticides. Sometimes, obtaining the right proportions of the active and inert materials can be challenging during formulation (Sesan *et al.*, 2015; Sales *et al.*, 2016; Shiberu and Getu, 2016). These are heightened as there are no standard preparation methods and guidelines for testing the efficacy of botanicals (Okunlola and Akinrinnola, 2014).

Today, *Azadirachta indica* A. Juss. (commonly known as Neem) stands out with documented efficacy on over 350 arthropod species, 12 nematode species, 15 fungal species, three viruses, two snails and one crustacean species (Elanchezhyan and Vinothkumar, 2015). Its efficacy has been attributed to its different secondary metabolites with varying mode of actions (Shannag *et al.*, 2015). Plant secondary metabolites and essential oils have largely been used in formulating biopesticides against insect pests due to their vast antifeedant, repellent, oviposition deterrent and insecticidal properties (Koul, 2016; Okrikata *et al.*, 2019).

The biopesticide technology is still considered nascent. In-depth studies are required in the areas of production, formulation, delivery and commercialization of the products. Some of the botanical biopesticides currently under development are potentially excellent alternatives to synthetic chemical pesticides and many of them are based on locally available plants which can be easily processed and made available to the users. More so, the continuous search for new active bioagents and improving efficiency of the known biopesticides via cutting edge technologies such as recombinant DNA technology and other innovative approaches will improve their acceptability as pest control options (Kumar, 2013). A few plants and parts having pesticidal properties used in African traditional agriculture are shown in Table 6.

Table 6. Plants used in traditional agriculture in some developing countries

Common names	Species	Families	Parts*
African mahogany	*Khaya senegalensis* (Desr.) A. Juss.	Meliaceae	S, B
African basil, Basilic, Basilic sauvage	*Ocimum gratissimum* L.	Liminaceae	L
African mesquite, Iron tree	*Prosopis africana* (Guill. & Perr.) Taub.	Leguminosae	S, B
Aztec marigold, Dwarf marigold	*Tagetes minuta* L.	Asteraceae	L
African custard-apple, Wild soursop	*Annona senegalensis* Pers.	Annonaceae	S, B
Bead vine, Coral bead plant, Coral bean	*Abrus precatorius* L.	Fabaceae	L, S
Black-jack, Cobbler's pegs	*Bidens pilosa* L.	Asteraceae	L
Barbados nut, Black vomit nut, Curcas bean or Physic nut	*Jatropha curcas* L.	Euphorbiaceae	sap, F, S, B
Button grass	*Mitracarpus scaber* Zucc.	Rubiaceae	S
Belhambra, Packalacca	*Phytolacca dodecandra* L'Herit.	Phytolaccaceae	L, F
Bitter leaf plant, little ironwood	*Vernonia amygdalina* Del.	Asteraceae	L
Chicken spike	*Spenoclea zeylanica* Gaertn.	Sphenocleaceae	S
Camel's foot	*Piliostigma thonningii* Schum.) Milne-Redh.	Leguminoceae	R, B
Clapperton's parkia	*Parkia clappentoniana* Keay.	Mimosaceae	S, B
Chinaberry, Persian lilac, Ale-laila	*Melia azadarach* L.	Meliaceae	L, R, B
Chinese mint, Hyptis or Mint Weed	*Hyptis sauvcolens* Poit.	Labiatae	S
Chili pepper, Bird pepper, Tabasco Pepper	*Capsicum frutescens* L.	Solanaceae	F
Cashew	*Anacardium occidentale* L.	Anarcadiaceae	L
Desert date	*Balanites aegyptiaca* (L.) Del.	Zygophyllaceae	R
Dalbergia	*Dalbergia saxatilis* Hook. f.	Fabaceae	L, B
Fish-poison bean, or Vogel's Tephrosia	*Tephrosia vogelii* Hook.f.	Fabaceae	L
Gamhar, Gmelina, Gumhar, Malay beechwood, Malay bush beech, Snapdragon	*Gmelina arborea* Roxb.	Verbenaceae	L
Garlic	*Allium sativum* L.	Alliaceae	L
Henna Plant	*Lawsonia inermis* L.	Lythraceae	L
Horsewood	*Clausena anisata* (Wild.) Hook. f. ex Benth.	Rutaceae	L, R
Hemp, Hashish, Mary-Jane pot, Marijuana	*Cannabis sativa* L.	Cannabaceae	L, S, F
Mums or Chrysanths	*Chrysanthemum coccineum* Wild.	Asteraceae	L, F
Neem, Nim tree or Indian Lilac	*Azadirachta indica* A. Juss.	Meliaceae	L, B, R,
Pawpaw, Papaya	*Carica papaya* L.	Caricaceae	R, B
Pepper fruit	*Dannettia tripetala* Barker F.	Annonaceae	L
Plum mango	*Lannea acida* Λ. Rich.	Anacardiaceae	B
Sweet wormwood, Sweet annie, Sweet sagewort, Annual mugwort or Annual worm	*Artemisia annua* L.	Asteraceae	L, B
Tobacco	*Nicotiana tabacum* L.	Solanaceae	L
Tasmanian Blue Gum, Eurabbie, Blue Gum	*Eucalyptus globules* Labill.	Myrtaceae	L, B
West African black pepper, Ashanti pepper,	*Piper guineense* Schum &Thonn.	Piperaceae	F

Source: (Adapted from Okwute, 2012; Ekefan and Eche, 2013).
*- L = Leaf; B = Bark; S = Seed; R = Root; F = Fruit

Researches on the potential of plant probiotic microbes in pest management are gaining interest largely within the last two decades. The microbes contribute to plant protection by indirectly or directly antagonizing pest through the release of compounds into the immediate environment (Menendez and Garcia-Fraile, 2017; Manchikanti, 2019). The organisms produce biocidal metabolites and volatile substances. *Bacillus, Mycobacterium, Paenibacillus, Pseudomonas, Rhizobium* and *Streptomyces* species are known to produce a range of plant probiotics (Li *et al.*, 2016) such as butanediol, HCN, certain hydrocarbons, insecticidal proteins, phenazines, and water-soluble compounds, such as phenolics, pyrrolnitrins and siderophores and, other cell wall degrading enzymes (Sivasakthi *et al.*, 2014; Anderson and Kim, 2018). It has been shown that a biopesticide formulated with plant growth promoting rhizobacterial strains of *Pseudomonas putida* and *Rothia* sp. induced defense against *Spodoptera litura* (Muqarab and Bano, 2016). In a recent study, a combination of Fluorescent *Pseudomonas jessenii* strains R62 and *Pseudomonas synxantha* strain R81 protected tomato plants against root-knot nematodes in laboratory trials (Sharma and Sharma, 2017). This area of research will no doubt identify many leads in the use of microbes in pest management.

The last decade has also experienced a scientific revolution in the application of nanoformulations in pest management technologies (Ganguli, 2019; Lade *et al.*, 2019). Nanobiopesticides are pesticidal chemical complexes of nanoparticles (NPs) with biological agents (Pestovsky and Agustino, 2017). Through nanoformulations and microencapsulation techniques which involve combining polymers, metal oxides, and/or active particles with micelles, the stability, bioavailability and residual action of biopesticidal agents can be enhanced thereby increasing field usage of biopesticdes. The antibacterial, antifungal, larvicidal, antiviral and insecticidal actions of nanobiopesticides have been established and efficacies have been achieved via a variety of mechanisms (Ganguli, 2019; Lade *et al.*, 2019). Nanoformulations are generally characterized by increased solubility and spreadability of active ingredients, precise and slow release of active ingredients, faster degradability in soil and slower degradability in plants with residue levels below the regulatory maximum in food and feeds (Bergeson, 2010; Ragaei and Sabry, 2014). Controlled release of nanobiopesticidal active ingredients is achieved by entrapping the bioagent with specific polymers such as carbohydrates or proteins that binds the active compound complex to the nanoparticles enabling prolonged pesticidal activity (Ragaei and Sabry, 2014).

Conventional spraying of pesticides leads to high pesticide loss from spray drifts and splashes, an issue that can hardly occur with encapsulated nanoparticles or nanopesticides with 'specialized sprayers' which can selectively release them in relation to heat, sunlight, or pH conditions (Hoffmann *et al.*, 2007). Studies are still ongoing with respect to delivery devices for nanobiopesticides such as nanospheres, nanoemulsions, nanocontainers, nanoencapsulates and nanocages. Some of which have been patented in various parts of the world (Bergeson, 2010; Prasad *et al.*, 2017; Ganguli, 2019; Lade *et al.*, 2019). These delivery devices have the potential of protecting the active pesticide ingredients from degradation, improving their effectiveness, and substantially reducing the quantity of pesticides used by at least 10 – 15 times when compared with application of conventional formulations (OECD and Allianz, 2008; Lade *et al.*, 2019).

In China, Sichuan Academy of Agricultural Science have developed plant-based nanopesticides (<100 nm) which are both environmentally friendly and highly efficient (Gao, 2006). A pheromone-based nanogel has also been successfully used against *Bactrocera dorsalis* infesting Guava fruits (Bhagat *et al.*, 2013; Routray *et al.*, 2016). Recent studies have also shown that *Azadirachta indica* (neem) extracts, *Annona squamosa, Acorus calamus, Argimone maxicana, Calotropis procera, Gnidia glauca, Toddalia asiatica* and *Vitex negundo* are pesticidal when formed as stable metallic nanoparticles (Lall *et al.*, 2014; Benelli, 2016; Nuruzzaman *et al.*, 2016; Lade *et al.*, 2019). A study has shown also that the use of nanoparticles effectively protects neem (*A. indica*) oil from rapid breakdown leading to prolonged effect on target pests (Damalas and Koutroubas, 2017). That the phytotoxicity of nanoparticles can be significantly suppressed by coating them with nanocoatings eg., biocompatible polyvinylpyrrole has also been demonstrated on Ag nanoparticles (Lu *et al.*, 2010). However, that nano-products can be harmful to the environment has been shown with respect to pyrethrin-based nanoparticles (30–100 nm) of novaluron (a water-insoluble insect growth regulator against *Spodoptera littoralis* larvae). Hence, environmental safety of NP-based formulations needs rigorous testing (Elek *et al.*, 2010).

Nanotechnology is therefore capable of developing less toxic biopesticides with higher safety profiles and increased stability of the active ingredients, thereby enhancing its pesticidal activity for easier adoption by the end-users (Agrawal and Rathore, 2014; Prasad *et al.*, 2014). There are rising interests in this pest management technology as some botanicals such as geraniol, eugenol, thymol and Malavone CS with yeast cells as matrix (<500 nm) have been formulated as nanobiopesticide by Eden Research UK for commercial use in greenhouse and field for kiwis and grapes (De Oliveira *et al.*, 2018).

iii. Biopesticide as a tool in integrated pest management
Biopesticides have been found to be cheaper and effective alternatives to synthetic chemical pesticides in integrated pest management programmes. They have a narrow pest range (suppressing only target pests and closely related organisms), faster biodegradability, potent with small quantities, natural in occurrence, capacity of replicating (microbial pesticides) in or on target pest ensuring sustainability of products, feasibility of delivery/application by conventional spray equipments, lower cost of development vis-à-

vis traditional pesticides, contains low volatile organic chemicals which reduces air pollution, and have a unique and complex mode of action which negates resistance development (Matanmi, 2011; O'Neal and Dara, 2017). They have high margins of safety for applicators/farm workers and generally short, if any pre-harvest and/or restricted-entry intervals (REIs) allowing for efficient agronomic practices and amenable to global crop market where there are stricter regulations on residue levels. Additionally, biopesticides are produced through environmentally friendly and sustainable production processes as wastes from fermentation processes (used in producing microbial pesticides) can be applied as fertilizers to promote the rapidly growing organic farming system. Biopesticides have also been shown to better yield when applied alone or in combination with other crop protection products with good return on investments (Chandler, 2011; Matanmi, 2011; Sporleder and Lacey, 2013; Eze *et al.*, 2016). Their advantages are triggering increased usage in farming, landscaping and home gardening.

Increasing awareness that the application of biopesticides is safer and more sustainable than conventional pesticides is promoting interest in their use as alternatives and particularly, as components of Integrated Pest Management (IPM) strategies. Experts have recommended IPM as the way out of the pesticide treadmill, and the EU has centrally placed it within its 2009 Pesticides Sustainable Use Directive (European Parliament, 2010). IPM is an approach that systemically combines different crop protection techniques, while monitoring carefully the dynamics of pests and their natural enemies (Chandler, 2011). It entails combining different pest control strategies to overcome the shortcomings of individual strategies. It is also targeted at suppressing pest populations below economic threshold levels. The University of California Integrated Pest Management Program (UCIPM) (2017) describes IPM as an ecologically-based technique that is focused at long-term prevention of pests or their damage through a combination of strategies which include biological control, modification of cultural practices, habitat manipulation and use of resistant varieties, and that pesticides are only used after monitoring proves that they are needed on the basis of established guidelines, and also that treatments are made with the aim of removing only the target pestiferous organism(s) using pest control materials that are selected and applied in a manner that minimizes risks to human health, non-target organisms, and the environment.

Since IPM encourages the development and use of less harmful substances, its success may depend partly on the adoption/use of biopesticides. Even though, most biopesticides are not as effectual as conventional pesticides, their relative higher safety and specificity to target pest shows that they can play a complementary role in pest management as indicative in the success of combining *B. bassiana* with invertebrate predators against two-spotted spider mites (Ullah and Lim, 2011). It is established for example, that Spider mites populations are routinely suppressed by regular releases of invertebrate predators in greenhouses. However, sometimes break down in control are observed. Hitherto, producers complemented with synthetic chemical pesticides which have triggered resistance aside suppressing insect natural enemies. Since, *B. bassiana* is efficacious against spider mites, has a short preharvest interval, and is compatible with predators; it is now recommended across Europe as a supplementary treatment for spider mite on greenhouse crops (Chandler *et al.*, 2011). However, it is worth mentioning that, the very high specificity of biopesticidal products might be disadvantageous when a pest complex is to be dealt with.

Many biopesticides have secondary benefits when used on crop plants making them good tools for IPM. For example, some microbial pesticides enhance the uptake of nutrients by plants while antagonizing the pathogens or pest that attack the same crop (Glare *et al.*, 2012; Manchikanti, 2019). This is seen in the case of the entomopathogenic fungal species; *B. bassiana* and *M. anisopliae* which were originally thought to be insect pathogens only. It is now however known that, they also function as plant endophytes, plant disease antagonists, rhizosphere colonizers and plant growth promoters (Vega *et al.*, 2009; Chandler *et al.*, 2011). They can thus be well used in IPM programmes.

In recent times, many progressive IPM professionals are integrating biopesticides with traditional pest management techniques. However, to ensure the use of best practices in the application and integration of biopesticides, and to facilitate better understanding of their distinct modes of action; there is need for more researches, education and training programmes (Kumar and Singh, 2015; https://www.epa.gov/pesticides/biopesticides#what, accessed November 11, 2019; O'Neal and Dara, 2017).

iv. Biopesticide market: current status and future projections
The biopesticide market experienced a rapid growth in the early 21st century as about 1400 biopesticidal products were reported to be sold world over accounting for about 2.5% of the global pesticide market (Marrone, 2007). A more recent report indicates that Biopesticides accounts for some US$3 billion of the US$61 billion global pesticide market (Marrone, 2019). It is projected that the global biopesticide market which was dominated by North America in 2018, will hit 6.4 billion US$ in 2023 as against the estimated 3.0 billion US$ of 2018 (Figure 1) with a Compound Annual Growth Rate (CAGR) of about 16% and that, biopesticide is expected to be the fastest growing sector of the crop protection market (Chandler *et al.*, 2012; https://www.marketsandmarkets.com/Market-Reports/biopesticides-267.html, accessed November 11, 2019). While it is expected that the market size of biopesticides will match synthetics between late 2040s and early 2050s (Figure 2), reservations mounts particularly, due to the slow rate of development and adoption of

biopesticides, particularly in Africa and Southeast Asia (Olson, 2015; Damalas and Koutroubas, 2018).

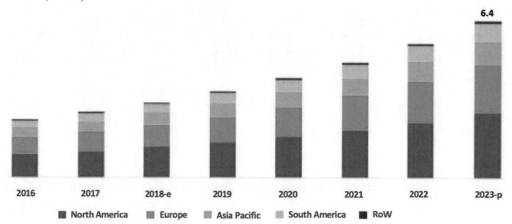

Figure 1. Global biopesticide market growth by regions in billions of USD
e - Estimated; **p** - Projected; **RoW** - Rest of the World
Source: https://www.marketsandmarkets.com/Market-Reports/biopesticides-267.html, accessed November 11, 2019

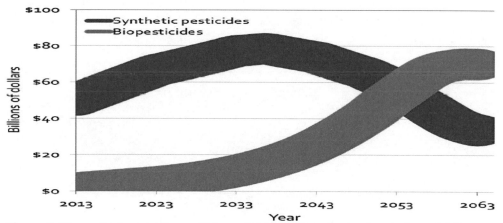

Figure 2: Theoretical projections of biopesticides and synthetic chemical pesticides
Source: Lux Research, Inc. www.luxresearchInc.com, cited in Olson, 2015

v. Factors influencing growth of biopesticide market
A recent analysis indicates that, the biopesticide market is experiencing a three times growth rate than that of conventional chemicals, and that a number of factors are responsible for this growth pattern (Chandler *et al.*, 2011; Balog *et al.*, 2017). The factors include; the exponential world's population growth which is placing a huge burden on agricultural production. The need to reduce the negative impact of traditional pesticides on the environment through sustainable pest control methods places biopesticides as a necessary

alternative (Shukla and Shukla, 2012; SEIPASA, 2018). Another factor is the growing awareness and market demand for residue-free food which is becoming more and more powerful. This has resulted to more organic products in the shelves of shops and super-markets in some developed nations (Shukla and Shukla, 2012; Eze *et al.*, 2016). Available statistics reveals that in 2016 for example, Spain organically cultivated over 2 million hec-tares. That figure was 8.5% higher than that of 2015 (La Vanguardia, 2017; SEIPASA, 2018). According to Eurostat data, organic farming rose by about 18.7% between 2012 and 2017 in the EU (Eurostat, 2017). The adoption of rotating biopesticide usage with conven-tional chemicals to reduce residue levels in food, faster approvals by regulatory bodies, lower cost of raw materials, banning of active ingredients of some synthetic chemical prod-ucts and implementation of IPM strategies are also all promoting biopesticide develop-ment, acceptance and usage (Yadav *et al.*, 2013; Dara, 2016; O'Neal and Dara, 2018b).

Additionally, higher pest infestation and development of resistance to existing chemical pesticides leading to heavy yield losses, increasing costs connected with outrageous use of synthetic chemicals coupled with the high and rising cost of developing chemical pesti-cides (biopesticides require up to 6 years and about $20 million for development, synthetic pesticides require up to 10 years and over $200 million) as shown in Figure 3 and gross reduction in the launchings of synthetic products have all favored biopesticide develop-ment and use (REBECA, 2007; Eze *et al.*, 2016; McDougall, 2016; O'Neal and Dara, 2018a).

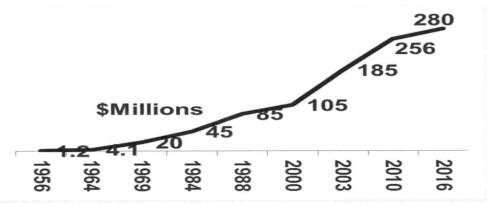

Figure 3: Costs of discovering and developing a synthetic agricultural chemical in Millions of US$
Source: O'Neal and Dara, 2018a

The introduction and implementation of regulatory processes appropriate for biopesticides in some countries plus stricter regulatory restrictions on synthetic chemical pesticides have promoted biopesticides against synthetic chemical pesticides (Chandler *et al.*, 2011;

Erbach, 2012; O'Neal and Dara, 2018b). In the US for example, the regulatory process for biotechnologically modified plants has developed into one in which regulations are dealt with on individual basis. As such, every new crop/gene combination requires separate submissions to the USDA and then to either the FDA and/or the EPA depending on the crop (feed or food) and the plant incorporated protectant, respectively (Hood et al., 2012). In the last two decades, a global meta-analysis of 147 studies showed that, aside reducing the use of synthetic chemical pesticides by 37%, adoption of GM technology has averagely increased crop yields and farmers' returns by 22 and 68%, respectively (Klumper and Qaim, 2014).

vi. Obstacles to the development and use of biopesticides

Researches on biopesticides were generally abysmal before 1995. This has however changed in recent times as more and newer biopesticides are being evaluated and introduced to the market (Sporleder and Lacey, 2013). However, despite appreciable increase in the use of biopesticides globally as outlined above, a number of factors impede its development and use. The factors include; negative perception attributed to insufficient awareness on its cost, efficacy and safety (particularly, in the case of GM crops) mainly among stake holders, small-scale growers and policy makers. Hence, a survey done in the USA suggests that biopesticide companies need to give more education to growers and the public on the products and their usage (Marrone, 2007; Marrone, 2019). The lack of trust among producers, marketers, buyers and users/potential users of biopesticides and particularly, microbial biopesticides vis-à-vis the perceived risk of importation and use (Kumar and Singh, 2015) also impedes the development and use of biopesticides.

In some countries, an important barrier to biopesticide development, use and commercialization is the tedious registration/regulatory process aimed at ensuring safety and consistency of the products. For example, unlike in the United States, Brazil, India and China, the EU and many other countries has fewer registered biopesticide-active substances, owing to the lengthy and complex process of registration which was modeled after those for conventional pesticides and scantily captures the peculiarities of 'biological products' (Bullion, 2018; Damalas and Koutroubas, 2018; Jampílek and Kráľová, 2019). The data required for registering new bioactive agents includes; mode of action, toxicological and eco-toxicological testings and host range evaluations. It has been shown that aside the high registration fees, collecting these data is time consuming and thus prohibitive, particularly to academics, small scale industries or developers of small market crops. This is again slowing down the innovation of new biopesticidal-based solutions (Chandler et al., 2011; Pavela, 2014). Simplifying the registration process while, ensuring safety and consistency of the products is therefore critical in the development and use of biopesticides.

The EPA generally requires fewer data at lower cost and timeframe to register a biopesticide than to register a conventional pesticide. This was made possible by modified test methodologies which require less data. Hence, biopesticides are often registered within a year in the US and for chemical pesticides, an average of more than 3 years (Sporleder and Lacey, 2013; Jampílek and Kráľová, 2019). Following the US pattern, Canada reviewed the process for the development and registration of biopesticides with huge successes (Bullion, 2018, Jampílek and Kráľová, 2019).

Some special provisions have been developed in the EU to favor products classified as 'low risk', which includes biopesticides, but the process is adjudged to be practically ineffective, as registration timeframes are still twice as long and double the cost of registration elsewhere, and this situation is discouraging developers from submitting products for regulatory reviews in Europe. However, member states, particularly France, Hungary, Netherlands and the UK have made giant strides in this regard (Bullion, 2018).

Determining whether a product can be classified a biopesticide has been difficult in some cases due to variations in the regulations and requirements for registration in different countries (Sporleder and Lacey, 2013; SEIPASA, 2018) hence, the need for global harmonization of the process and procedures for biopesticide approval is brought to the fore. For instance, species of the fungus *Trichoderma* are used as biopesticides against soil-borne plant pathogenic fungi. However, they also promote plant growth by producing auxin-like compounds (Verma *et al.*, 2007; Vinale *et al.*, 2008). Some *Trichoderma* products were sold as plant growth promoters instead of as plant protection products, thereby escaping the required scrutiny by regulators with respect to their efficacy and safety (Chandler *et al.*, 2011).

At a global level, the OECD's Expert Group on BioPesticides is taking the lead to promote a harmonized and proportionate process of biopesticide regulation, and to develop guidance documents and smoothen information and communication exchange among the critical stakeholders (Bullion, 2018). However, some regions and countries are still lagging in this respect.

Approval or otherwise of a biopesticide product is the prerogative of the experts within the regulatory institutions. Rising numbers of biopesticidal agents seeking registrations vis-à-vis fewer staff and resources in the regulatory institutions are also slowing the review processes (Bullion, 2018). When the regulators are lacking expertise on biopesticides, they may delay the approval, or request the applicants to provide more data, some of which may not be appropriate for biopesticides or may be primarily for synthetic chemical pesticide registration. A number of regulatory institutions, particularly in the UK have realized that regulating biopesticides with the same model for chemical pesticides has been a barrier to its development, and use and have done much to effect review on the data requirements (Chandler *et al.*, 2011; Isman, 2015).

The capital intensive and highly competitive marketplace dominated by multibillion-dollar synthetic pesticide companies with their well developed and organized market are obstacles for biopesticide companies. Similarly, well established biopesticide companies like Monsanto, Syngenta, Bayer, DuPont, BASF and Dow and their products makes it difficult for small and/or medium biopesticide companies and their products to make inroads into the market mainly because of the huge organizational, financial and technological requirements for developing, commercializing and marketing biopesticides (Fountain and Wratten, 2013). Since biopesticides are largely niche products and may be targeted on non commodity crops coupled with the relatively expensive and tasking nature of biopesticide registration and production in some countries, commercialization can be deterred (Chandler *et al.*, 2011). Also, owing to limited commercial use of biopesticides, they are often developed by research institutes and other less organized setups rather than by the conventional industries. This may be responsible for some of the inappropriate formulations which are less efficient particularly in the developing countries (Sporleder and Lacey, 2013).

Despite substantial progress in the global acceptance and utilization of biopesticides, popularity and usage have remained low in agrarian developing countries, particularly in Africa and Southeast Asia. This has largely been hinged on high cost due to insufficient indigenous industrial production, low efficacy in some cases, inconsistent performance in the field, aside weak enforcement as clearly seen with respect to Nigeria's Pesticide Regulation (Glare *et al.*, 2012; Ekefan and Eche, 2013; McDougall, 2016; Ivase *et al.*, 2017). Additionally, insufficient awareness, mostly among rural farmers, and poor advocacy have hampered the development and use of biopesticides in Africa and Nigeria, in particular (Ivase *et al.*, 2017). Insufficient knowledge and technological skills tied to poor funding of research and development infrastructure for the development and promotion of biopesticides in developing countries is another critical factor. Weak synergy between the academia/research institutes and the pesticide industries is also a key factor (Oruonye and Okrikata, 2010; Ivase *et al.*, 2017). Other militating factors influencing the current status of biopesticide use in Nigeria and Africa in general are: slow mode of action of biopesticides vis-à-vis synthetic chemicals pesticides, shorter shelf life and persistence of biopesticidal agents leading to low profit potential due to repeated application, higher specificity and unavailability of biopesticides in the market, to mention but a few (Koul, 2011; Ekefan and Eche, 2013; Ivase *et al.*, 2017).

Conclusion and the Way Forward to Enhancing Sustainable Development and Utilization of Biopesticides

The provision of an almost unlimited bioactive pesticidal molecules by nature and the scientific advancements in isolation and characterization of these compounds is changing the over reliance on synthetic chemicals. That biopesticides are advantageous as they ensure safe environment and provide healthy food for human consumption is largely not in doubt. However, some factors previously mentioned such as slow action, 'high cost', production difficulties, inappropriate formulations and negative perception due to previous poor performance amidst others limit their full adoption in pest and pathogen management across the world. In some cases, higher quantities of the active compounds are required for effectiveness under field conditions, and regulatory procedures continues to be tedious in many countries as processes designed for evaluating synthetic chemical pesticides are unjustifiably applied on biopesticides.

To enhance sustainable development and utilization of biopesticides, a number of technological and policy gaps have been identified at national and international levels that need to be addressed. Researchers for example, are currently using molecular-based technologies to reorganize the phylogeny of microbial natural enemies and segregate the molecular basis for their pathogenicity. The technology is also used to decipher crop-weeds and weeds–herbicides interactions, and for insects; to characterize the receptor proteins used for detecting semiochemicals. Such technology if well harnessed, will give information that will provide further insights into biopesticides–pests interactions with the possibilities for improving the efficacy of biopesticides. The application and improvement of high throughput screening (HTS) in assaying potential biopesticides should also be encouraged. The need for Agrochemical companies to change their market and, research and developments strategies in favor of biopesticides due to the build-up of resistance to conventional chemical pesticides, and the evolving legislations for the protection of man and the environment are so obvious.

The potential of nanobiopesticide is great. However, future research need to focus on ways of overcoming the risk factors associated with the use of nanoparticles as comprehensive understanding of their risk factors on the ecosystem is currently lacking. While there are certainly ongoing studies aimed at unraveling this, the technology is still far from being well developed as there are still basic issues pertaining to release rates, storage stability and cost effectiveness that need to be attended to.

Policy measures need to be strengthened to minimize indiscriminate use of chemical pesticides while promoting the use of biopesticides. To achieve this, a public-private sector approach to policy making and implementation with respect to the development, registration, sales and utilization of biopesticides particularly, in developing countries should be given due attention. This is because many times critical stakeholders particularly, farmers (the main users of the products) are sidelined in policy making and reduced to 'policy takers'.

Owing to the slow efficacy of biopesticides, it may be difficult for users to drop the widely accepted and adopted 'fast acting' conventional pesticides. However, the adoption of IPM practices will enhance the use of biopesticides as conventional pesticides will only be used as an alternative pest management tool to minimize the development of resistance in pest populations thereby promoting the use of biopesticides and other environmental friendly strategies.

The regulatory and registration process should ensure quality and safety of biopesticidal products. However, the regulatory agencies should fast-track the process on the basis of justified regulations. Additionally, the regulatory system should be supportive of the development and growth of small and medium-sized biopesticide firms to enable them provide reliable, available, effective and economical pest control tools. These can be achieved by constituting dedicated specialist regulatory team with expertise in different types of biopesticide. Developing and harmonizing guidance documents specifically for biopesticides and running training programs for regulators and applicants is also very important. Pre-submission interactions between regulators and applicants should also be encouraged to identify, in good time, potential grey areas in the review process.

Governments, particularly, in developing countries should as a matter of policy encourage commercial investors and pesticide industries to take up biopesticide enterprises and stringently enforce policies on chemical pesticide minimization. The noticeable inadequate research funds, personnel, cutting edge equipment/technologies and, other bureaucratic impediments coupled with insufficient synergy between industry and academia should be sufficiently addressed to promote research and developments for the purpose of enhancing the development and commercialization of home-grown biopesticide technologies, particularly, in the developing countries of Africa and Southeast Asia.

References

Anderson, A. and Kim, Y, C. (2018) Biopesticides produced by plant-probiotic Pseudomonas chlororaphis isolates. *Crop protection*, 105, 62-69. doi: 10.1016/j.cropro.2017.11.009

Agrawal, S. and Rathore, P. (2014) Nanotechnology pros and cons to agriculture: A review. *International journal of current microbiology and applied sciences*, 3, 43–55.

Agrow (2018) Biologicals 2018: an analysis of corporate, product and regulatory news in 2017/2018. Retrieved 20 October 2019 from https://agrow.agribusinessintelligence.informa.com/-/media/agri/agrow/ag-marketreviews-pdfs/supplements/agrow_biologicals_2018_online_v2.pdf

Aiuchi, D., Inami, K., Kuramochi, K., Koike, M., Sugimoto, M., Tani, M. and Shinya, R. (2008) A new method for producing hybrid strains of the entomopathogenic fungus *Verticillium lecanii* (*Lecanicillium* spp.) through protoplast fusion by using nitrate non-utilizing (nit) mutants. *Micologia aplicada international*, 20, 1–16.

Aktar, M. W., Sengupta, D., and Chowdhury, A. (2009) Impact of pesticides use in agriculture: their benefits and hazards. *Interdisciplinary toxicology*, 2(1), 1–12. doi: 10.2478/v10102-009-0001-7

Arjjumend, H. and Koutouki, K. (2018) Science of biopesticides and critical analysis of indian legal frameworks regulating biocontrol agents. *International journal of agriculture, environment and biotechnology*, 11(3), 563-571. doi: 10.30954/0974-1712.06.2018.20

Balog, A., Hartel, T., Loxdale, H. D. and Wilson, K. (2017) Differences in the progress of the biopesticide revolution between the EU and other major crop-growing regions. *Pest management science*, 73(11), 2203–2208. doi: 10.1002/ps.4596

Benelli, G. (2016) Plant-mediated biosynthesis of nanoparticles as an emerging tool against mosquitoes of medical and veterinary importance: a review. *Parasitology research*, 115(1), 23-34. doi: 10.1007/s00436-015-4800-9

Bergeson, L. L. (2010) Nanosilver: US EPA's pesticide office considers how best to proceed. *Environmental quality management*, 19(3). 79-85. doi: 10.1002/tqem.20255

Bhagat, D., Samanta, S. K. and Bhattacharya, S. (2013) Efficient management of fruit pests by pheromone nanogels. Scientific Reports. 3, 1294. doi: 10.1038/srep01294

BPIA (2014) Biopesticides Offer Multiple Benefits for Agricultural Dealers and Consultants Retrieved 12 October 2019 from http://www.bpia.org/wp-content/uploads/2014/01/dealer_consultant-final.pdf

BPIA (2019) History of biopesticides. Retrieved 10 November 2019 from http://www.bpia.org/history-of-biopesticides/

Bullion, A. (2018) US still leads the way in biopesticides, as EU rules remain complex. *In:* Biologicals 2018: analysis of corporate, product and regulatory news in 2017/2018.

Retrieved 14 October 2019 from https://agrow.agribusinessintelligence.in-forma.com/-/media/agri/agrow/ag-market-reviews-pdfs/supplements/agrow_bio-logicals_2018_online.pdf

Chandler, D., Bailey, A. S., Tatchell, G. M., Davidson, G., Greaves. J. and Grant, W. P. (2011) The development, regulation and use of biopesticides for integrated pest management. *Philosophical transactions of the royal society B,* 366, 1987–1998. doi: 10.1098/rstb.2010.0390

Damalas, C. A. and Koutroubas, S. D. (2018) Current status and recent developments in biopesticide use. Agriculture, 8, 13, doi: 10.3390/agriculture8010013 www.mdpi.com/journal/agriculture

Dara, S. K. (2016) IPM solutions for insect pests in California strawberries: efficacy of botanical, chemical, mechanical, and microbial options. CAPCA Adviser 19(2), 40-46.

De Oliveira, J. L., Campos, E. V. R., Bakshi, M., Abhilash, P. C. and Fraceto, L. F. (2014) Application of nanotechnology for the encapsulation of botanical insecticides for sustainable agriculture: prospects and promises. *Biotechnology advances,* 32, 1550–1561. doi: 10.1016/j.biotechadv.2014.10.010

Erbach, G. (2012) Pesticide legislation in the EU: towards sustainable use of plant protection products. Library briefing, Library of the European Parliament 120291REV1: 1-6. Retrieved 13 November 2019 from http://www.europarl.europa.eu/Reg-Data/bibliotheque/brief-ing/2012/120291/LDM_BRI%282012%29120291_REV1_EN.pdf

Ekefan, E.J. and Eche, C.O. (2013) Indigenous biopesticides use and biodiversity management. Proceedings of the 9th National Conference on Organic Agriculture in Nigeria, 11-15 November, 2013. pp. 45–63.

EFSA (2018) Guidance on the characterisation of microorganisms used as feed additives or as production organisms. doi: 10.2903/j.efsa.2018.5206

Elanchezhyan K. and Vinothkumar B. (2015) Neem: an ecofriendly botanical insecticide in pest management. *The journal of insect science. photon,* 116, 207-217

Elek, N., Hoffman, R., Ravi, U., Resh, R., Ishaaya, I. and Magdassia, S. (2010) Novaluron nanoparticles: formation and potential use in controlling agricultural insect pests. *Colloids and surfaces A: Physicochemical engineering aspects,* 372, 66–72. doi: 10.1016.j.colsurfa.2010.09.034

El-Wakeil, N. E. (2013) Botanical pesticides and their mode of action. *Gesunde pflanzen,* 65, 125-149. doi: 10.1007/s10343-013-0308-3

EPA (2018) Biopesticide active ingredients. Retrieved 14 November 2019 from https://www.epa.gov/ingredients-used-pesticide-products/biopesticide-active-in-gredients

EPA (2019) Overview of plant incorporated protectants. Retrieved 10 November 2019 from https://www.epa.gov/regulation-biotechnology-under-tsca-and-fifra/overview-plant-incorporated-protectants

Escobar, M. A. and Dandekar, A. M. (2003) *Agrobacterium tumefaciens* as an agent of disease. *Trends in plant science*, 8(8), 380-396.

European Parliament (2010) Pesticides: framework for community action to achieve a sustainable use of pesticides. Retrieved 11 October 2019 from http://www.europarl.europa.eu/oeil/file.jsp?id=5372322

Eurostat (2017) Organic farming statistics. Retrieved 11 October 2019 from https://ec.europa.eu/eurostat/statistics-explained/index.php/Organic_farming_statistics

Eze, S. C., Mba, C. L. and Ezeaku, P. I. (2016) Analytical review of pesticide formulation trends and application: the effects on the target organisms and environment. *International journal of science and environmental technology*, 5(1), 253-266

FAO (2009) How to feed the world in 2050. *In:* World Agricultural Summit on Food Security 16 – 18 November 2009. Food and Agriculture Organization of the United Nations, Rome. Retrieved 17 November 2019 from http://www.fao.org/fileadmin/templates/wsfs/docs/expert_paper/How_to_Feed_the_World_in_2050.pdf

Fountain, E. D. and Wratten, S. D. (2013) Conservation biological control and biopesticides in agricultural. *In:* Elias, S. A. (ed). Reference module in earth systems and environmental sciences. Elsevier. Amsterdam. doi: 10.1016/B978-0-12-409548-9.00539-X

Ganguli, P. (2019) Patenting issues in the development of nanobiopesticides. *In:* Koul, O. (ed). Nano-biopesticides today and future perspectives. Pp. 367-395. doi: 10.1016/B978-0-12-815829-6.00017-6

Gao, S. H. (2006) Advent of nano-bio-pesticides. Pesticides Market News, pp. 2-3.

Glare, T., Caradus, J., Gelernter, W., Jackson, T., Keyhani, N., Köhl, J., Marrone, P., Morin, L. and Stewart, A. (2012) Have biopesticides come of age? *Trends in biotechnology*, 30(5), 250–258.

Gupta, S. and Dikshit, A. (2010), Biopesticides: an eco-friendly approach for pest control. *Journal of biopesticides*, 3(1), 186–188.

Hikal, W. M., Baeshen, R. S. and Said-Al Ahl, H. A. H. (2017) Botanical insecticide as simple extractives for pest control. *Cogent biology*, 3:1, doi: 10.1080/23312025.2017.1404274

Hood, E. E., Requesens, D. V. and Eversole, K. A. (2012) Regulatory issues of biotechnologically-improved plants. *In:* Altman, A. and Hasegawa, M. (eds). Plant biotechnology and agriculture, prospects for the 21st Century; 2011; Academic Press, Elsevier, Amsterdam; pp. 541-550. doi: 10.1016/B978-0-12-381466-1.00034-1

Hoffmann, M., Holtze, E. M. and Wiesner, M. R. (2007) Reactive oxygen species generation on nanoparticulate material, *In:* M.Wiesner, R. and Bottero, J. Y. (ed).

Environmental nanotechnology: applications and impacts of nanomaterials. New York, NY: McGraw Hill, pp 155–203.

Hubbard, M., Hynes, R. K., Erlandson, M. and Bailey, K. L. (2014) The biochemistry behind biopesticide efficacy. *Sustainable chemical processes*, 2, 18, doi: 10.1186/s40508-014-0018-x

ISAAA (2017) Global Status of Commercialized Biotech/GM Crops: 2017. ISAAA Brief No. 53. ISAAA: Ithaca, NY.

Isman, M.B. (2015) A renaissance for botanical insecticides? *Pest management science*, 71(12), 1587–1590. doi: 10.1002/ps.4088

Ivase, T. J. P., Nyakuma, B. B., Ogenyi, B. U., Balogun, A. D. and Hassan, M. N. (2017) Current status, challenges, and prospects of biopesticide utilization in Nigeria. *Acta universitatis sapientiae agriculture and environment*, 9, 95-106. doi: 10.1515/ausae-2017-0009

Jampílek, J. and Kráľová, K. (2019) Nanobiopesticides in agriculture: state of the art and future opportunities. *In:* Koul, O (ed). Nano-biopesticides today and future perspectives, pp. 397–447. doi: 10.1016/B978-0-12-815829-6.00018-8

Kachhawa, D. (2017) Microorganisms as a biopesticides. *Journal of entomology and zoology studies*, 3, 468-473.

Kiewnick, S. (2007) Practicalities of developing and registering microbial biological control agents. *CAB Reviews: Perspectives in agriculture, veterinary science, nutrition and natural resources,* 2(13), 1–11.

Klumper, W. and Qaim, M. (2014) A meta-analysis of the impacts of genetically modified crops. *PLoS One* 9:e111629. doi: 10.1371/journal.pone.011162

Koul, O. (2011) Microbial biopesticides: opportunities and challenges. *CAB Reviews: Perspectives in agriculture, veterinary science, nutrition and natural resources*, 6, 56. doi: 10.1079/PAVSNNR20116056

Koul, O. (2012) Plant biodiversity as a resource for natural products for insect pest management. *In:* Gurr, G. M., Wratten, S. D., Snyder, W. E. and Read, D. M. Y. (eds). Biodiversity and insect pests: key issues for sustainable management. John Wiley and Sons Ltd., West Sussex, pp. 85–105.

Koul, O. (2016) Naturally occurring insecticidal toxins. CAB International, Wallingford. 850 pp.

Kumar, S. (2013) The role of biopesticides in sustainably feeding the nine billion global populations. *Journal biofertilizer biopesticide*, 4(2), e114. doi: 10.4172/2155-6202.1000e114

Kumar, D. and Kalita, P. (2017) Reducing postharvest losses during storage of grain crops to strengthen food security in developing countries. *Foods*, 6 (8), doi: 10.3390/foods6010008

Kumar, S. and Singh, A. (2015) Biopesticides: present status and the future prospects. *Journal of fertilizers and pesticides*, 6, e129. doi: 10.4172/2471-2728.1000e129Kumar, K. K, Sridhar, J., Murali-Baskaran, R. K., Senthil-Nathan, S., Kaushal, P., Dara, S. K. and Arthurs, S. (2019) Microbial biopesticides for insect pest management in India: current status and future prospects. *Journal of invertebrate pathology*, 165, 74-81. doi: 10.1016/j.jip.2018.10.008

Lade, B. D., Gogle, D. P., Lade, D. B, Moon, G. M., Nandeshwar, S. B. and Kumbhare, S. D. (2019) Nanobiopesticide formulations: application strategies today and future perspectives. *In:* Koul O. (ed.). Nano-biopesticides today and future perspectives. Pp. 179 – 206. doi: 10.1016/B978-0-12-815829-6.00007-3

Lall, D., Summerwar, S. and Pandey, J. (2014) Bioefficacy of plant extract against larvae of American bollworm, *Helicoverpa armigera* (Noctuidae: Lepidoptera) special reference to the effect on peritrophic membrane. *In:* International Conference on Chemical, Civil and Environmental Engineering, Nov 18-19, 2014 Singapore, pp. 21–23. doi: 10.15242/IICBE.C1114026

La Vanguardia (2017) Spain maintains the European leadership in organic farming. Retrieved 14 October 2019 from https://www.lavanguardia.com/natural/20171116/432922649829/espana-lider-ue-hectareas-agricultura-ecologica.html

Lengai, G. M. W. and Muthomi, J. W. (2018) Biopesticides and their role in sustainable agricultural production. *Journal of biosciences and medicines*, 6, 7-41. doi: 10.4236/jbm.2018.66002

Li, Z., Alves, S. B., Roberts, D. W., Fan, M., Delalibera, I., Tang, J., Lopes, R. B., Faria, M. and Rangel, D. E. M. (2010) Biological control of insects in Brazil and China: history, current programs and reasons for their success using entomopathogenic fungi. *Biocontrol science and technology*, 20, 117–136.

Li, H., Ding, X., Wang, C., Ke, H., WU, Z., Wang, Y., Liu, H. and Guo, J. (2016) Control of Tomato yellow leaf curl virus disease by *Enterobacter asburiae* BQ9 as a result of priming plant resistance in tomatoes. *Turkish journal of biology*, 40, 150–159.

Lu, W., Senapati, D., Wang, S., Tovmachenko, O., Singh, A.K., Yu, H. and Ray, P. C. (2010) Effect of surface coating on the toxicity of silver nanomaterials on human skin keratinocytes. *Chemical physics letters*, 487, 92–96. doi: 10.1016/j.cplett.2010.01.027

Marrone, P. G. (2007) Barriers to adoption of biological control agents and biological pesticides. *CAB Reviews: Perspectives in agriculture, veterinary science, nutrition and natural resources*, 2(15), doi: 10.1079/PAVSNNR2007205

Marrone, P. G. (2019) Pesticidal natural products – status and future potential. *Pest management science*, 75(9), 2325-2340. doi: 10.1002/ps.5433

Matanmi, B.A. (2011) Use of biopesticides in agriculture. *Nigerian journal of plant protection*, 25(1). 1-12.

Manchikanti, P. (2019) Bioavailability and environmental safety of nanobiopesticides. *In:* Koul O. (ed). Nano-biopesticides today and future perspectives. Pp. 207 – 222. doi: 10.1016/B978-0-12-815829-6-00008-5

McDougall, P. (2016) The cost of new agrochemical product discovery, development and registration in 1995, 2000, 2005-8 and 2010-2014. R&D expenditure in 2014 and expectations for 2019. A consultancy study for CropLife International, CropLife America and the European Crop Protection Association, pp 41. Retrieved 11 November 2019 from https://croplife.org/wp-content/uploads/2016/04/Cost-of-CP-report-FINAL.pdf

Menendez, E. and Garcia-Fraile, P. (2017) Plant probiotic bacteria: solutions to feed the world. *AIMS Microbiology*, 3(3), 502-524. doi: 10.3934/microbiol.2017.3.502

Mondal, K and Parween, S. (2000) Insect growth regulators and their potential in the management of stored-product insect pests. *Integrated pest management reviews*, 5(4), 255–295. doi: 10.1023/A:1012901832162

Muqarab, R. and Bano, A. (2016) Plant defence induced by PGPR against *Spodoptera litura* in tomato. *Plant biology*, 19, 406–412.

Nuruzzaman, M., Rahman, M. M., Liu, Y. and Naidu, R. (2016) Nanoencapsulation, nanoguard for pesticides: a new window for safe application. *Journal of agricultural and food chemistry*, 64, 1447-1483. doi: 10.1021/acs.jafc.5b05214

OECD and Allianz (2008) Opportunities and risks of nanotechnologies. OECD International Futures Programme, pp. 1-44. Retrieved 14 November from https://www.oecd.org/science/nanosafety/44108334.pdf

Okrikata, E. and Ogunwolu, E. O. (2017) Farmers' perceptions on arthropod pests of watermelon and their management practices in the Nigerian Southern Guinea Savanna. *International journal of agricultural research,* 12(4), 146–155. doi: 10.3923/ijar.2017.146.155

Okrikata, E. and Ogunwolu, E. O. (2019) Determination of the critical period of cyperdiforce® treatment against arthropod fauna and productivity of watermelon. *Iraqi journal of science*, 60(9), 1904- 919. doi: 10.24996/ijs.2019.60.9.3

Okrikata, E., Ogunwolu, E. O. and Ukwela, M. U. (2019) Efficiency and economic viability of neem seed oil emulsion and cyper-diforce® insecticides in watermelon production within the Nigerian Southern Guinea Savanna Zone. *Journal of crop protection*, 8(1), 81–101. doi: 10.13140/RG.2.2.20031.36001

Okunlola, A. I. and Akinrinnola, O. (2014) Effectiveness of botanical formulations in vegetable production and bio-diversity preservation in Ondo State, Nigeria. *Journal of horticulture and forestry*, 1, 6-13.

Okwute, S. K. (2012), Plants as potential sources of pesticidal agents: a review. *In:* Soundararajan, R. P. (ed). Pesticides – advances in chemical and botanical pesticides. InTech: USA. 207–323. doi: 10.5772/46225

Olson, S. (2015) An analysis of the biopesticide market now and where it is going. *Outlooks on pest management*, 26, 203-206. doi: 10.1564/v26_oct_04

Oluwatobi, O. (2018) Progress needed to characterize the active agents in biologicals. *In:* Biologicals 2018: analysis of corporate, product and regulatory news in 2017/2018. Retrieved 14 October 2019 from https://agrow.agribusinessintelligence.in-forma.com/-/media/agri/agrow/ag-market-reviews-pdfs/supplements/agrow_bio-logicals_2018_online.pdf

O'Neal, M. and Dara, S. K. (2017) Biopesticides and integrated pest management. *E-Journal of entomology and biologicals*. Retrieved 13 November 2019 from https://ucanr.edu/blogs/blogcore/postdetail.cfm?postnum=25912

O'Neal, M. and Dara, S. K. (2018a) Brief history of botanical and microbial pesticides and their current market. *E-Journal of entomology and biologicals*. Retrieved 10 October 2019 from https://ucanr.edu/blogs/blogcore/postdetail.cfm?postnum=26249

O'Neal, M. and Dara, S. (2018b) Biopesticide development, registration, and commercial formulations. *eJournal on production and pest management practices for strawberries and vegetables*. Retrieved 27 October 2019 from https://ucanr.edu/blogs/blog-core/postdetail.cfm?postnum=26135

Oruonye, E. D. and Okrikata, E. (2010), Sustainable use of plant protection products in Nigeria and challenges. *Journal of plant breeding and crop science*, 2(9), 267–272. Retrieved 14 November 2019 from https://academicjournals.org/journal/JPBCS/article-full-text-pdf/B3CE4C24692

Pavela, R. (2014) Limitation of plant biopesticides. *In:* Singh, D. (ed). Advances in plant biopesticides. Springer Publishing: New Delhi, India, pp. 347–359.

Peace O. O., Uche M. N. and Mariano J. B. (2019) Nigeria approves the commercial release of Bt. pod-borer resistant cowpea. USDA Foreign Agricultural Service, Global Agricultural Information Network (GAIN) Report Number: NG-19003. 5/20/2019

Pestovsky, Y. S. and Agustino, M. (2017) The use of nanoparticles and nanoformulations in agriculture. *Journal of nanoscience and nanotechnology*, 17(12), 8699-8730. doi: 10.1166/jnn.2017.15041

Prasad, R., Bhattacharyya, A. and Nguyen, Q. D. (2017) Nanotechnology in sustainable agriculture: recent developments, challenges, and perspectives. *Frontiers in microbiology*, 8,1014, doi: 10.3389/fmicb.2017.01014

Prasad, R., Kumar, V. and Prasad, K. S. (2014) Nanotechnology in sustainable agriculture: present concerns and future aspects. *African journal of biotechnology*, 13, 705–713. doi: 10.5897/AJBX2013.13554

Ragaei, M. and Sabry, A. H. (2014) Nanotechnology for insect pest control. *International journal of science, environment and technology*, 3(2), 528–545.

Regulation of Biological Control Agents (REBECA) (2007) Balancing the benefits and costs of regulating biological plant protection products, WS 6 Synthesis, Deliverable No. 25, Regulation of Biological Control Agents (REBECA)

Roach, D. R., Castle, A. J., Svircev, A. M. and Tumini, F.A. (2008) Phage-based biopesticides: characterization of phage resistance and host range for sustainability. *Acta horticulturae,* 793, 397–401.

Roh, J. Y., Choi, J. Y., Li, M. S., Jin, B. R. and Je, Y. H. (2007) *Bacillus thuringiensis* as a specific, safe, and effective tool for insect pest control. *Journal of microbiology and biotechnology,* 17(4), 547–559

Rosell, G., Quero, C., Coll, J. and Guerrero, A. (2008) Biorational insecticides in pest management. *Journal of pesticide science,* 33, 103–121.

Routray, S., Dey, D., Baral, S., Das, A. P. and Patil, V. (2016) Potential of nanotechnology in insect pest control. *Progressive research,* 11 (Special-II), 903–906.

Ruiu, L. (2018) Microbial biopesticides in agroecosystems. *Agronomy,* 8, 235. doi:10.3390/agronomy8110235

Sales, M. D. C., Costa, H. B., Fernandes, P. M. B., Ventura, J. A. and Meira, D. D. (2016) Antifungal activity of plant extracts with potential to control plant pathogens in Pineapple. *Asian pacific journal of tropical biomedicine,* 1, 26-31. doi: 10.1016/j.apjtb.2015.09.026

Seiber, J. N., Coats, J., Duke, S. O. and Gross, A. D. (2014) Biopesticides: State of the art and future opportunities. *Journal of agriculture and food chemistry,* 62(48), 11613-11619. doi: 10.1021/jf504252

SEIPASA (2018) Biopesticides: 5 reasons to understand their growth in the global market. Retrieved 28 October 2019 from https://www.seipasa.com/en/blog/biopesticides-growth-global-market/

Sesan, T. E., Enache, E., Iacomi, M., Oprea, M., Oancea, F. and Iacomi, C. (2015) Antifungal Activity of some plant extract against *Botrytis cinerea* Pers. in the Blackcurrant Crop (*Ribes nigrum* L.). *Acta scientiarum polonorum technologia alimentaria,* 1, 29-43.

Sharma, I. P. and Sharma, A. K. (2017) Effective control of root-knot nematode disease with Pseudomonad rhizobacteria filtrate. *Rhizosphere,* 3, 123–125.

Sivasakthi, S., Usharani, G. and Saranraj, P. (2014) Biocontrol potentiality of plant growth promoting bacteria (PGPR)-*Pseudomonas fluorescens* and *Bacillus subtilis*: A review. *African journal of agricultural research,* 9, 1265–1277.

Shiberu, T. and Getu, E. (2016) Assessment of selected botanical extracts against *Liriomyza* Species (Diptera: Agromyzidae) on Tomato under glasshouse condition. *International journal of fauna and biological studies,* 1, 87-90.

Shukla, R. and Shukla, A. (2012) Market potential for biopesticides: a green product for agricultural applications. *International journal of management research and review*, 2(1), 91-99.

Sporleder, M. and Lacey, L. A. (2013) Biopesticides. *In:* Giordanengo, P., Vincent, C. and Alyokhin, A. (eds). Insect pests of potato: global perspectives on biology and management. Academic Press, Amsterdam, pp. 463–497.

Stockwell, V. O., Johnson, K. B., Sugar, D. and Loper, J. E. (2010) Control of fire blight by *Pseudomonas fluorescens* A506 and *Pantoea vagans* C9-1 applied as single strains and mixed inocula. *Phytopathology*, 100(12), 1330-1339

Sundh, I. and Goettel, M. S. (2013) Regulating biocontrol agents: a historical perspective and a critical examination comparing microbial and macrobial agents. *Biocontrol*, 58, 575–593.

UCIPM (2017) What is integrated pest management (IPM)? Retrieved 18 November 2019 from http://www2.ipm.ucanr.edu/WhatIsIPM/

Ullah, M. S. and Lim, U. T. (2011) Synergism of *Beauveria bassiana* and *Phytoseiulus persimilis* in control of *Tetranychus urticae* on bean plants. *Systematic and applied acarology*, 22(11), 1924-1936. doi: 10.11158/saa.22.11.11

Vega, F. E., Goettel, M. S., Blackwell, M., Chandler, D., Jackson, M. A., Keller, S., Koike, M., Maniania, N. K., Monzo´Ni, A., Ownley, B. H., Pell, J. K., Rangel, D. E. N and Roy, H. E. (2009) Fungal entomopathogens: new insights into their ecology. *Fungal ecology*, 2, 149–159. doi:10.1016/j.funeco.2009.05.001

Verma, M., Brar, S. K., Tyagi, R. D., Surampalli, R. Y. and Valero, J. R. (2007) Antagonistic fungi, *Trichoderma* spp.: panoply of biological control. *Biochemical engineering journal*, 37, 1-20. doi:10.1016/j.bej.2007.05.012

Vinale, F., Sivasithamparam, K., Ghisalberti, E. L., Marra, R., Woo, S. L. and Lorito, M. (2008) *Trichoderma*—plant–pathogen interactions. *Soil biology and biochemistry*, 40, 1–10. doi:10.1016/j.soilbio.2007.07.002

Yadav, S. K., Babu, S., Yadav, M. K., Singh, K., Yadav, G. S. and S. Pal, S. (2013) A review of organic farming for sustainable agriculture in northern India. *International journal of agronomy*, ID 718145, 8 pages, doi: 10.1155/2013/718145.

CHAPTER 12

Challenges and Prospects of Biopesticides

by
Johanna Bitzenhofer* and Ralf Thomas Voegele
University of Hohenheim, Faculty of Agricultural Sciences,
Institute of Phytomedicine, Department of Plant Pathology,
Otto-Sander-Straße 5, 70599 Stuttgart, Germany.
*Corresponding author: bitzenhoferjohanna@gmail.com

Abstract

Chemical crop protection has contributed tremendously to enhance agricultural productivity since the 1960s. In spite of this achievement, pesticides leave a long term mark in the environment. As a result, the desire to combine productivity and environmental compatibility in agriculture has increasingly become the focus of social and scientific attention. In this regard, the use of biopesticides appears to be particularly promising, although their application currently still faces major challenges. Biopesticides often have a limited shelf life, they are very heterogeneous in terms of their effects, and generally have a narrower application timeframe than synthetic pesticides. In addition, biopesticides are not harmless *per se*. Using microbial biocontrol agents (BCAs) can lead to a release of toxic secondary metabolites into the environment, the effects of which are difficult to predict. Furthermore, lengthy approval procedures and low acceptance in the farming sector itself limit the rise of biopesticides. Nevertheless, biopesticides have a great potential, above all, the trend towards more organic food and the increasing amount of organically farmed land worldwide offer many opportunities to the biopesticide market. Moreover, biopesticides do not infringe upon non-target organisms due to their specificity. As opposed to synthetic pesticides, it is also rather unlikely that BCAs induce resistance in the pathogens/parasites. The authors conclude that biopesticides are an important alternative to conventional crop protection, but cannot, at least not in the near future, replace it completely. With regard to avoiding resistance and practicing integrated pest management (IPM), biopesticides are crucial components. However, for biopesticides to be broadly accepted, they must be easy to handle and reliable in their effect.

Keywords: biopesticides, agrochemicals, sustainability, integrated pest management, resistance management

Introduction

The use of pesticides has undoubtedly contributed to increasing food production and feeding a rapidly growing world population (Carvalho, 2017; Damalas and Koutroubas, 2018). Nevertheless, there is a strong call on the society for a change in agricultural mindset (Harvey, 2020). In particular, the worldwide use of pesticides is currently under frequent criticism. As numerous studies have shown, the downside of extensive pesticide use is the pollution of ecosystems and toxic effects on humans (Carvalho, 2017). In addition, the use of pesticides is seen as one of the main reasons for the global decline in insect species diversity (Zaller, 2020). In contrast to many other chemicals from industrial cycles, pesticides are openly released into the environment; this is their specific problem (Zaller, 2020). Accordingly, the role of pesticides is ambivalent: On the one hand, they secure agricultural yields, on the other hand, their use can damage the environment. Due to this ambivalence, the political and social pressure to find alternative solutions is enormous (Liu *et al.*, 2019; Mishra *et al.*, 2015). Dillard (2019) described increase in yields while at the same time ensuring ecological sustainability as one of the greatest challenges for food production. This is where biopesticides entered the stage. As already described in details in previous chapters, they are increasingly moving into the focus of science. However, there are quite some controversial issues with regard to biopesticides, as can be seen, for example, from the open discourse between Deising *et al.* (2017) and Koch *et al.* (2018). In the present chapter, both challenges and perspectives of biopesticides are elaborated. In Chapter 1, biopesticides were defined according to the criteria of the Environmental Protection Agency (EPA). This chapter is also based on the EPA definition, which distinguishes the three main classes of biopesticides: biochemical, microbial, and what it calls Plant-Incorporated Protectants (PIPs) (Kachhawa, 2017; Seiber *et al.*, 2014). Finally, the authors addressed the question whether biopesticides can be a supplement to or even a substitute for synthetic plant protection products or not.

Challenges

One of the greatest challenges in the use of biopesticides is their strong dependence on environmental conditions. Lengai and Muthomi (2018) argued that *in vitro* tests for the effectiveness of biopesticides often produce excellent results. In contrast, biopesticides often perform insufficiently during field trials. Under field conditions, high doses have to be used to reach the effectiveness of synthetic pesticides (Lengai and Muthomi, 2018). This discrepancy between *in vitro* and *in planta* is due to the limited shelf life of natural products and to their sensitive reaction to environmental influences (Lengai and Muthomi, 2018; Copping and Menn, 2000). Their strong dependence on environmental conditions is also one of the reasons why biopesticides have a narrower application timeframe. As biopesticides are generally used for protective purposes, they have to be applied at an earlier stage than synthetic pesticides. Prognostic models are increasingly being used to optimize the timing of plant protection measures. However, these methods are based on certain damage thresholds where the use of conventional pesticides is still possible while the application of biopesticides would be too late. A better adaptation of forecasting systems to plant protection products of organic origin is therefore necessary to boost biopesticides (Marrone, 2019).

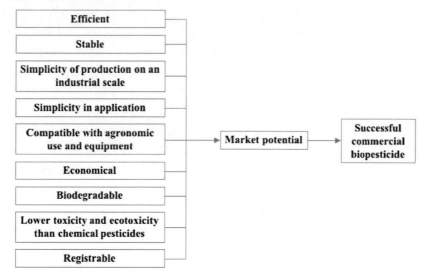

Figure 1: The most important characteristics of successful commercial biopesticides (Zaki *et al.*, 2020).

As already noted with regards to dependence on environmental conditions, the reduced effectiveness of biopesticides is a major problem, especially under field conditions (Lengai and Muthomi, 2018). According to Zaki *et al.* (2020), Figure 1 shows the most important properties a biopesticide should have in order to be successfully marketed. Among these nine properties, Zaki *et al.* (2020) saw effectiveness and stability as the two most important factors for success. When Zaki *et al.* (2020) speak of stability, they mean long shelf life during transport and storage. Many biopesticides only have a limited shelf life, which reduces not only their efficiency but also their competitiveness. Synthetic pesticides can be stored for a longer period of time without their effects decreasing significantly (De la Cruz Quiroz *et al.,* 2019). Like Zaki *et al.* (2020), De la Cruz Quiroz *et al.* (2019) therefore plead for an increased focus on efficacy and durability in biopesticide research.

Constantine *et al.* (2020) also attribute particular importance to the effectiveness of biopesticides for their market success. In their study, they investigated the question why only 10 % of smallholder farmers in Kenya use biopesticides, even though the country has a high number of registered biopesticide products, and even though the farmers interviewed regarded the use of chemical synthetic agents as generally dangerous for humans and the environment. The survey revealed that only 23 % of smallholders considered biopesticides to be effective, while synthetic pesticides were perceived as effective by 80 %. The speed of action, too, played a decisive role for the respondents (Constantine et al., 2020). 45 % of the farmers interviewed attributed a rapid effect to synthetic pesticides, while only 5 % did so to biopesticides (Constantine et al., 2020). In their Ghana study, Coulibaly et al. (2008) also confirm the importance of the speed of action of a crop protection product for choosing a pesticide. As a consequence, the use of biopesticides could be increased by ensuring that they perform in field trials as well as conventional pesticides do, and that the farmers perceive their efficiency as being equivalent to synthetic analogues. Dhakal and Singh (2019) rightly mentioned acceptance by farmers as one of the key tasks in driving biopesticides forward.

Furthermore, lengthy approval procedures are certainly an obstacle to a greater market success of biopesticides (Lengai and Muthomi, 2018). The market for biopesticides in the USA is larger than in the EU (Damalas and Koutroubas, 2018). This could be due, among other things, to the different legal framework for the registration of pesticides' active ingredients. In their article, Frederiks and Wesseler (2019) pointed out that the EU does not differentiate between synthetic and organic active substances in its registration procedures. This is why the approval of microbial BCAs by the EU takes on average 1.62 years longer than in the USA. Many researchers see the complicated registration procedures in some countries as a major obstacle for a wider spread of biopesticides (Lengai and Muthomi, 2018; Damalas and Koutroubas, 2018; Glare et al., 2012; Leng et al., 2011; Copping and Menn, 2000).

In their opinion paper, Deising et al. (2017) expressed their concerns about BCAs. Although they believed that biopesticides could prevent the development of fungicide-resistant pathogen populations, however, the authors pointed out that possible environmental burden could be caused by microbial toxins (Deising et al., 2017). Deising et al. (2017) argued that 'minute amounts' of low-toxic fungicides in our food are the subject of emotional debate, while society underestimates the toxicity of microbial secondary metabolites. Copping and Menn (2000) also noted that the natural origin of biopesticides does not automatically make them harmless. At this point, the phenomenon of cross-feeding should be mentioned: This is a biological interaction within microbial communities where one species uses the metabolites of another species as a source of energy or food (Estrela et al., 2012). The application of microbial pesticides could lead to the input of substances into the environment, which might enable undesirable species to develop. So far, the ecological result of a metabolic interaction cannot be predicted in detail (Estrela et al., 2012). In the USA, the registration of biopesticides is proceeding more quickly, based on the argument, among other things, that this type of pesticide tends to be less risky than synthetic pesticides (Mishra et al., 2015). However, a mere tendency must not be an argument for a faster, and thus less thorough, approach. Therefore, when assessing micro-bial pesticides, it should also be seen to it that negative interactions with other organisms as well as risks to humans are excluded (Egbuna *et al.*, 2020). Taking into account that microbial BCAs make up the bulk of biopesticides (Zaki *et al.*, 2020; Dunham and Trimmer, 2018), assessing them with special accuracy is particularly important.

One of the great advantages of biopesticides is their high specificity, which considerably reduces the potential damage to non-target organisms (Kachhawa, 2017; Leng *et al.*, 2011). Paradoxically, it is precisely this specificity which is simultaneously interpreted as a disadvantage of biopesticides (Copping and Menn, 2000), especially from the perspective of their users. While the application of a broad-spectrum synthetic pesticide combats several pests at once, the use of a product which acts specifically requires a tar-geted approach and thus an increased workload. The work of Constantine *et al.* (2020) makes it clear that the high specificity of biopesticides reduces their attractiveness in the eyes of farmers. The change of perspective from food production to environmental protec-tion reveals a special dilemma confronting biopesticides.

Prospects

Where there are challenges, there is always potential for development: Although, biopesti-cides currently account for only 5 % of the total crop protection market (Damalas and Koutroubas, 2018; Dunham and Trimmer, 2018), the biopesticide market is growing quite rapidly. Ramírez-Guzmán *et al.* (2020) gave annual growth rates from 7 to 45 % for se-lected countries. On the average, the compound annual growth rate (CAGR) for biopesti-cides is 17 % worldwide (Dunham and Trimmer, 2018). In the EU, too, the relevant figures speak for themselves, despite long registration procedures (Frederiks and Wesseler, 2019): In a press release dated 28 February 2019, the German Federal Research Centre for Culti-vated Plants, *Julius Kühn-Institut* (JKI), stated that biopesticides account for approx. 50 % of all approval requests for new active substances in the EU (JKI, 2019). According to Pelaez and Mizukawa (2017), the leading agrochemical companies are also increasingly investing in the development of biopesticides as part of a diversification strategy. In addi-tion, social and political perceptions of what constitutes an ideal pesticide have changed. In their review article, Villaverde *et al.* (2016) named three conditions a pesticide should meet today so as to avoid contamination and resistance: (1) minimal toxicity to non-target organisms, (2) high effectiveness at low application rates and (3) low persistence. At least points (1) and (3) are met by the majority of biopesticides (Egbuna *et al.*, 2020).

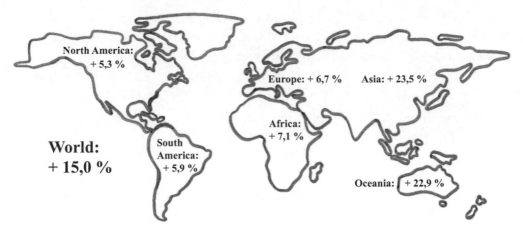

Figure 2: Growth of Organic agricultural land (including in-conversion areas) from 2015 to 2016 itemized by continent (Lernoud and Willer, 2018).

Biopesticides are potentially acceptable for use in organic farming (Liu *et al.*, 2019; Marrone, 2019). Accordingly, the trend towards more biopesticides is supported by the increase in organically farmed areas (Lernoud and Willer, 2018), although the share of organic farming is declining in some countries, there is an overall increase in organic areas worldwide and in all continents (see Figure 2). In that respect, Asia has the highest growth rate, followed by Oceania (Lernoud and Willer, 2018). Closely linked to organic agriculture, and thus to the use of organic pesticides, is the associated market for organically produced food and beverages, which is also growing (Sahota, 2018). In contrast to the global distribution of organic farming, around 90 % of sales of organically produced products are currently concentrated in North America and Europe; in large parts of the world, organic food and beverages are produced for export only (Sahota, 2018). In any case, the increased demand for organically produced food and the associated global increase in organic farmland offer great opportunities for the biopesticide market.

On top of the positive market trends, there are several properties that make biopesticides stand out in the eyes of scientists. According to Bhagat *et al.* (2014), BCAs could be particularly useful for developing countries, as they are easily available and relatively inexpensive. Kachhawa (2017) confirmed lower costs for the development of microbial insecticides compared to synthetic ones. For industrialized countries, Bhagat *et al.* (2014) argued that biopesticides could have a direct impact on the production of high-quality food. Their specificity and thus their harmlessness to non-target organisms are also properties that researchers consider to be major advantages (Liu *et al.*, 2019; Lengai and Muthomi, 2018; Kachhawa, 2017). In addition, the use of biopesticides leads to lower unwanted residues in food, which averts risks from end consumers (Egbuna *et al.*, 2020; Liu *et al.,* 2019).

Moreover, biopesticides can be used within the framework of IPM, which is becoming increasingly important (Kachhawa, 2017; Mishra *et al.*, 2015). This also pushes the BCAs' own market position. By practicing IPM, various methods are combined with one another, so that the incompleteness of isolated methods is overcome (Chandler *et al.*, 2011). The combination of pesticides (microbial and conventional) can lead to a reduction in the use of synthetic agrochemicals and, simultaneously, to a greater success in pest control (Mishra *et al.*, 2015). In alternation with conventional pesticides, BCAs can delay pest re-sistance (Glare *et al.,* 2012). Biopesticides, especially the microbial ones, play an im-portant role in avoiding resistance due to their multiple action sites (Zaki *et al.* 2020; De la Cruz Quiroz *et al.*, 2019). Therefore, researchers such as Villaverde *et al.* (2016) argued that biopesticides should be used as a cornerstone in the development and implementation of resistance management programs.

In spite of the obstacles currently impeding a broad success and acceptance of BCAs, there are voices in the scientific community predicting a great future for biopesticides. De la Cruz Quiroz *et al.* (2019) are convinced that biopesticides will compete with conventional agrochemicals in the future. This forecast is based on the assumption that the preference of end consumers of agricultural products for green technologies will continue to increase (De la Cruz Quiroz *et al.*, 2019). Furthermore, De la Cruz Quiroz *et al.* (2019) argued that the development of formulations of fungus-based biopesticides with a focus on efficacy and stability is promising. Mishra *et al.* (2015) also consider biopesticides to be competi-tive. The authors believed that biopesticides could gradually overtake the market of syn-thetic pesticides, provided that they show good performance under field conditions (Mishra *et al.,* 2015).

Conclusion

As pointed out at the beginning, the central question dealt with in this chapter is whether biopesticides can be an alternative to or even a substitute for chemical synthetic pesticides or not. The market data clearly show that biopesticides are on the rise. How long this trend will continue mainly depends on whether biopesticides, similar to their conventional ana-logues, are easy to use and reliable in their effects in the long term. As far as shorter ap-proval procedures specifically designated for biopesticides are concerned (as, e.g., in the USA), the authors dismiss such a procedure for the following reasons: (1) pesticides of natural origin are not automatically harmless and (2) the use of microbial preparations can lead to cross-feeding effects. The authors therefore advocate equal treatment of all pesti-cides in their assessment.

Copping and Menn (2000) already considered it unlikely 20 years ago that biopesticides were going to replace synthetic plant protection within the next two decades. Nevertheless, they believed that the amount and quality of biopesticides would increase and that their market would grow (Copping and Menn, 2000). Incidentally, the authors were right with both forecasts. Scientists like Villaverde *et al.* (2016), considered it unrealistic that the use of biopesticides could completely replace synthetic pesticides, except in organic food production. The authors also believe that biopesticides will not completely replace conventional agrochemicals. Although BCAs are of great importance, especially with regards to resistance management and IPM, their widespread use is still confronting too many obstacles. Environment-friendly crop protection should not be a question of faith, no 'yes or no' fight. What is needed is a pragmatic approach based on the best of our knowledge. Accordingly, all promising possibilities should be exhausted so as to secure global food production and to avert risks from the environment and people. Finally, the authors welcome the global progression of biopesticides as a significant alternative to the preponderant use of synthetic pesticides.

References

Bhagat, S., Birah, A., Kumar, R., Yadav, M. S. and Chattopadhyay, C. (2014). Plant disease management: prospects of pesticides of plant origin. In: Singh, D. (ed) Advances in plant biopesticides. Springer, New Delhi, 119-129.

Carvalho, F. P. (2017). Pesticides, environment, and food safety. *Food and Energy Security* 6, 48-60.

Chandler, D., Bailey, A. S., Tatchell, G. M., Davidson, G., Greaves, J. and Grant, W. P. (2011). The development, regulation and use of biopesticides for integrated pest management. *Philosophical Transactions of the Royal Society B: Biological Sciences* 366, 1987-1998.

Constantine, K. L., Kansiime, M. K., Mugambi, I., Nunda, W., Chacha, D., Rware, H., Makale, F., Mulema, J., Lamontagne-Godwin, J., Williams, F., Edgington, S. and Day, R. (2020). Why don'tsmallholder farmers in Kenya use more biopesticides? *Pest Management Science*. https://doi.org/10.1002/ps.5896.

Copping, L. G. and Menn, J. J. (2000). Biopesticides: a review of their action, applications and efficacy. *Pest Management Science* 56, 651-676.

Coulibaly, O., Cherry, A. J., Nouhoheflin, T., Aitchedji, C. C. and Al-Hassan, R. (2008). Vegetable producer perceptions and willingness to pay for biopesticides. *Journal of Vegetable Science* 12, 27-42.

Damalas, C. A. and Koutroubas, S. D. (2018). Current status and recent developments in biopesticide use. *Agriculture* 8, 13.

Deising, H. B., Gase, I. and Kubo, Y. (2017). The unpredictable risk imposed by microbial secondary metabolites: how safe is biological control of plant diseases? *Journal of Plant Diseases and Protection* 124, 413-419.

De la Cruz Quiroz, R., Cruz Maldonado, J. J., de Jesús Rostro-Alanis, M., Torres, J. A. and Saldívar, R. P. (2019). Fungi-based biopesticides: shelf-life preservation technologies used in commercial products. *Journal of Pest Science* 92, 1003-1015.

Dhakal, R. and Singh, D. N. (2019). Biopesticides: a key to sustainable agriculture. *International Journal of Pure & Applied Bioscience* 7, 391-396.

Dillard, H. R. (2019). Global food and nutrition security: from challenges to solutions. *Food Security* 11, 249-252.

Dunham, W. and Trimmer, M. (2018). Biological products around the world. URL: http://www.bpia.org/wp-content/uploads/2018/03/Biological-Products-Markets-Around-The-World.pdf (13.07.2020).

Egbuna, C., Sawicka, B., Tijjani, H., Kryeziu, T. L., Ifemeje, J. C., Skiba, D. and Lukong, C. B. (2020). Biopesticides, safety issues and market trends. In: Egbuna, C. and Sawicka, B. (eds) Natural remedies for pest, disease and weed control. Academic Press, 43-53.

Estrela, S., Trisos, C. H. and Brown, S. P. (2012). From metabolism to ecology: cross-feeding interactions shape the balance between polymicrobial conflict and mutualism. *The American Naturalist* 180, 566-576.

Frederiks, C. and Wesseler, J. H. H. (2019). A comparison of the EU and US regulatory frameworks for the active substance registration of microbial biological control agents. *Pest Management Science* 75, 87-103.

Glare, T., Caradus, J., Gelernter, W., Jackson, T., Keyhani, N., Köhl, J., Marrone, P., Morin, L. and Stewart, A. (2012). Have biopesticides come of age? *Trends in Biotechnology* 30, 250-258.

Harvey, F. (2020). Young climate activists call for EU to radically reform farming sector. The Guardian. URL: https://www.theguardian.com/environment/2020/may/22/young-climate-activists-call-for-eu-to-radically-reform-farming-sector (04.07.2020).

JKI (Julius Kühn-Institut) (2019). PI Nr. 8: Julius Kühn-Institut publiziert Statusbericht 2018 zum Stand des biologischen Pflanzenschutzes in Deutschland. URL: https://www.julius-kuehn.de/presse/pressemeldung/news/pi-nr-8-julius-kuehn-institut-publiziert-statusbericht-2018-zum-stand-des-biologischen-pflanzenschu/ (02.07.2020).

Kachhawa, D. (2017). Microorganisms as a biopesticides. *Journal of Entomology and Zoology Studies* 5, 468-473.

Koch, E., Becker, J. O., Berg, G., Hauschild, R., Jehle, J., Köhl, J. and Smalla, K. (2018). Biocontrol of plant diseases is not an unsafe technology! *Journal of Plant Diseases and Protection* 125, 121-125.

Leng, P., Zhang, Z., Pan, G. and Zhao, M. (2011). Applications and development trends in biopesticides. *African Journal of Biotechnology* 10, 19864-19873.

Lengai, G. M. W. and Muthomi, J. W. (2018). Biopesticides and their role in sustainable agricultural production. *Journal of Biosciences and Medicines* 6, 7-41. Lernoud, J. and Willer, H. (2018). Current statistics on organic agriculture worldwide: area, operators, and market. In: Willer, H. and Lernoud, J. (eds) The world of organic agriculture: statistics and emerging trends. FiBL & IFOAM – Organics International (2017): Frick and Bonn, 2017-02-20.

Liu, X., Cao, A., Yan, D., Ouyang, C., Wang, Q. and Li, Y. (2019). Overview of mechanisms and uses of biopesticides. *International Journal of Pest Management*. https://doi.org/10. 1080/09670874.2019.1664789.

Marrone, P. G. (2019). Pesticidal natural products – status and future potential. *Pest Management Science* 75, 2325-2340.

Mishra, J., Tewari, S., Singh, S. and Arora, N. K. (2015). Biopesticides: where we stand? In: Arora, N. K. (ed) Plant Microbes Symbiosis: Applied facets. Springer, New Delhi, 37-75.

Pelaez, V. and Mizukawa, G. (2017). Diversification strategies in the pesticide industry: from seeds to biopesticides. *Ciência Rural* 47. https://doi.org/10.1590/0103-8478cr20160007.

Ramírez-Guzmán, N., Chávez-González, M., Sepúlveda-Torre, L., Torres-León, C., Cintra, A., Angulo-López, J., Martínez-Hernández, J. L. and Aguilar, C. N. (2020). Chapter 1 - Significant advances in biopesticide production: strategies for high-density bioinoculant cultivation. In: Singh, J. S. and Vimal, S. R. (eds) Microbial services in restoration ecology. Elsevier, 1-11.

Sahota, A. (2018). The global market for organic food & drink. In: Willer, H. and Lernoud, J. (eds) The world of organic agriculture: statistics and emerging trends. FiBL & IFOAM – Organics International (2017): Frick and Bonn, 2017-02-20.

Seiber, J. N., Coats, J., Duke, S. O. and Gross, A. D. (2014). Biopesticides: state of the art and future opportunities. *Journal of Agricultural and Food Chemistry* 62, 11613-11619.

Villaverde, J. J., Sandín-España, P., Sevilla-Morán, B., López-Goti, C. and Alonso-Prados, J. L. (2016). Biopesticides from natural products: current development, legislative framework, and future trends. *BioResources* 11, 5618-5640.

Zaki, O., Weekers, F., Thonart, P., Tesch, E. and Kuenemann, P. (2020). Limiting factors of mycopesticide development. *Biological Control* 144, 104220.

Zaller J. G. (2020). Insektensterben – inwiefern sind Pestizide dafür verantwortlich. *Entomologia Austriaca* 27, 285-295.